ANDREW WELBURN is a wide-ranging author of works exploring the history and evolution of consciousness. Starting out with a thesis on William Blake and the Gnostics, his main concern has been bridging academic and spiritual perspectives. He has been particularly engaged with Rudolf Steiner's philosophy and esotericism, through which he has striven to illumine the Christian imagination in poetry and art, highlighting the Romantic writers. In his books *The Beginnings of Christianity* and *From a Virgin Womb* he has studied archaic traditions concerning the 'transfiguration of the world' in their relevance to Christian origins and a vision of future reality. In his most recent work he is extending the bounds of his researches into evolution and prehistory. He has been a British Academy Fellow and Frances Yates Fellow in the universities of London and Oxford.

The researcher David Lewis-Williams has pointed out that the animal depictions in Upper Palaeolithic art evidently do not stand, but hover in space—indicating their visionary character. It is now widely acknowledged that the prehistoric cave paintings discovered in abundance since the celebrated find at Lascaux depict shamanic journeys into the realm of other-worldly images. Upper Palaeolithic painting, Lascaux, France.

NEANDERTHALS AND ATLANTIS

Rudolf Steiner and the New Prehistory

Andrew Welburn

Clairview Books Ltd.,
Russet, Sandy Lane,
West Hoathly,
W. Sussex RH19 4QQ

www.clairviewbooks.com

Published by Clairview Books 2024

A CIP catalogue record for this book is available from the British Library

ISBN 978 1 912992 69 0

Cover by Morgan Creative. Montage includes Commons image of soft tissue reconstruction of a male homo neanderthalensis with child from Natural History Museum, Vienna, by Wolfgang Sauber
Typeset by Symbiosys Technologies, Visakhapatnam, India
Printed and bound by 4Edge Ltd, Essex

CONTENTS

PROLOGUE: WHERE DO WE COME FROM?

Ideas about prehistory are currently being rocked by a disturbing new realisation: we were not alone. For long we have piqued ourselves on being the only intelligent, creative, the only genuinely human species. But now we know it is not, or at least was not, so. There were other versions of the human, who were not just forerunners but richly developed embodiments of human qualities with language, memory and music. And they actually helped us to become what we are. Many have desperately fought off the realisation but now it is clear that a major reorientation in our thinking is required. Our own prehistory becomes different as well. Yet we may derive unexpected assistance from a brilliant though frequently challenging thinker to whom such matters were by no means alien.

'For a long time,' wrote Rudolf Steiner in the definitive outline that he wrote of his ideas, 'the human race consisted of two separate kinds of human being.' (One might equally translate: 'two different species …'.) One of them had evolved in close connection with the earthly environment; the other experienced itself as related to the heavens, and to the several cosmic influences which differentiated them and pointed toward the emergence of intelligence and individuality. Even today some so-called primitive peoples still share that cosmic perspective. Steiner is talking, however, of human prehistory—long before the earliest civilisations. Both of these types merged in subsequent human evolution, a process which led to ourselves, and to modern consciousness. However, as Steiner elaborated on his findings elsewhere, before that could happen 'the great mass of mankind, amongst whom thought was only dully present (though they possessed natural abilities which modern people have lost) … was doomed to gradual extinction'. Out of the older population it was only one particular line of human beings which combined the two natures and engendered man as known today.

In evolutionary terms it was actually quite late, he notes, that a specific group of the second human type played a critical role in developing Modern Human intelligence. Still, even this was during the period following the end of the last Ice Age (remembered mythologically as the Great Flood, or End of Atlantis), long prior to the emergence of human lifestyles that shaped the earliest high cultures, based on agriculture, or the great urban civilisations such as Egypt and Mesopotamia. Human beings in those prehistoric times, during or close to the

Ice Age, interacted with their environment and each other through highly developed memory, in an intense symbiotic relationship with the world that is fitfully preserved nowadays in faculties of clairvoyance and psychokinesis.

Rudolf Steiner's descriptions frequently challenge our usual preconceptions. Yet to those who follow recent research into early humans, these descriptions may strike a mysteriously pertinent note. Until recently conventional scientific research into human origins, it is true, had backed a radically different framework, but discoveries and new ideas have brought major shifts in perspective, notably just in the last few years. Let us note a few. In the time following Darwin, nearly every older reconstruction shared among its presuppositions the assumption that *Homo sapiens* (ourselves) must be regarded as a unique success-story in evolution. We did have, of course, remote predecessors of an ape-like character, like the endearingly nicknamed fossil called 'Lucy' found in Ethiopia, and then there were various kinds of early tool-fashioning near-humans.* But from these, true humans emerged and steadily outstripped all previous evolutionary achievements with a combination of uprightness, manual dexterity (making tools and weapons) and intelligence. These features Darwin had already linked together on our species' unique trajectory to world-domination. It was *Homo sapiens* alone which, as mankind spread all over the world, ascended multi-regionally to intellectual superiority, technological advance and socialisation. No room here for other human types or species whose interaction was needed for our own emergence.

There was though one thorn in the side of the theory and its happy assumption of our unique superiority: the strangely 'other' human, whose fossilised bones had been discovered in the Neander Valley near Düsseldorf in 1856. He was not just an ape, nor certainly a human

---

* I refer broadly for the present to prehistoric humans, to some of whom more technical descriptions will later be applied, e.g. hominins. I have tried as far as is reasonable to avoid the over use of specialist parlance which, though technically justified, is confusing to general readers. I draw justification for having have done so especially from the fact that some important terms have in the last decade or so been re-allocated so that, for instance, what are now called hominins were once called hominids (who are now not humans at all). Readers must therefore take note that when they read even quite modern books on the subject of prehistory, older meanings of words may still pertain. Similarly, for our purposes I have not thought it necessary to differentiate a cranium from a skull—nor for that matter to refer to the paintings found in prehistoric caves as 'parietal art'.

like ourselves. The remains were soon shown to be those of a prehistoric type of man who did not fit the usual assumptions—i.e. that all human ancestors must lead towards us. As a result, this hapless fossil was subjected to an extraordinary scientific campaign of vilification. He was reconstructed as low-browed, ape-like, stooping, hairy and thuggish—all of which interpretations of the fossil evidence are now admitted to have been completely unjustified. Nevertheless he and the others like him discovered later are still regularly denigrated both anatomically and culturally, called brutal or even accused of cannibalism on thoroughly concocted grounds, and all the achievements evidenced in associated sites dismissed as attained by plagiarising 'true' humans. Most outrage was stirred up by the unforgivable fact that this prehistoric threat to our special status actually had a greater brain-capacity than we do. The most inventive reasons nevertheless often continue to be given for considering the Neanderthal to be just an inferior species. Howsoever he was to be fitted into the family-tree of humans, he either was not in the line which produced genuine human beings, or if he was, he represented a crude stage which we very soon outgrew. But alas (for the conventional theory), every step in further knowledge of the Neanderthals has shown them to have been more intelligent, socially complex, and technically advanced than previously thought. The cave-man brute has now yielded to reconstructions of a race clothed rather than hairy, large of countenance and touched by nobility (and sometimes striking beauty), possessed of language and of abstract symbols, not to mention music as well as (controversially) art! No one nowadays can avoid knowing that *sapiens* did not single-handedly surge forward in evolution. There was *Homo neanderthalensis* the 'smart Neanderthal', a second human kind. But how do they fit into the picture—and is Steiner relevant?

To crown all, geneticists recently showed that Neanderthals were not animal-like forerunners of mankind but interbred with 'Anatomically Modern Humans' like ourselves in the Middle East, to leave their trace in everybody's current genetic make-up, including in our large brains. Shortly after their intermarriage with humans, however, the Neanderthals who had been numerous all over Europe and parts of Asia, quite rapidly (in evolutionary terms, at least) died out in a fashion about which there are many theories but as yet not one convincing explanation.

To compound the problem, back in biology things have meanwhile been getting steadily less clear-cut for humans even as the Neanderthals have emerged from the shadows to stand revealed in new glory.

One major shock came from improvements in the relative dating of fossil remains where humans and Neanderthals were preserved in close proximity. It had been assumed that the Neanderthals were the older, more 'primitive' form, and that they were 'superseded' by the advent of humans. But no. It turned out that the Anatomically Modern remains were much older. And that was not only a problem for the general shape of the human story. For it highlighted the falling-apart of the entire reconstruction instigated by Darwin, according to which the special anatomical features of humans were intrinsically interwoven with their 'higher' achievements, as they had slowly but surely evolved into us. Alas (for the theory) even 'Lucy', who initially had seemed to many to exhibit a stage of animal-human transition, finally proved the exact opposite. The best recent researchers into early *Homo*, such as the late Richard Leakey, a great scientist from a dynasty of human-fossil finders, were struck instead by the lack of human features despite Lucy's uprightness, making him see in humans a 'biological new beginning' rather than a continuation of the apes.

Darwinian gradualism has failed on all fronts: the final 'leap' to modernity and the spread of humans over the globe is no longer understood as a slow but sure emergence from animality, but has now come to be seen as a single 'explosive' event. A specific history now accounts for the emergence of Modern Humans, based on a particular line of descent coming 'out of Africa'—and on that remarkable encounter between them and the Neanderthals. In the prehistoric migration which formed early Modern Humans, as Steiner long ago described it, 'one stream ... was spread across Europe and over to Asia and the region of the Caspian Sea, while another came through the land we now call Africa. In the Middle East a kind of confluence of these two streams took place, as when two currents meet and a vortex is generated.' In the most recent couple of decades a remarkably similar view has become almost universally accepted, in which context the strong advocacy of Chris Stringer, of the London Natural History Museum must be mentioned with honour. The theory is normally termed 'Out of Africa *II*' because of previous waves of early human dispersion.

The shortly following later phase of the Stone Age bears witness, not to a slow 'multi-regional' human emergence from animality, but to the outburst of human culture in startling sudden richness. The phenomenon of the 'Upper Palaeolithic explosion' now celebrated in many studies has turned human prehistory upside down even more than the chronological dislocation caused by new dating techniques. And it is no longer possible to suppose that the associated breathtaking cave-art is just fertility images or hunting fantasies: we now know that it possesses a spiritual focus that points to the guiding ideas of archaic culture, involving a hierarchical spiritual order and cyclical patterns of renewal. Long before Plato, we have to do not just with life in caves but with 'the mind in the cave' (so David Lewis-Williams, one of its expert interpreters).

The question of how humans became human in this dramatic fashion has suddenly started to be asked in a quite new and urgent way. Many brilliant ideas have surfaced in response, making it an issue with a resonance and importance far beyond the special subject of palaeo-anthropology (the study of ancient humans). Nor is evaluating early humans any longer just a matter of estimating their IQ. Researchers are clear that more is involved, and are trying specifically to understand the *consciousness* of early humans. More than just being bright, humans saw how to use their intelligence to change themselves and their situation. The crucial issue for evolution is not now man's animal affinities but the riddle of our Humanness. It is to the appreciation of that emerging question in its wider ramifications that I hope to contribute something in this book—which aims to draw attention to Rudolf Steiner's renewed relevance. Through the original work of biologists like Wolfgang Schad, among others, Steiner's evolutionism has recently been highlighted anew and won high praise.[*] The meaning of humanness is of course the crux of his anthroposophy—a wisdom of man—whether expressed in educational theory, medicine, religion or understanding our origins. I believe that Steiner's approach may contribute most richly to the growing realisation that human emergence is fundamentally an evolution of consciousness. It will be possible to trace the roots of much that is still vital to us today, and for our future, in the resources of thought, memory, social connections and imagination (myths, or their modern equivalents by which we structure our world) from this perspective. We will learn that there have been other versions of being human, and

---

[*] I will repeatedly mention Schad's work in what follows, especially his study on *Mammals* listed below. It is of course much more than a restatement of Steiner's views. It has a special place here, since its praise was sung (not of course literally but very inspiringly) by Martin Lockley, a professional palaeontologist from the University of Colorado when I met him in Oxford—which encouraged me to write this account, albeit my approach is from a rather different direction, and I recommend to all readers Lockley's own fascinating book for its use of Steiner's ideas along with much original thought (again, listed below).

there are further transformations yet to come. Evolution is not a finger of approval pointing reassuringly at ourselves. Indeed, many 'strange' former perspectives turn out to underlie what we take for granted in our lives. Steiner enables us to bridge our prehistoric past with our still-to-be-created future.

The elements I mentioned at the outset—the several human varieties, their interbreeding, the great extinction, the leap forward that does not correspond to gradual animal development—are they mere scattered resemblances in current thinking and Steiner's thought? I shall try to show not. Though quite independent, the two accounts, from Steiner and modern science, run deeply parallel and strikingly illuminate one another. At this point, however, a disclaimer. I certainly offer no 'new synthesis', nor is the moment ripe for one even if I would. I am painfully aware how little that is original I can contribute beside the wealth of actual research. I stand dumb with admiration before those investigators who can interpret the tiniest tendency in the shape of a fossil fragment, to place it in the stream of human evolution. Nor am I among those who have learned to chip constructively at flint axe-heads so as to feel myself into the very attitudes and skills of our ancestors. But I respect their realisation that what has often been dismissed as 'primitive' workmanship in fact demands a subtle degree of professionalism, from assessing the right-weight stone, to judging the exact angle and force of the strokes. I can only express wonderment at the way that through long familiarity with the art and industries of different phases and species of mankind, experts have so convincingly been able to intuit their meaning and the cultural gestures which lie behind them, to estimate the range and structure of the language they must have needed to co-operate on them—even to uncover the musicality which makes sense of them when images were not the primary mental tool (Steven Mithen); or share the sensitivity which reconstructs the compartmentalised consciousness of Neanderthal humans who did not yet find a centre in themselves, and the realisation that their unchanging and uncorrelated activities imply language without a future tense, and all that that implies. In the contrast between their musical organisation of experience, and the use of representation in the image-based consciousness which from the cave-artists onwards has dominated human cultural expansion, we come back to Steiner's contrast of the 'two types [or species] of mankind'. Among so much that is new I have found his indications a continuing source of illumination precisely in these most modern of approaches to understanding human origins.

There are of course also profound gulfs to be crossed. Strange to say, his background allusions to a continent of 'Atlantis' do not actually pose great geographical difficulties (as we shall see from modern 'paleomaps'!); but to put human evolution rightly into the framework he offers will undoubtedly demand some detailed efforts of readjustment and insight. And yet—there are such striking advances in the way that biological evolution is understood in a cosmic context in recent science, which speaks of astrobiology, or of 'Anthropic' science based on the special meaning of human existence for understanding the universe, that the ideas come remarkably to hand. I can merely try to sketch how they might be relevant, specifically with Steiner's thought in mind. Much in modern science is already starting to bridge that gulf.

At this point in Introductions, with due humility, it is customary to express a pious hope that more adequate books on the subject will be written in the future. I can do better. Marvellous books have already been written by scientists who are not only front-line investigators but authors with a gift, not for any crude 'popularisation' but simply for putting their ideas lucidly across. In lieu of a conventional 'Further Reading' book-list, here may be a better place for signalling the main sources of whatever this work can offer. Of course I trust this book explains itself as it goes along, but in some notable scientific works is explained the background of thought on which it gratefully draws along with Steiner's ideas. It may be worth adding that I consider it no paradox that such front-line thinkers will be most apt to indicate where Steiner's contribution can most relevantly be brought to bear. First, four books which illumine the nature of the cosmos and the place of life within it: as for life's origins, all of them unapologetically jettison the supposed accidental fusion in a chemical witch's-kitchen, trying to explain instead the 'biological imperative', or the laws of complexity behind life which are written into the universe. Paul Davies' brilliant *The Origin of Life* (London 1999), is especially valuable for its approach in terms of the 'biological information forces' which organise living things; Lee Smolin's *The Life of the Cosmos* (London and Oxford 1997) beautifully explains the way that life is not mere local organisation but is membered into the 'nested hierarchies' of form reaching up to stellar level; Stuart Kauffman's *At Home in the Universe: The Search for Laws of Self-Organization and Complexity* (Oxford and New York 1995) shares with us his extensive ruminations on the deeper meanings that arise as substances and organisms grow more complex, revealing the foundations of a world-order in which human conscious existence profoundly *belongs;* Walther Cloos, *The Living Earth* (Sussex 1977) unfolds the stages of the Earth's development and the phases of its life, in a presentation already shaped by contact with Rudolf Steiner's ideas. It will be apparent to any who have studied them that many of the ideas about science, life and development found in many of these books in contemporary form have roots in the way of thinking pioneered by figures popularly associated more with poetry than science like Goethe in his work on *Plant-Metamorphosis* or Coleridge in his *Essay on Life*. Steiner, editor and commentator on Goethe's scientific writings, was fully aware of this rich history, but even more of the potential still to be pursued in the approach they adumbrated. See Rudolf Steiner, *Goethe the Scientist* (New York). For an impressive formulation of evolution growing out of Rudolf Steiner's approach there is Wolfgang Schad's *Man and Mammals: Towards a Biology of Form* (New York 1977).

My unofficial prize for the best accessible introduction to the evidences of human evolution still goes to Richard Leakey: *The Origin of Humankind* (London 1994), not least for its judiciously selected 'Further Reading' sections to each chapter. For the more inward and even spiritual side there is Martin Lockley (with R. Morimoto), *How Humanity Came into Being: The Evolution of Consciousness* (Edinburgh 2010). Any work on early humans must nowadays be supplemented by studies reflecting the on-going discoveries and re-evaluations, but still invaluable are: Chris Stringer and Clive Gamble's *In Search of the Neanderthals: Solving the Puzzle of Human Origins* (London 1993); Erik Trinkaus and Pat Shipman's *The Neandertals: Changing the Image of Mankind* (London 1993); James Shreeve's *The Neandertal Enigma: Solving the Mystery of Modern Human Origins* (London 1995) shrewdly surveys many of the views taken by those in the field; Steven Mithen's wonderful *The Singing Neanderthals: The Origins of Music, Language, Mind and Body* (London 2006) opens many doors to understanding the presence of prehistory in our own consciousness; Clive Finlayson, *The Smart Neanderthal: Bird Catching, Cave Art and the Cognitive Revolution* (Oxford 2019) catches up with physical-genetic conclusions, the emerging relationship to birds and the advanced thinking-consciousness for which there is now overwhelming evidence. David Lewis-Williams' *The Mind in the Cave: Consciousness and the Origins of Art* (London 2002) grapples with the nature of early spirituality and its formative presence behind much that we think of as modern in our consciousness—though he might say he struggles to evade many of the conclusions accepted in the approach adopted here! He confronts the issues unflinchingly.

Coming to the dispersion-phase, and the qualities associated with different streams: Rudolf Steiner's account of the 'Northern ecstatic' i.e. shamanistic spirituality, interacting with a 'Southern' (e.g. Middle Eastern, North African) spirituality focussed on inner-bodily states has found an echo in many of Mircea Eliade's influential works on the history of religion. A good 'taster' is Eliade's *Myths, Dreams and Mysteries* (London 1968), while the foundation-study for his ideas is still Eliade, *Shamanism* (London 1989). I try to show how Steiner's threefold model of society could benefit the understanding of threefoldness in the Indo-European mythologies which has been brilliantly uncovered through the comparative-religious studies of Georges Dumézil, and (I shall suggest) help trace it back into prehistory. A classic introduction to Dumézil's 'tripartite' ideas is C. Scott Littleton, *The New Comparative Mythology* (California 1966), though there is nothing quite like exploring the demonstration of the method's scope in Dumézil's magisterial *Mythe et Epopéé* I.II.III (Paris 1995).

The work of Rudolf Steiner has come down to us in a form at once congenial and problematic. The large body of work in lecture-form gives a wonderfully living impression, but finding one's way into the major themes and connections is not always easy. Commentary has its limitations. A fine introductory selection (with commentary) is Robert MacDermott, *The Essential Steiner: Basic Writings of Rudolf Steiner* (San Francisco 1984). On the aspect of the evolution of consciousness, the profoundly original work of Owen Barfield, *Saving the Appearances: A Study in Idolatry* (London 1957) is still an inspiration.

Obviously the reader who tackles this book is under no obligation to delve further than he or she wishes. I merely gesture toward the riches available.

Chapter 1

EXPLOSION TO MODERNITY

Ourselves in the Mirror of Time

No doubt everyone has heard of the Stone Age. We all know vaguely
that our ancestors thousands of years ago made tools out of stone—
also wood and bone (ivory), but stone was their starting-point (hence
the term). They very effectively used sharply fractured flints for blades,
either mounting them on cut and trimmed shafts for use as spears, or
employing them as knives to prepare and carve their food, or in tailoring clothes from skins and other materials. In the days before cultivation, they lived the life of foragers as do certain scattered peoples today,
doubtless with many similarities though we should be wary of assuming
complete identity of lifestyle.

Whereas even a few decades ago it was usual to talk about them as
'primitives' or 'savages', mere forerunners of ourselves, the progress of
science has led us to recognise a surprising similarity between ourselves
and even very archaic people. In fact we can say identity. In-depth studies of artefacts and artworks from early people, including some of the
Old Stone Age human beings and their culture, and still more those of
the Middle and New Stone Age times, indicate that those who made
them could speak, think and co-operate just as we can. We will hardly
be able to avoid learning the modern jargon for the periods when they
lived, namely the Palaeolithic, Mesolithic and Neolithic, from the Greek
lithos = stone and the Greek terms for Old, Middle and New. Further
subdivisions produce a scheme as follows:

STONE AGE →

PALAEOLITHIC (OLD STONE AGE) →	MESOLITHIC →	NEOLITHIC →

LOWER PALAEOLITHIC	MIDDLE PALAEOLITHIC	UPPER PALAEOLITHIC

Even in the Palaeolithic, almost every modern discovery made
about people's life and activities increases our awareness of their
sophistication, though admittedly they were living under what seem to us
the simplest material conditions. By the Upper Palaeolithic, blade-tools

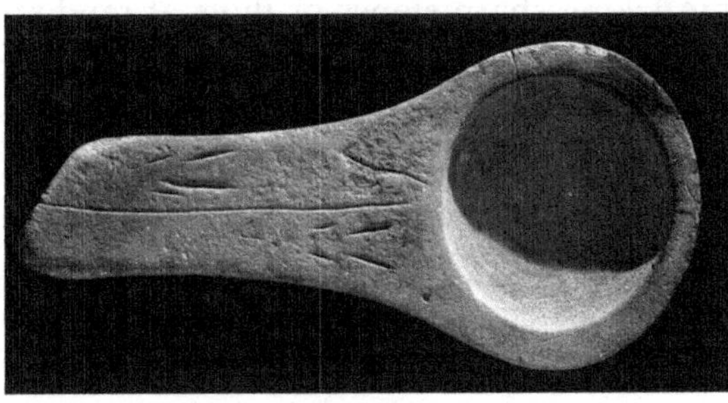

The idea of cave-dwellers still tends to evoke semi-animal notions of our ancestors' life, huddling together and warmed at best by a smoky fire. Perhaps a telling clue to the reality is the realisation that they were lit by oil-lamps, many of consummate artistic craftsmanship, such as the famous Brûloir de Lascaux shown here.

Oil Lamps of the Palaeolithic

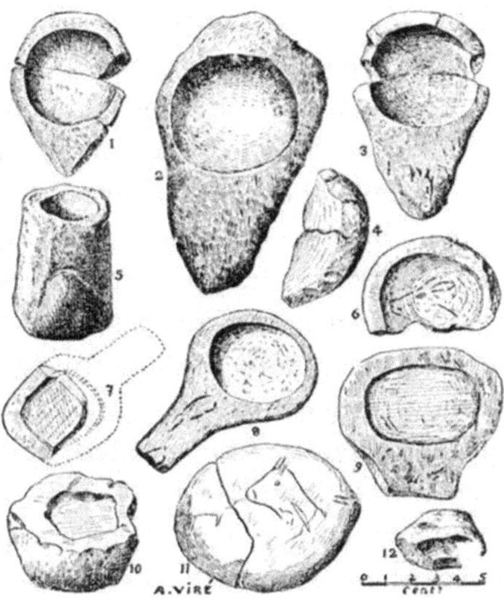

Other lamps from the Upper Palaeolithic show similarly fine workmanship, and are often decorated on the underside with animal engravings, ibex-horn patternings, etc.

1 La Mouthe (Dordogne) Emile Rivière. Sandstone. The reverse carries an engraving of the head of an Ibex. Height 170 mm, width 120 mm.

2 Grotte du Coual (Lot). Félix Bergougnous. 250 mm x 150 mm. Sandstone.

3 Grotte des Scilles, at Lespugue (Haute Garonne) Comte de Saint-Perrier. Sandstone. 190 mm. On the bottom a rudimentary horse's head.

4 and 5 Anval (Puy de Dôme), Dr Baudon. Trachyte, 140 mm and 130 mm.

6 Grotte de Thévenard (Corrèze). Abbés Bardon and Bousyssonnie. Red sandstone. 130 mm. Lightly engraved ruminant.

7 Grotte des Fadets (Charente). Collection Maret. Red sandstone. 50 mm.

8 Bois du Roc (Charente). Fermont. Engraving of a stylised fish. Sandstone. 170 mm.

9 Mouthiers (Charente). Trémeau de Roche-Brune. Sandstone. 140 mm.

10 Pair-non-Pair (Gironde). Daleau. Limestone 110 mm.

11 Grotte de la Mairie, à Teyjat (Dordogne). Bourrinet. Bottom of a lamp of sandstone with the head of a reindeer. 130 mm.

12 Grotte des Harpons, at Lespuge. Comte de Saint-Perrier. Limestone. 60 mm.

Photo and text: *A. Viré, in* Bulletin de la Société préhistorique française, 1934, *tome 31, N. 11. pp. 517-520.*

were no longer just made by flaking sharp stones or flints at random but follow a carefully thought-through plan that produces a well-balanced, functional and beautiful implement, in which the material is pre-selected for size and shape.

From the later Palaeolithic onward, clothing becomes strikingly well-designed and fitted, including tunics with belts, matching boots and headgear, and clearly bore social significance as well as serving to keep warm. Leather was treated by smoking it to render it pliable before use. Plant fibres were also already being woven into fine, linen-like cloths for lighter or perhaps undergarments. Women wore a bandeau round the breasts (the brassiere was not invented until 1913 AD). Evidence of pigments for colouring, and the use of pendants, jewellery, beads, and other accessories tells an obvious human tale. Obviously we must not read modern fashions into the situation, but clothing for status and concern for design clearly indicate reflexive consciousness—an awareness of others' awareness of ourselves. If life still took place partly in caves, finely made and decorated stone lamps were available, sometimes of great beauty. Significantly large numbers of artefacts were produced overall and we may even speak of Stone Age 'factories'; there must have been correspondingly large numbers of skilled and trained craftsmen who manufactured and used these tools and taught others the techniques in 'schools'. They created 'portable art' in ivory, stone and wood: beautifully carved horse-heads, figurines, as well as more utilitarian kitchen equipment.[*] Women apparently designed and made their own clothing, stretched upon tailoring 'dummies', and must similarly have handed down skills and stylistic-ideas to the new generation.[†]

From the later or Upper Palaeolithic, too, burials are increasingly elaborate, involving clear religious symbolism, deliberate positioning of the deceased (expressing *post-mortem* expectation?) and sometimes stunningly rich grave-goods. Indeed on occasion there is a lavish-

---

[*] Anthropologists now admit that in the first pioneering times, after Darwin and especially around the time of the great Wars, far too much of the evidence was interpreted in terms of 'man the mighty hunter'—blades were considered as spears or weapons which in fact belong to preparing and sharing food (i.e. kitchenware): cf. Richard Leakey, *The Origin of Humankind* (London 1995) pp.75ff.

[†] Olga Soffer of the University of Illinois has been a leading researcher of Stone Age womenswear—a subject which she jokingly refers to as fashion B.V. ('before *Vogue*').

ness that will not be surpassed in spirit until the Pharaohs of Egypt, thousands of years later. An Upper Palaeolithic site at Sungir near Moscow revealed the bodies of two evidently exalted individuals, both adolescents, laid out with hands folded. 'One ... that of a boy, was covered in strands of beads, 4,903 in all; he also had a beaded cap. ... Around his waist was the remains of a decorated belt with more than 250 canine teeth of the polar fox; a simple calculation shows the minimum number of foxes (63)—animals that must be individually trapped or hunted—required to supply so many teeth. In addition, the boy was buried with a carved ivory pendant in the form of an animal, an ivory statuette of a mammoth, an ivory lance made from a straightened mammoth tusk, a carved ivory disc with a central perforation, and other items. The lance was probably too heavy to have served a practical purpose. The adjacent burial ... that of a girl, was accompanied by no fewer than 5,274 beads and other objects.'* The archaeologist assessing the site estimates that it took more than 45 minutes to fashion each single bead, so that the beads for the female burial alone took 3,500 hours to create!

It is impossible to doubt that the people who devoted themselves to the making of the Sungir grave-artefacts were expressing the world-view of a whole society, making a statement about their values, the meaning of life and death, and about the status of the young people who had died. To understand the burial, we must assume the use of complex language and complex social structures and loyalties, symbolic conceptions of ultimate realities beyond the here-and-now and the sense of artistic beauty (already divorced from practical utility). 'There is no doubt in any researchers' minds that Upper Palaeolithic people had fully modern language,' says palaeoanthropologist David Lewis-Williams from this and extensive further evidence; they were able 'to manipulate complex grammatical constructions, to speak about the past and future, to convey abstract notions, and to utter intelligible sentences which had never before been put together'—in short to express themselves creatively in thought and word just as human beings do today.†

The impression of a creative human mentality at work is still more deeply reinforced when we turn to cave-art. Discoveries at cave-sites

---

* As reported by Randall White, cited in David Lewis-Williams, *The Mind in the Cave* (London 2002) p.80.

† Lewis-Williams, *Mind in the Cave* p.88.

such as Lascaux in France, and subsequently across Europe and Asia, have literally brought back to light astonishing images of animals such as bovines, horses, reindeer, woolly mammoths or predators such as lions. Though located sometimes far underground, they shine out from the rock face or are cunningly contrived to use natural edges or planes in the rock; they are linked in conception too by the presence of abstract designs, so that we must suppose the whole impression made by the pictures *in situ* was conceived as an aesthetic whole—whether or not it had other purposes. The images themselves are still artistically stunning, and are achieved by technically adept methods of using pigment, burnt-on marks, outlines (made by interposition of the hand, for instance). Picasso felt humbled: 'We cannot paint like that!' he famously said—but he clearly meant to indicate the affinity he felt with those archaic artists who do paint like us, not just as forerunners but as equals or more.

The amazing images seem almost to float off the walls. Their interpretation was long disputed, and fall-back positions were adopted by many: psychologists diagnosed in them Freudian-sexual symbolism, Marxists found in them evidence of the prehistoric class-struggle, more down-to-earth researchers declared they were primitive hunting-magic to ensure those hungry generations caught enough food to survive. But none of these knee-jerk explanations rang true. It is nowadays established beyond any doubt that their meaning is spiritual. The bovines which hover in immaterial space, the abstract patterns which connect and interpret them, the combined animal-human figures which remind us of shamans who take into themselves the animal-powers, and commune with the cosmic beings or the ancestors, are still known in some archaising societies today. They belong to a primordial religion which also played a part in giving early humans a sense of their special powers and dignities, a consciousness of their place in the world, and of God.

Recent researchers have been made sensitive—sometimes even acutely troubled—by the realisation that we are untangling here the threads which are woven together in our own modern cultural understanding of the mind, of our individual value, of our belief in ultimate truths. The inner world which these remote shamans first explored may have undergone many layers of abstraction, of deeper interiorisation, of analytic assessment—but we recognise in it the authentic beginnings of our own mental life. Here we touch on the paradox of similarity and otherness which is the crux of contemporary research into the Upper Palaeolithic, where to our

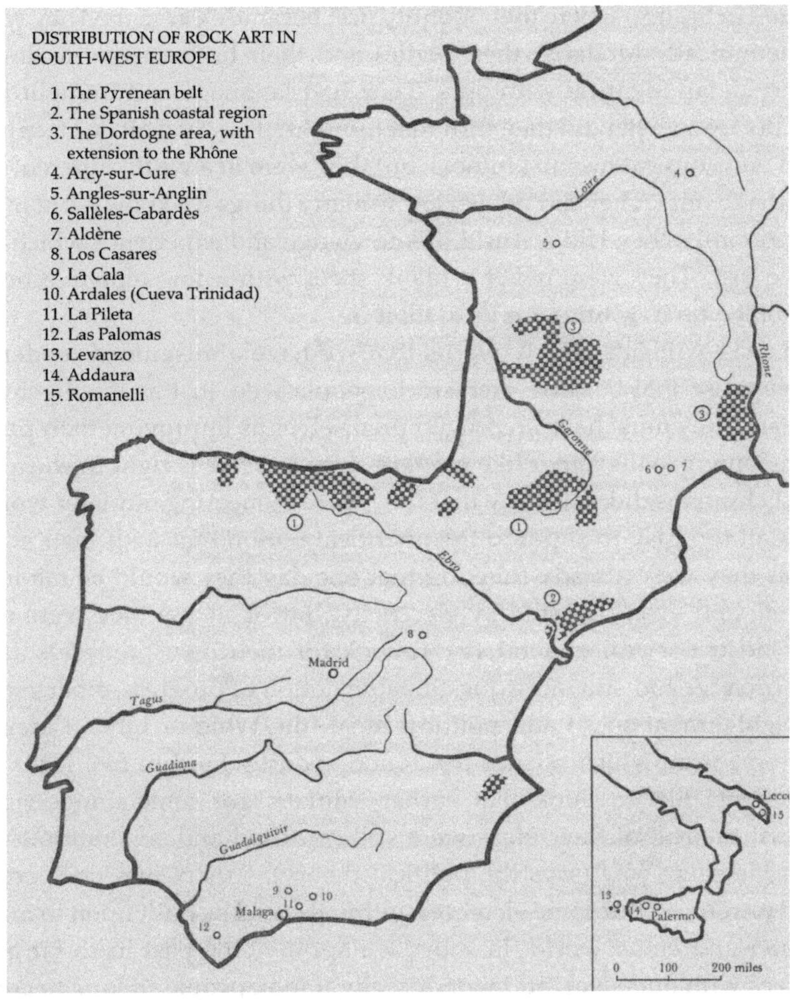

Cave paintings are the most spectacular evidence of early Modern Human culture in the Upper Palaeolithic. This rough map shows their distribution across much of south-west Europe. The Pyrenean belt contains well-known sites such as Altamira and the Grotte de Niaux. The Spanish coastal region near Barcelona is also very rich in locations. The Dordogne area, with extension to the Rhône, includes some of the oldest known and important sites such as the Grottes de Lascaux. 1. The Pyrenean belt. 2. The Spanish coastal region. 3. The Dordogne area, with extension to the Rhône.

ever-increasing surprise we find that we are looking at ourselves in the vast mirror of time. Yet its modality is not always what we would expect, and many domains of experience alien to most of us today turn out to have been instrumental in shaping our own mental worlds. In effect they are redefining the human.

The culturally advanced people of the time, as the world emerged from the Ice Age, are sometimes called Cro-Magnon after the place (again in France) where their identity first became clear to us. They were 'Anatomically Modern': their bodies and their brains were in almost every detail identical with ours. They had language and thought like us. But upon what did they turn that thought? They had all our faculties and our physical resourcefulness, but they were in a vastly different situation to ours. They did not yet know many things we know—but more importantly, they had a world of knowledge and experience which we have lost. They were never a blank sheet with a few tentative notes scribbled on it, waiting for us to fill it in.

Lewis-Williams rightly warns that we have a misguided tendency to suppose that because prehistoric people were just as intelligent as we are, they must have envisaged themselves as improving their problem-solving skills rather like modern scientists. He is right to warn us. Early humans did not know that they were pioneering our later world. It is all too easy to slip into the uncritical notion that with their stone tools they were already thinking that one day they would be made of metal, and even powered by electricity. But of course they were not. Behind the scientists' tendency to look for their own predecessors is the broader and still more questionable notion of progress, a persistent Enlightenment social and political ideal (the Whig or Liberal idea of history) from which science has found it oddly hard to free itself. Its adherents like to think that earlier thinkers got some things right, whilst the rest of their ideas were still confused and superstitious. In just the same way (says the political theory), society was once crude and unrefined, but some elements within it could be built upon to make a later and better world. In short, we approve the past in so far as it agrees with ourselves, and wonder why it took people so long to come round to our point-of-view. Previous ages could not have had such a perspective, however, and did not know they were supposed to 'make progress'; they made sense of their world in a manner that belonged to them and their time and place, which presented themselves as true and real. It is certainly true that a continuous development links them with ourselves, but we need to let go of our privileged position. It is not so much ideas or intelligence that is at stake here, as an understanding of their different consciousness.

Assessments of human evolution based on prehistoric intelligence, recent researchers suggest, have distracted attention from this real issue of *consciousness*. 'The emphasis on intelligence has marginalized the importance of the full range of human consciousness in human behaviour.'* It is simply not the case that an anthropoid ape which surpasses a certain level of intelligence is at once human. Some animals show considerable intelligence. The way human beings have come to manipulate awareness, of themselves and objects, e.g. through images, is more crucial to understanding them, however. (Some animals are even clever enough to work out that their reflection in the mirror is themselves—but they do not make that the starting-point for a way of placing themselves consciously into the world.) If we think about early humans in this way, there is a possibility we can then gauge the different direction in which they utilised their thinking—which may surprise us. Rudolf Steiner offers here an invaluable further dimension to the picture of early man, with his groundbreaking ideas about the evolution of consciousness, on which we shall draw extensively in shaping our presentation. He boldly broke with that simplistic view of former times and their world-pictures, according to which they possessed some small nuggets of genuine knowledge surrounded by a mental fog, which piece-by-piece grew into the extensive knowledge we enjoy today. He grasped the truth that former times could think as well as we, but within different frameworks of consciousness.

These did indeed evolve into ours, but at every stage each culture had its own point-of-view and its own validity. The child is forerunner to the grown-up; yet the child is never incomplete, but in its own way harmonious and complete. The child's way of seeing is also harmonious and self-consistent (again Steiner's pioneering view) as was the consciousness of archaic peoples. And Steiner was able to deal equably with a component in ancient consciousness which still causes the modern palaeoanthropologist much soul-searching, namely the important role within it of what may be called visionary or ecstatic (clairvoyant) states.

Enlightened scientific thinking is supposed to distance itself from content that derives in any way from states other than rational consciousness; but the evidence of their importance is pressing itself upon research into early humanity with ever increasing force. Of course, such states can be denigrated because they are sometimes based upon psychotropic substances. But in fact they can be evoked in a number of contexts (initiatory ordeals, spiritual exercises, techniques of sensory deprivation or just strong expectation), but all of them far removed from the vacuous drug-culture of the affluent West; the availability

---

* Lewis-Williams, op. cit., p.111; and see pp.92f; 180ff.

of psychedelic indulgences is essentially a modern-day phenomenon.* In antiquity intoxicants might sometimes play an ancillary role. However, it was the more heightened ecstatic and systematically cultivated states which were valued in the ancient world—for a number of reasons, but fundamentally because in the charged situations in which they were employed (perhaps divinatory or oracular, seeking visions) they facilitate extraordinarily strong impressions, even to the point of full identification. When in an ecstatic condition the mind is without many of the psychological barriers which we develop either as we grow into adulthood or through the increasing rationalisation of our culture. It is enabled on the other hand to receive the stamp of powerful experiences, and also to assimilate large amounts of material or memories, in a way beyond the resources of the rationally conscious ego. It seems also that the mind expresses itself then in potent visual or other symbols.† Ancient cultures made significant use of such possibilities, and we may certainly trace them back into remote prehistory. In ages before there were written records or even oral stories, it was necessary for at least some individuals to identify intensively, even totally with the body of knowledge (practical as well as social and spiritual) which enabled a special pattern of life in a landscape, climate which perhaps formed the basis of a group's shared self-understanding.

Rudolf Steiner has enabled us to understand most profoundly how these states of consciousness played a part in early humanity's development. They are related to childhood modes of pre-self-conscious awareness—especially infantile states, in which of course we are imbib-

---

* See the critical discussion in Violet MacDermot, *The Cult of the Seer in the Ancient Near East* (London 1971); on the modern drug-aspect, pp.234ff.

† Lewis-Williams, op. cit., pp.129-30. A rare direct comment on the working of the Greek Mystery-cults of ecstasy comes from Proclus the Neoplatonist, who was well-informed. The rites, he says, activate an affinity with the things done 'which is beyond our understanding, and divine, so that some of the novices are stricken with panic, being filled with divine awe; others assimilate themselves to the holy symbols, leave their own identity, become at home with the gods, and experience divine possession'—cited in W. Burkert, *Ancient Mystery Cults* (Harvard 1987) p.114. It is also increasingly clear that the emergence of reason was not a negation of these archaic mental states, but a further stage in the same process of utilising them. Steiner argues this for pre-Socratic philosophy in his *Christianity as Mystical Fact* (New York 1997) pp.22-39; cf. now M.L. Morgan, *Platonic Piety* (Yale 1990) who argues similarly for Socrates as effecting 'a rational transposition of ecstatic rites'. This became what we call philosophy.

ing and assimilating huge amounts of basic knowledge and adapting to them far more effectively and fluidly than later in life. How much more difficult it is to master a language when we are grown-up, or even as an older child! Knowledge and psychological, or actual biological growth here still stand closely connected. Steiner is far removed from the superficiality of considering consciousness to be like a camera looking on: we are rooted behaviourally and biologically in the world we grow out of, and we relate sensually and emotionally to it, all at a lower level of awareness, before we refine and bring into focus our intellectual consciousness.* And only very gradually have we evolved to the highly structured ego-awareness we now so easily take for granted. In prehistory, such ego-awareness can only have been in its flimsiest beginnings. It is the relative weakness of the ego in humanity's early evolution which gives the keynote to understanding archaic consciousness.

Steiner wrote:

> Men of earlier times do not as yet separate their own soul-experience from the life of nature. They do not feel that they stand as a special entity over against nature. They experience *themselves* in nature as they experience lightning and thunder in it, the drifting of the clouds, the course of the stars or the growth of plants. What moves man's hand on his own body, or places his foot on the ground to make him walk, for prehistoric man belongs to the same sphere of cosmic forces that brings lightning, cloud-formations and other outer happenings.†

---

* Steiner, *The Spiritual Guidance of Humanity* (London New York 1970) pp.3ff stresses the intensity of the development of the child in the three years or so before memory, relating to self, begins to be dominant. In this phase, mental development is closely bound up with organic growth. Indeed the two only separate fundamentally with the change of teeth around the age of seven.

† Steiner, *The Riddles of Philosophy* (New York 1973) p.13. Steiner has further characterised ancient man's mode of perception, which was less analytical and still strongly bound up with a more undefined self-awareness: 'On approaching a lake, a feeling arose in him, merely through looking at it, that was like a taste of what lay in front of him physically, without his having drink the water. Simply through looking at it he would have felt whether the water was sweet or salty. It was not at all like our seeing water today. We see only the surface and do not penetrate into the inner qualities. But while a dim clairvoyance still prevailed, the man who approached water had no alien feeling toward it. He felt himself as being within the properties of the water; he did not stand over against the object as we do; it was as though he could penetrate into the water. ... Man was, as it were, within the whole, and he perceived things as though they were ensouled. ... Every-

We can illustrate his point from many known archaic cultures which have survived into historical times, where it manifests itself quite practically. Traditional societies and individuals do not act out of their own decision-making or set their own goals for life, but characteristically align themselves with a cosmic moment particularly when they undertake some new initiative—on whatever scale, whether it be a dawn, a solstice/equinox, or the observation of a heavenly body that marks a rhythm which carries some significance in their world. As a result, they do not feel that their own resources alone are the basis of their action, but they act with the force of the cosmic event.[*] Survivals of similar attitudes in advanced cultures are not hard to find: the Romans believed their general Camillus won his victories when he attacked the enemy with the rising sun and the dawn wind aiding him—features that reveal his story to be 'mythical', yet a genuine expression of a world-conquering culture's self-understanding.[†] Steiner points out that elements of such experience still persist, and in fact reappear in imaginative literature's response to nature, for example. One might equally mention those 'genius' moments in science which facilitate discovery, when everything 'comes together'

---

thing revealed something to him. He could feel himself into the interior of objects; he experienced their inner being. Nothing appeared to him as a soulless object in the modern way' (Steiner, *Egyptian Myths and Mysteries* (New York 1971) p.30). Modern children echo that archaic experience, even in simple ways such as running or blowing with the wind, turning with the bus, etc. Psychologists still often speak of them projecting their self into things in a confused 'animism', but Steiner insists quite rightly that it is because the child has not yet developed a separate self. We have a vestige of a primary mode of experience here, not a secondary mix-up: 'Modern thinking has produced the concept of "animism". A child bumps against a table, and strikes it in anger. We say today that the child credits the table with life, imagines it to be alive, dreams life in to it, and strikes it. Now this is not the case.' The child does not yet limit the pain as applying only to itself as a living thing. 'Since everything is without life for the child, he treats the living and the lifeless in exactly the same way' (Steiner, *A Modern Art of Education* (London 1972) p.116).

[*] Cf. Mircea Eliade, *Myths, Dreams and Mysteries* (London 1968) who speaks especially of alignment with a 'primordial situation', pp.158ff; pp.180-1; etc.

[†] Georges Dumézil, *Mythe et Épopée* (Paris 1995) pp.1165-87. It is well known that the Romans divided their calendar days into favourable and unfavourable, *fas* and *nefas*, on which various works could or could not validly be done. Original astrological religion, before it became systematised, must also be understood as an archaic mode of experience, based on living out of the changing patterns of the sky.

rather than seeming to be worked out (even if the Newton-and-the apple story is apocryphal).*

The farthest intensification of archaic consciousness (ecstasy), however, was undoubtedly cultivated already in prehistoric times among prototypical shamans and 'charismatic' leaders, who merged themselves mystically with the cosmic order in initiation-practices. They were enabled to take the impressions of these forces deeply into themselves in what today we should probably think overpowering and quite likely traumatic experiences.† When they subsequently translated these

---

* Inevitably there is Goethe; but we might also mention Wordsworth's 'mountings of the mind' which 'be they what they may ... are yet the fountain-light of all our day, the master-light of all our seeing'; Wordsworth's way of looking at nature has become almost universal in Western culture today. Scientific discoveries may come to a discoverer as dream-like images (the benzene ring of Kekulé being a famous instance), or as breakthrough insight like Einstein's relativity, arrived at holistically, outstripping any step-by-step argumentation. Evolution itself is a case in point.

† Traditional ecstatics typically describe their calling as 'being pursued by the spirits', to which their human identity has finally to yield, allowing them to enter in and take possession of the psyche. But these spirits are none other than the powers which shape and direct the life of nature and the social order nested into it: cf. I.M. Lewis, *Ecstatic Religion* (Harmondsworth 1971) pp.45-57; and, for the cosmological dimension, pp.149ff. Popular notions of 'possession' preserve a dim recollection of a time when mental life did not stem from one's own effort or ego. The impress of cosmic patterns from stars or landscape or animal movements had the quality of 'ideas'. Steiner rightly points to their cognitive and spiritual character, yet by our standards they are factual, objective, i.e. not arising out of ego activity (*Riddles of Philosophy* pp.14–15). We moderns are trained to say: well of course they smuggled in their own subjective ideas (superstition). But why? Out of fear?—but we shall see that anthropologists characterise archaic hunter-gatherer communities as 'fearless societies'. At any rate it is simply true that, until early Greek times, and frequently even then, the experience of knowledge belongs to the gods and is manifested to human beings only at their pleasure, and *ab extra*; it is for the first time with Xenophanes, an early Greek thinker in the sixth century BC that 'we have the notion that men acquire their knowledge through their own striving'—Bruno Snell, *The Discovery of the Mind* (New York 1982) pp.139-40; Robert Lamberton, *Homer the Theologian* (London and California 1989) pp. 2-12. In modern psychology that moment counts as 'the end of hallucination' (discussed by Lewis-Williams p.287) but that term covers so many sorts of images from dreams, visions, revelations, symbols. Is it not we who are lumping together everything that is not *our* experience, as if ancient humanity should have known our viewpoint but somehow did not? Our strict (or is it perhaps philosophically dubious) attempt to draw a strict line of demarcation between mind and world is no older than Descartes (seventeenth century AD).

into mythical pictures and symbols, they give us an inkling of the way that prehistoric humanity probably experienced their place in nature and their struggles—less a little patch of human control in an alien, hostile world, and more like participating in a titanomachy or battle and victory-celebration of cosmic powers. They struggled and triumphed along with superhuman presences. If we follow recent interpretations of the famous Lascaux cave and those who contemplated its images deep underground, mentally they were travelling among the stars and the cycles of time.* The cave-images, as we shall see, are to be seen as pressing through from the Beyond. Even the later mythologies that have come down to us generally convey an unparalleled sense of the richness and greatness of life, with an immediacy which may well be the envy of modern individuals and our whole civilisation, where the possibility of such full participation is largely lost, leaving us isolated and impoverished, despite our intellectual achievements. We shall explore in more detail later the evidence for the role of ecstatic participation in Upper Palaeolithic culture. We must assume at least that the experiences of early humans resembled those which produced the surviving archaic mythologies, or that they lived them perhaps even more intensely.

Rudolf Steiner, then, did not agree that prehistoric man possessed a tenuous patch of realism amid mental confusion. It was rather that in the absence of developed subjectivity like ours, they allowed environmental experiences and cosmic patterns to determine their lives. The cave-symbols with their shamanic implications indicate that rather than trying to anticipate our form of consciousness, they worked systematically to intensify states in which cosmic patternings could penetrate deeply into them—much more deeply than we would tolerate, since we protect our own ego and like to picture our self as facing the world-out-there as if on equal terms. We may indeed have gained much in so doing, on a

---

'Hallucination' is best reserved for instances of a mistake concerning our own categories—e.g. where we think we saw a person physically present who was not, etc. Rudolf Steiner, most significantly, integrates clairvoyance into the history of our own mentality.

*See the brilliant (and stunningly photographed) work of Norbert Aujoulat, *Lascaux: Movement, Space and Time* (New York 2005). The ability of early humans and of very young children to identify with the breadth of the sky is well documented, and deeply grounded in the psyche. Karl König, *The First Three Years of the Child* (New York 1969) p.49 observes: 'the smaller the child the larger he experiences himself in relation to the world of space.' Fitting oneself into a structured finite earthly space is a later achievement—and a construct!

personal and historical plane, in terms of knowledge and power over the world; even so, ours is but one way of relating to the world, evolved from the earlier ones. The intense patternings of early man's primary experience have been tamed as 'ideas' in the mind, domiciled within us. For ultimately, Lewis-Williams and others have reminded us, our inner world is itself a descendant of the one they discovered. It is just that their consciousness was turned another way. Modern science makes it conceivable that they were thereby in touch with a 'cosmic consciousness'(so-called 'panpsychism'). Rudolf Steiner took the radical step of proposing that if we are to comprehend the factors and the elements of the world which determined the emergence of humanity and its earliest development, we may have to transform our own consciousness again in order to comprehend concretely the features which really affected our ancestors' life. … But we are perhaps running ahead of our story.

Yet of one thing we may be clear. As the last Ice Age waned, we find human beings in Europe, and to some degree in Africa and Asia, who are living an impressively civilised life, quite different from the vulgar notion of cavemen. We find people who are physically and mentally equipped like ourselves, if differently focussed. The decipherment of the cave-symbols indicates the source of their conceptions in Mystery-rites and visionary states. We ask all the more where they and all this come from, and how does it fit into the broader picture of man's evolution?

It appears these 'Anatomically Modern Humans' emerged out of Africa, but the sudden cultural maturity we encounter in the Upper Palaeolithic, a whole world 'venerable and articulate and complete' (as the poet said), was a blossoming that defies any easy explanation. One still widespread current view claims it is the expression of that new type of human, *Homo sapiens* first showing what he/she could achieve: with their physical dexterity, large brains and a sophisticated ability to deploy consciousness, they developed much we still recognise in ourselves today, and through their trance-explorations they mapped out an inner world. They did so in a 'cognitive leap' or 'revolution', an 'explosion' to modernity.* Then when the last Ice Age went into recession and

---

* The view is enthusiastically summed up recently in Y.N. Hariri, *Sapiens: A Brief History of Humankind* (London 2011) pp.1-69. Criticisms, suggesting the explosion really had more in the way of preparatory developments, in Lewis-Williams, op. cit., pp.69ff; C. Finlayson, *The Smart Neanderthal: bird catching, cave art and the cognitive revolution* (Oxford 2019) pp.7ff, p.62, etc. But one must beware of the 'academic fallacy'. Take a case in recent cultural history, when the eighteenth-nineteenth century Romantic Movement revolutionised art,

our present climatic phase began, they began an extraordinary journey from Europe right across Asia and on into the farthest recesses of the inhabited world. We are all their descendants. Was it really quite so simple? How did they come by their astonishing modernity? And why are there so few signs of it in the making?

These ancestors of ours are extraordinarily engaging, and to make them unique fits many of our cultural assumptions both from the Bible and from older science which we have inherited. There have long been, however, other lines of thought which have referred to very ancient civilisations—on some of them, indeed, thinkers such as Rudolf Steiner drew, though he always pursued his own critical line of thought. Such things are evoked in Plato's myth of Atlantis, and other pre-scientific ideas. But modern discoveries indicate there could be something valid there also. After all, we have quite recently come to understand that continents change and break up, indeed come and go, over geological ages of time. Also, we have realised that there were other humans before, and different from, ourselves.

Before we can even start to join up the dots, therefore, we must make the acquaintance of another startling prehistoric character from the same, or even a somewhat earlier epoch. Modern research into the earliest human beings did not start by finding the Cro-Magnon type of people from the Upper Palaeolithic, in what has turned out to be so like an encounter with ourselves in a misty mirror. It had previously come up against a disturbing, even potentially terrifying remnant of early humanity, who is still hard to assimilate today. It encountered a man indeed, or his fossil remains in an unsuspected cave—but he was emphatically not one of us.

The Neanderthal's Gift

It is 1856, and on a sleepy estate in a district near Düsseldorf the workmen are clearing a small cave of debris, when they pull out some unexpectedly large bones including part of a sizeable skull, along with hefty thigh and rib-bones—doubtless, as they remark, the remains of some cave-bear who spent his final hibernation there and never woke up again. They set

---

poetry and music. It swept Europe (and beyond). However, academic research can find every specific feature in the time before it was supposed to happen, and some conclude it never did. Yet events and the mood of the French Revolutionary period caught it all up and overwhelmed people with its new potency.

them roughly aside and continue working. But they know that the local schoolteacher is a keen amateur naturalist with a cabinet of curiosities, and in due course they offer their find to the startled young man, who knows instantly that the oddly disturbing remains are not those of a cave-bear. They are hard to recognise, looking rather human in some ways but in others decidedly not. He notes the thick mineral deposits partially encrusting the bones and realises with equal rapidity that they are not from a recent inhabitant of the cave at all, but are extremely, perhaps fabulously old. He asks to be shown exactly where they were found, and enquires (vainly) about any other bones, objects or artefacts that might have associated with the discovery. Feeling somewhat out of his depth, he takes them to the University of Bonn to have them properly identified, and the mystery of their significance only becomes deeper. Professor Schaaffhausen gives his opinion that these fossil remains exhibit 'a natural conformation hitherto not known to exist'. At a joint conference with the schoolteacher, he announces to his colleagues that the specimen shows no sign of pathology, except for a broken left arm; otherwise it exhibited no hint of deformation but represented a normal human being of a type not previously seen. It is extremely ancient, 'antediluvian'. Its anatomy has no exact parallels in any living human being.

Sadly his colleagues were not in a mood to be persuaded by what we now know to be his fundamentally sound conclusions. Instead, all kinds of far-fetched explanations were immediately put forward, and long continued to dominate the arguments over this most spectacular find in palaeo-human archaeology. After all the discovery was staggering. The remains were those of a man, but such a man as no one alive had ever seen and who challenged everything we thought we knew, with our biblical and scientific culture, about our own uniqueness and our special place in the universe. Scientists are no more exempt from anxieties and troubled feelings than the rest of us, and the innocent remains of a prehistoric man soon became the vent for all kinds of denigration, disgust, horror, resentment and distorted human pride. This figure who was emphatically human but sensationally not one of us had to be dealt with somehow, and our own claim to be alone truly human, rational, inwardly and outwardly beautiful, must at all costs be reasserted. By a perverse logic, therefore, this upsetting and unwanted fellow-human who had emerged from the cave must have been everything we are not (or claim not to be): brutish, violent, ape-like, lumbering, stooping, ignorant, clumsy, selfish, hairy, earth-bound in gaze and thought, low-browed because low-spirited, grunting and inarticulate.

The famous image of the 1856 cave-man created by Marcellin Boule is a masterpiece of irrational propaganda posing as scientific reconstruction. But politicians (and other people) are aware how difficult it can be, once an argument is tripped onto the wrong foot, to regain once more the elusive trick of sounding plausible. If once caught off balance and brought down, even the victim's most positive achievements will be turned into embarrassments, bold assertions will sound like insecure half-truths, and can all be effortlessly exploited in the hands of a skilled opponent. The mute remains of the man from the cave had no chance to make retort. Everything was turned against him: if he had a human form, it must have been a slouching and simian parody of ours; if he had, as it turned out, a bigger brain than we do—that only showed what a big brute he was. Popular and sensational literature has recently added the charge of cannibalism; a discussion of Neanderthal attitudes to women confides to us damningly that not every encounter need have been rape!* *Homo sapiens* has kept the theoretical upper hand ever since, but it is remarkable how the evidence perversely might seem to suggest the opposite. No put-down has ever proved quite final. Popular and technical studies on Neanderthals alike continue to use terms such as 'mystery' and 'enigma', or speak of the 'search' for the truth even as decades of research turn into centuries. Two recent experts comment drily that there is simply no evidence for a stooping, ape-like posture. And the fossil's embarrassingly large brain keeps on putting the question mark over our superiority, since evidence of his prehistoric subtlety and advanced cultural level constantly accumulates. Indeed, as we shall soon see, the man from the cave is finally staking his claim.

The estate where he was found had long been owned by the Neumann family. But affecting a more elevated, cultured lifestyle and notably

---

*Cannibalism is of course another stock-in-trade of paranoia. Here the vacuousness of the supposed evidence is exposed by E. Trinkaus and Pat Shipman, *Neandertal. Changing the Image of Mankind* (London 1994) pp.254-5. A broken Neanderthal skull found at Monte Circeo suggested to one person's fantasy a ritual murder; the stones supposedly placed significantly ringing the skull are just the stones which the illustrator chose to show, not a distinct pattern at all. A site in the Ardèche with human remains is also unconvincing on close examination, since they are not edible parts. We do not know their significance, but Ludovic Slimak, *The Naked Neanderthal* (London and elsewhere 2023) p.96 concludes: 'The more we explore the exact context of these … events, the less *raw* evidence of subsistence cannibalism we find.' See his Ch.3 for the whole story. Treatment of women is another free-for-all. Even where the aim is supposedly to exonerate, we find terms which are culturally loaded and inappropriate, i.e. say more about ourselves than past reality. There is absolutely no evidence Neanderthal men (or women) were sexually violent.

H.G. Wells helped to perpetuate the idea of the Neanderthal as bestial, hairy and stooping, a reconstruction for which experts now agree there is no evidence. This drawing by J.F. Horrabin was included in his Outline of History *(1920). That the interpretation even of the early fossil finds could have led to a different view is shown by the impressively human, intelligent Neanderthal inferred by the respected anatomist Sir Arthur Keith (1911).*

earning recognition in musical circles, they had classicised their German name into Neander, Greek for 'New Man', and the lands in which the discovery had been made were styled the Neander Valley (*Neanderthal*), and the specimen received a name that still resonates down nearly two centuries of study and speculation: *Homo neanderthalensis*, Neanderthal Man.[*]

Since that first discovery, many extraordinary advances have been made. A few even earlier but enigmatic finds could now be identified as Neanderthals. And many, many more were uncovered, predominantly in Western Europe (France, Gibraltar, Germany, Croatia) but then increasingly in Asia from Israel and the Crimea to Teshik Tash beyond the Aral Sea. The finds in France were historically amongst the most important, not least the almost complete skull and skeleton of the Neanderthal unearthed at La Chapelle-aux-Saints in 1908, and a discovery at Le Moustier in the same year which brought to light some of the typical artefacts made and used by the Neanderthals, their industry being given the name 'Mousterian' which is still employed today. Subsequently, the evidence from the Near East was to raise many questions; but it reinforces a now well-established pattern of discovery demonstrating that these people were widespread inhabitants of Europe and Asia during the Middle Palaeolithic, i.e. the time-period prior to the creative 'explosion'; and the settings in which they were found have provided many clues to the environment in which they lived and the way they adapted to it.

---

[*]The full story of the discovery is told very circumstantially in Erik Trinkaus and Pat Shipman, *The Neandertals. Changing the Image of Mankind* (London 1994) pp.3-8, 48-54.

The time falls within the periods of European glaciation, or Ice Ages; but that was not a mere negative. Herded by the advancing glaciers into the habitable zones, an astonishingly rich array of animal and especially bird-species created an environment crowded with wildlife almost beyond what we can imagine in our own devastated world.

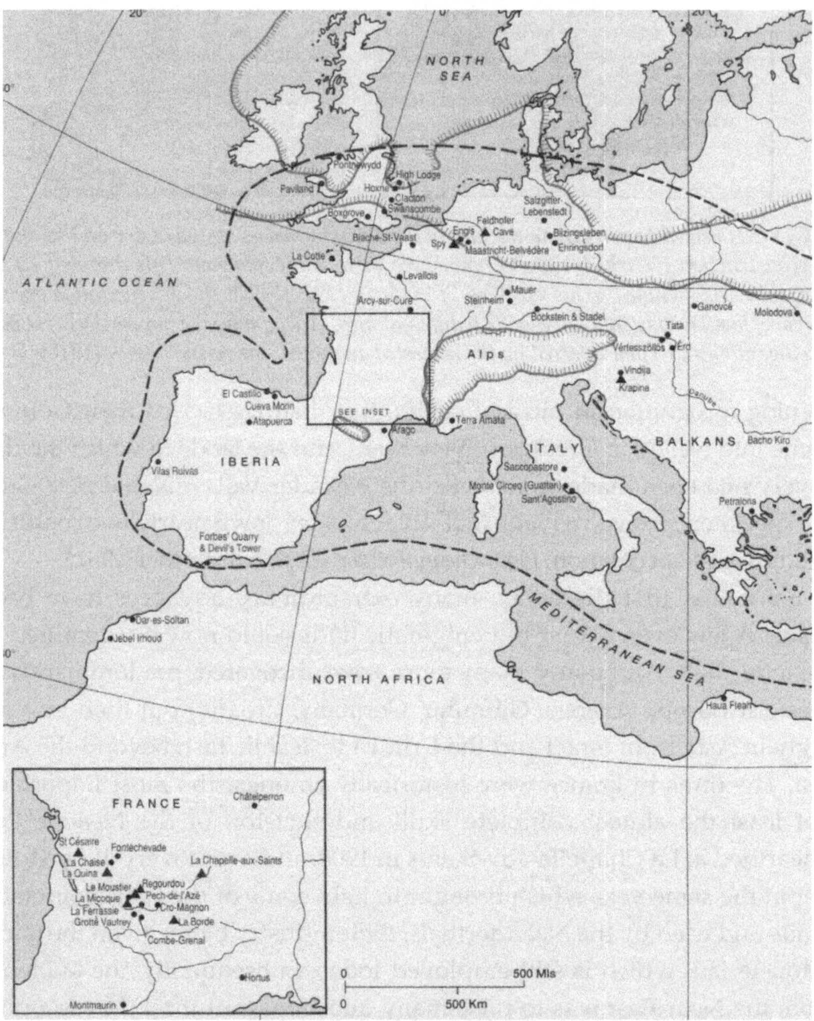

The major Neanderthal sites in Europe. Although Neanderthals take their name from the valley in Germany where their first identification as a species of extinct human was made, subsequent discoveries have shown that they lived all over Europe. Later they expanded into the Middle East and even spread farther afield. No true Neanderthal has been found, however, outside Eurasia—whereas it has become clear that Anatomically Modern Humans reached the rest of the world after migrating out of Africa. Chris Stringer and Clive Gamble remark on the convergence between Neanderthal territory and the later Graeco-Roman world. Drawing Annick Peterson © Thames and Hudson from Stringer and Gamble, In Search of the Neanderthals *1993.*

Recent advances in comparative anatomy and forensic techniques of reconstruction have yielded stunning results: in place of the caricature-images of Boule and his like, we now have expertly built-up three-dimensional recreations of Neanderthal men, women and children. The fossil now in the Vienna Museum of Natural History can be encountered, life-size, as an engaging, keen little boy.* In place of the grunting ape-men who stooped and scratched, we now know about human beings whose upright bearing matches our own, with a physiognomy which, while distinctive, has famously been claimed as within the normal range of faces on the New York subway.† Forensic restoration has also given us back at least one wide-eyed, beautiful girl, and adults with a serious and reflective look quite at odds with the early projections.‡

Early excavators usually laid waste the site from which the Neanderthal remains were recovered, but scientific procedures which sift every detail of evidence from the field-location have progressively refined our understanding of what these people did, ate and planned, how they socialised. We know that they enjoyed shellfish 'barbecues', but also that they went after selected sources of food in their hunting, at pre-decided seasons, rather than being opportunists who killed and ate whatever they came across. They lived in small cohesive communities but must have interacted with other groups too. Before the advent of Anatomically Modern Humans they successfully occupied much of Europe and spread to its southern limits and into the Middle East.

We must naturally presuppose, behind this, significantly developed language-skills. Even so, their speech must have sounded very different from ours, for their larynx was positioned rather differently—

---

* Cf. Clive Finlayson, *The Smart Neanderthal* (Oxford 2020) p.2 fig.1 and pl.1. Though salutary in comparison to the denigrating treatment of Neanderthals previously, Finlayson's giving of homely names to the individuals and linking them in an imaginary family group seems methodologically dubious, however. We do not know how they related to each other personally or if they used familiar names. Empathy should not become assimilation to ourselves.

† See Christopher Stringer and Clive Gamble, *In Search of the Neanderthals* (London 1993) pp.26ff.

‡ There is, however, one memorable exception to the caricatures: reproduced in our illustration on p. 27 which shows a pensive, almost Rodinesque Neanderthal, reconstructed by Sir Arthur Keith in 1911 from his examination (as a professional anatomist) of the Chapelle-aux-Saints discovery.

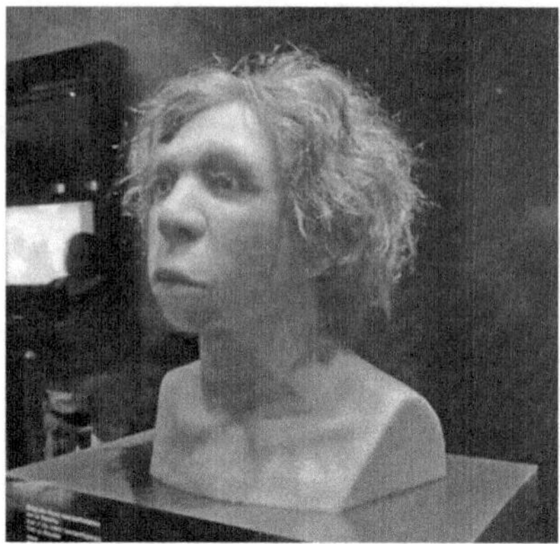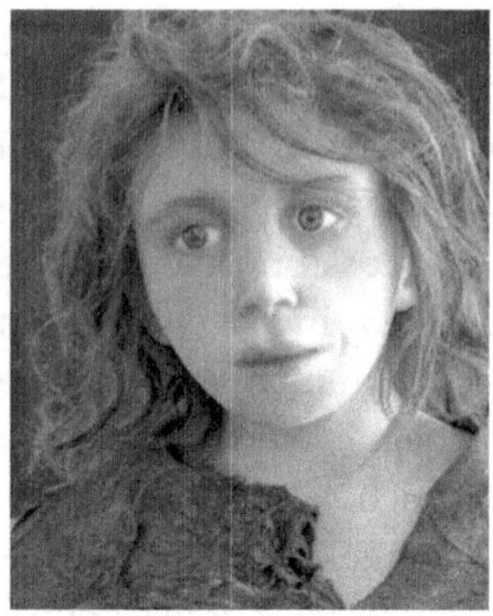

Quasi-forensic methods of reconstruction that are now used to recreate prehistoric human features have brought the Neanderthals vividly before us, enabling us to meet their fellow-human gaze. Even their range of expressions can be accurately assessed through detailed anatomical research. Their often large and expressive faces confront us with manifold emotions.
(Right) The wide-eyed innocence of a Neanderthal girl.

though there are strong reasons to suppose their voices were neverthe-less highly evolved. It is no longer claimed that they were inarticulate. (A brilliant new theory about their speech will be taken up later.) They trapped birds, clearly not just for food but for the harvesting of their feathers which they cut and trimmed. Not that they were naked feathered savages. Their clothing was simpler, it is true, than that of 'post-Explosion' sartorial developments in the Upper Palaeolithic—but after all, it was still the Ice Age. Though evidence is hard to interpret definitively at such distance in time, it is demonstrable that they looked after their frail or injured, and that when they died they buried them with religious ritual, sometimes holding objects which must have had symbolic, spiritual meaning. Indeed it is widely acknowledged nowa-days that the Neanderthals were the first humans to bury their dead. 'One of the most poignant burials,' attests Richard Leakey out of his wide range of knowledge, was that of a Neanderthal: 'a mature male was buried in the entrance to a cave; his body had apparently been placed on a bed of flowers of medicinal potential, judging by the pollen that was found in the soil around the fossilized skeleton.'* The moving

---

* The evidence of the burials was earlier treated with exaggerated scepticism. (Neanderthals should not be so advanced, and the evidence must be 'acci-dental'.) But nowadays all the main authorities consider that the Neander-thal evidence is clearly burial and indicates ritual and symbolic behaviour: Leakey, *The Origin of Humankind* (London 1994) pp.198-9. For details of

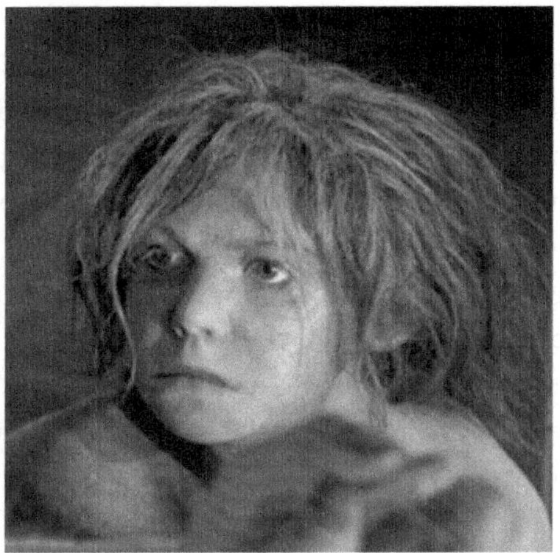

(Left) An adult Neanderthal male, as reconstructed for the 'Human Story' exhibition (Commonwealth Institute) , exudes a warm intelligence.

(Right) Interpretation of the Neanderthals and their lifestyle has recently moved beyond the old minimalist view of their human characteristics, resulting partly from better forensic reconstructions which nowadays emphasise their complex expressions. Not only highly intelligent, argues Steven Mithen, 'they were also intensely emotional beings: happy Neanderthals, sad Neanderthals, angry Neanderthals, disgusted Neanderthals, envious Neanderthals ... grief-stricken Neanderthals, Neanderthals in love.' Shown here is a Neanderthal boy in a pose of fascinated interest (Museum of Natural History, Vienna).

practice of laying flowers at a grave is thus very old indeed. Archaeologists have recently become refreshingly open to recognising deeper meanings in such evidence. It has been acknowledged that among present-day archaic peoples even elaborate ceremonial events, festivals or burials with mythic recitals, ritual gatherings and powerful social import leave behind only the flimsiest of material traces. Resistance, however, has also been long and hard and persists. A whole literature exists which argues that whenever Neanderthals did anything which transcended the well-known 'Mousterian' cultural level, they cannot have created it themselves but must have copied from their superior *sapiens* neighbours. The anthropologists then explain that they simply could never have fully understood it. Yet the momentum of discovery has become inexorable, and hardly a month passes without further evidence of Neanderthal sophistication. There have been some

---

the 'flower-burial' at Shanidar in Iraq, see Trinkaus and Shipman, op. cit., pp.338ff. The pollen-evidence was analysed by Arlette Leroi-Gourhan, but the popularisation of the Neanderthals as the first 'flower-people' belongs to R. Solecki, another archaeologist at the site: see his *Shanidar. The First Flower People* (New York 1971).

outstanding contributions which we shall mention repeatedly, as increasingly we find ourselves ever more on the verge of uncovering what has rightly been called 'the Neanderthal world'. There seems less and less to be a real gap in culture and intelligence between humans and their unsettling cousins. We have been dragged unwillingly to the recognition that we (*sapiens*) were not, after all, the only protagonists in human evolution. A new world of human significances is starting to appear.

'A glance at the Neanderthal world,' assert two of its most notable exponents, 'reveals a size and shape that is reminiscent of the so-called civilized world of Classical Greece and Rome.… The Neanderthal world is therefore very much our own world.' They urge us to engage with our predecessors, or as they more figuratively say, 'enter that "stone with an opening", and an opportunity emerges to undertake a sociological analysis that is normally regarded only as the preserve of the imagination.'[*] Their words have proved prophetic, and in the last couple of decades the Neanderthal world has increasingly taken on concrete reality for us—not only in scholarly and popular books but even in the columns of the daily press. No longer just fossils, Neanderthals have become people we know, at least through a mist of strangeness which we should also not forget.

For there has been since the first momentary shudder of discovery a disturbing difference about the Neanderthals which we do not find when we turn to the Upper Palaeolithic *Homo sapiens* people of Europe with whom we began. With them we can often feel even exaggeratedly at home. Experts who have worked strenuously to empathise with the Neanderthals still usually feel that they were different. The fact of such strenuous empathy is nonetheless an impressive feature of modern research. One of the best researchers on Neanderthals, Steven Mithen, comments on a strangely compartmentalised mentality, which he likens to a mental cathedral with several chapels of thought, but no linking passageways. The 'cognitive revolution' people, on the other hand, seem to be able to hover over one or another field of insight and make it fertile for the whole of their experience.[†] That does not mean the Neanderthals were not just as intelligent; it is again an issue best understood in terms of consciousness. Mithen's brilliant observation is one to which we shall return in a later context, when we return to the insights of Rudolf Steiner.

---

[*] Chris Stringer and Clive Gamble, *In Search of the Neanderthals* pp.8-9.

[†] See Steven Mithen, *The Prehistory of the Mind* (London 1996).

Is that otherness just our phobia about an alien species—and will the continued trend of discovery and revision upwards of the Neanderthals' abilities and mindset bring us to accept them as human? Will we make friends of them? The question has taken a new turn since 2010 when genome researchers compared *sapiens* and *neanderthalensis* and found that, despite decades of scientific attempts to relegate the Neanderthals to a secondary branch-line, unconnected with human development or separate from ours, the evidence was clear: Neanderthals and humans had interbred extensively in the region of the Middle East, and Neanderthal genes are part of our own inheritance.[*] Outside of Africa, i.e. in most of Europe and Asia, most of us have something like 4% Neanderthal genetic make-up. Within its new setting, it is estimated that up to 20% of the Neanderthal genome survives. In other words, we are different—but we are also the recipient of the Neanderthal's gift. And his contribution is in a domain which we long thought was all our own. For it has been established that the Neanderthal strain shows itself specifically in our skull and brain morphology. Recent studies have concluded:

> The associations between Neanderthal sequence variation and co-localized skull and brain morphology in modern humans engender an enduring, living footprint of *H. neanderthalensis*—a residual echo of shared intimate history.[†]

Evidently when our Anatomically Modern Human predecessors were moving into the Middle East and Europe, there was friendly contact and interbreeding, which despite our basically *sapiens* origins was also crucial in making us precisely what we now are: specifically, in our brains. That was in the period immediately prior to the cultural 'Explosion', which we also call our 'cognitive revolution'. All people living today have emerged from that development. Did that awakening itself have something to do with the Neanderthal encounter that affected the shape of our skull, and our brain? Were they our poor relations, or co-partners in creating modern humanity?

For just recently, in 2018, the once unthinkable has happened. Art was supposed to be the dividing line between full humans like ourselves and those (i.e. Neanderthals) who somehow did not quite

---

[*] See Steven Mithen, *The Prehistory of the Mind* (London 1996).

[†] In Clive Finlayson, *The Smart Neanderthal* pp.17-18 citing various technical studies (detail on p.211).

make it. Only the Cro-Magnon forerunners of our anatomy and civilisation could symbolise their inner states, could combine images into meaningful statements and make art. But in Malaga researchers uncovered some of the oldest cave paintings ever found, already showing some of the central symbolism that combines geometry with animal forms—dating from long before the Upper Palaeolithic expansion of Modern Humans. The setting indicates that the paintings were done by Neanderthals, using pigments manufactured from marine shells and minerals, and that there was no way they could have been aping Modern Humans.

The re-dating of some images on the walls of the cave at La Pasiega in Spain shows that they are much older than the Upper Palaeolithic—so much older that, in conjunction with other indications, they must be attributed to the Neanderthals who once occupied that region. In view of the shamanic content in later paintings, it is tempting to connect this ladder-form with conceptions of multiple levels, etc. which are so important there.

Meanwhile in Gibraltar a Neanderthal site produced an abstract symbol carved into the rock. Its engraving had required more than skill and experience. 'It showed, above all else, that the person had cognitive skills comparable to … a Modern Human. That Neanderthal was in every way human.'[*] The wheel has finally come full circle. Still more recently it has been concluded that far from the Neanderthals trailing our forebears in the intelligence stakes, 'in the fields of creativity, *Sapiens* was probably no match for Neanderthal populations and was in all likelihood intellectually inferior.'[†]

---

[*] Finlayson, p.193 and for the Neanderthal cave-art and its dating, his n4 and p.220.

[†] Slimak, *The Naked Neanderthal* p.184.

Though its meaning is enigmatic, the discovery of this inscribed sign at a Neanderthal site in Gibraltar helped force the recognition of their capacity for abstract thought, when their mental abilities have so often been denigrated in comparison with Homo sapiens.

To deepen the mystery further, in the time following the explosion, as the great trek of Modern Humans over to the far reaches of Asia and beyond got under way, the European Neanderthals rather surprisingly and rapidly died out. What really happened? Just when the 'creative Explosion' seemed to have polished our credentials as *Homo sapiens*, the need to fit together our history with these prehistoric people who are decidedly not us has suddenly grown sharper, and the picture once more grows problematic. If it was not that the Neanderthals just could not keep up, what did wipe them out? Is there perhaps a sinister history of genocide concealed in their unexplained extinction? Or do we simply need a new and better understanding of human evolution? One may well be becoming available, as an older established one is crumbling into dust.

An Idea of Darwin's Goes Extinct

A cognitive revolution in our own scientific thinking has been going on since the 1980s, but is perhaps only now coming to a head as these new data challenge us once more. Since then, much that was previously central to understanding human evolution has fallen unceremoniously apart. New questions, as well as new answers, have been needed—with some fascinating results and a large space for new ideas.

Darwin was notoriously reticent about applying his theories to human origins (though many of his followers were not). But he set the stage for a century or so with his idea that man was just a further step on from the anthropoid apes. Put briefly, Darwin suggested that the great African apes included species which had already begun to develop great dexterity in handling food, and even manipulating their environment using simple instruments—such as sticks to catch termites. As they used their 'hands' more and more, they left the trees to seek greater variety in the open savannas, they spent more time walking upright. They abandoned the African jungles for the plains of Europe and Asia. Stone weapons ('technology') were the critical marker of progress, establishing humans in the role of 'the noble hunter' then so important to many scientists. To deal with their increasingly complex and varied activities, their brains grew larger and more intelligent. The result was man. Such was the theory. One day, surely, the precise link between ape and man would be discovered. And then, in 1911 (or was it earlier?) the discovery was made—one that would surely eclipse what many regarded as a wrong turn in understanding the unique being, man, which had taken place in the Neander Valley. This one took place in Sussex, England, under the hands of a little known but somewhat-more-than-amateur researcher (Charles Dawson) who immediately (or more-or-less so) brought it to the attention of his friend in high places: Arthur Smith Woodward of the British Natural History Museum, London. Several other prominent researchers of the day also soon became involved.

The find was indeed stunning. It exhibited just what the scientists had been desperately looking for. The cranium was human, seeming almost modern, but the jawbone was much closer to that of an ape than anything found in man today, except that the great chewing teeth of the ape were much reduced. A number of other bones and prehistoric objects found in the vicinity of the fossil confirmed its great antiquity. The exhumed individual was christened *Eoanthropus*—'Dawning Humanity'—and a wave of quiet but enormous satisfaction ran through the palaeoarchaeological Establishment which did not abate for over forty years. Man's animal ancestry according to the theory had been irrefutably demonstrated, and his emergence located in Europe, in England, in the Home Counties, at Piltdown.

Unfortunately Piltdown was a complete fraud. The cranium looked modern and human because it was; the jaw looked more ape-like because it had belonged to an orang-utan, although the teeth had been cleverly filed down, and the broken bone cunningly stained to match the local gravel like the skull. The prehistoric animal remains had come from Africa. The artefacts later found to be intruded had come from a quite different archaeological stratum. Everybody nowadays knows Piltdown was a hoax, and many assume it was some kind of joke. It was nothing of the sort, but a warped and technically brilliant though insidious deception. 'The fact is that the original fossils were selected and broken with tremendous insight, betraying a keen understanding of exactly which anatomical features needed to be present and which absent to convince.... Indeed the forger covered his tracks so well that virtually everyone significantly involved with the fossil has been accused at one time or another' of the fraud—not excepting even Teilhard de Chardin!* It was meant to establish a view of man's animality, and for almost half a century it held back better understanding of human origins.

Only in 1953 did new scientific dating techniques show that none of the Piltdown component parts could possibly have belonged together. It had given academic respectability, however, to a view which turned attention away from the real questions, including that of the Neanderthals, and furnished much needed evidence for Darwin's favoured supposition. Of course it proved nothing. And none of the real evidence has confirmed Darwin's approach either. Indeed quite the reverse.

The judgement of a mere observer of developments like myself has little weight, but the case against Darwin here is that most forcibly presented by Richard Leakey, one of the best established and fully informed of the professionals in this sphere. The evidence he summarises has carried most present-day palaeo-anthropologists with it.† Darwin's 'package' of ideas on human origins predicted that the

---

* Trinkaus and Shipman, *Neandertals* p.206 and generally pp.199ff; 290ff.

† I summarise here, with acknowledgement of his authoritative position, Richard Leakey's presentation of recent developments in his *The Origin of Humankind* (London 1995). Leakey was another brilliant member of the human fossil-hunting dynasty which began with Mary and Louis Leakey, the explorer-researchers who first showed that Africa could furnish evidence of human beginnings. Richard Leakey in turn has uncovered several important remains, including the oldest complete cranium of the genus *Homo* (op. cit., p.xiii).

qualities which divide us from animals could be explained as evolving from an ancestral ape. The several factors we have mentioned would reinforce and 'leap-frog' one another: as the ape stood upright more, it would use its freed hands more, think more about the range of things it could do with them especially by extending their use through instruments, mainly spears, and get a larger brain to cope with those needs. Walking upright (bipedalism), use of technology and increased brain size should happen all together to bring about a new kind of animal—essentially, modern man. Their linked emergence should show us an ape turning into a human. But the fossil record, mostly uncovered since Darwin's day, shows a very different picture as regards those species most obviously related to us.

The exciting discovery in 1974 of the largely complete fossil skeleton popularly christened 'Lucy' has revealed, not a continuity but a divergence. She was an prehistoric African ape or *Australopithecus afarensis*, of the kind that anthropologists had long assumed would show extreme closeness to the earliest humans. Piltdown had proved a false dawn, but here was a bona fide discovery inviting the application of Darwin's specific evolutionary ideas. Analyses and reconstructions were undertaken. But all did not go as expected, and by the 1980s theoretical horizons were starting to look very different. On the one hand, Lucy had a number of human-looking features, and had clearly been able to hold herself upright. The researcher Peter Schmid set out hopefully to reconstruct the complete skeleton in fibreglass, but was startled by what he found:

> Schmid began assembling Lucy's body, with the full expectation that it would be essentially human in shape. He was surprised with what he saw: Lucy's rib-cage turned out to be conical in shape, like an ape's, not barrel-shaped, as would be seen in humans. Lucy's shoulders, trunk, and waist also turned out to have a strong ape-like aspect to them....*

So an ape that been able to hold itself erect had not, after all, shown any signs of turning into a human, but remained an ape.

Other anatomical studies pointed to the same conclusion. The jawbones of the australopithecines were ape-like, not at all like the ones of the genus *Homo*. (The Piltdown fantasy of ape-jaw synthesised with human cranium was simply not realised.) Comparison of estimated body-weights again highlighted the differences between

---

* Leakey, op. cit., p.71.

the heavy-weight apes and the freely active movement of *Homo*. Most significant of all, a comparative investigation of the inner-ear which reconstructed the little semi-circular canals which are necessary to keep balance when standing or walking upright, only reinforced the conclusion: 'In all species of the genus *Homo*, the inner ear structure is indistinguishable from modern humans. Similarly, in all species of *Australopithecus*, the semi-circular canals look just like those of apes.'[*] Lucy's were no different. The desire of many anthropologists to find prehistoric apes which started to move around in a way just like that of people today had not found anything to support it.

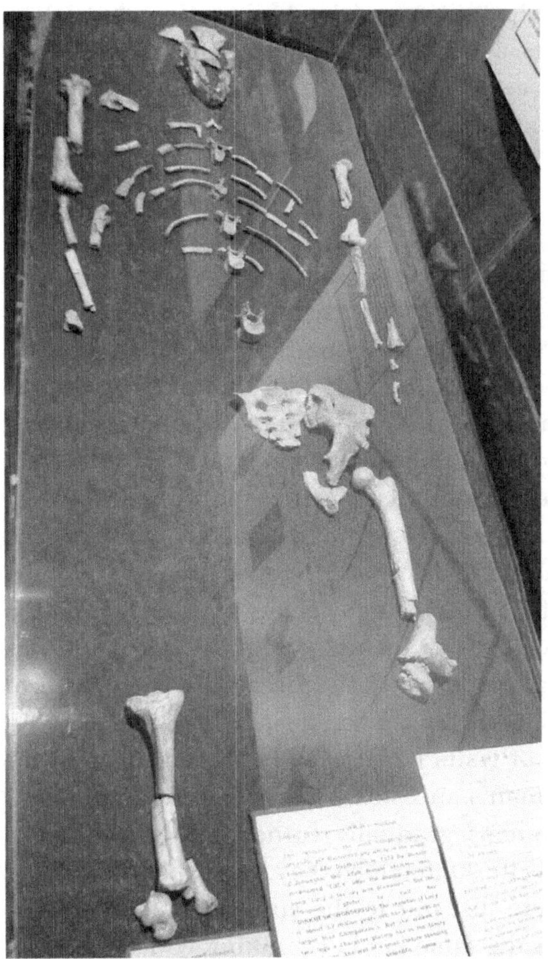

Lucy
Found in 1974, the fossil of an extinct australopithecine caught public imagination as 'Lucy'. At the time her skeleton was interpreted as showing virtually modern upright locomotion, making her a 'missing link' in the transition from animal to human. But later investigations in the 1990s proved discontinuity rather than closeness: her uprightness was not a delicate human balance, but powered by strong leg muscles so that, finally, her evidence is actually 'one more argument for the magnitude of the Homo *adaptation' (Richard Leakey). The only transitional form between ape and human that has ever been found was Piltdown Man—subsequently exposed as an extremely clever fraud mixing human and animal remains. 'Lucy' skeleton (replica), National Museum of Ethiopia.*

---

[*] Leakey, p.47: he summarises here the original research of Fred Spoor.

The evidence actually reinforced the contrast between a lifestyle of heavy engagement with the environment in apes (climbing and swinging in trees, albeit with some upright movement), while in humans we find the ability to move freely and often rapidly in a clear open space. There is nothing to confirm, however, that this stemmed from the throwing of stone spears for hunting, nor that it has to do with a large brain. The great African apes, Darwin's best hope, have now been assigned a separate, diverging branch on the evolutionary tree—not the one leading to humans. Their 'commitment to uprightness' reflected in their different anatomy, freeing movement in humans, seems to be not so much a by-product of activities nudged by technology, or dexterity requiring brain-power, but rather as Richard Leakey suggests, a primary biological phenomenon—from which almost everything else in human evolution will follow. 'We have come to see that *Homo* was a different kind of human right from its first appearance. The discovery of the biological discontinuity between *Australopithecus* and *Homo* has fundamentally changed our understanding of human prehistory.'[*]

The great prehistoric apes and the first appearance of *Homo* take us very far back indeed, of course, long before the mid-Stone Age Explosion, long even before the time of the Neanderthals who preceded it. But in the 1980s the same kind of older thinking which now underwent extinction was starting to give way in the later context too, where similar conceptions had previously reigned. For after long deliberations over whether to admit the Neanderthals into the family of human ancestors, scientists had eventually decided to acknowledge our kinship: but only to stress that Modern Humans had now outgrown their Neanderthal phase, thereby achieving the domination which we still uniquely enjoy. The theory as regards this stage was often called the 'multi-regional hypothesis', because it hypothesised that humans did not originate in a once-for-all development, but in a gradual assimilation of traits which slowly spread through the scattered populations of man's ancestors. Neanderthals and similar near-humans or proto-humans gradually turned into full humans across the range of habitats through which they had expanded. Those that remained as they were lost out to the superior intelligence of the more advanced. It was the same kind of thinking which had supposed that the great bipedal apes had gradually lifted themselves to

---

[*] Leakey,op. cit., p.55.

stooping, then to full uprightness, fiddling with primitive weapons then making spears to hunt with, acquiring brains to think out cleverer strategies to waylay their prey, gradually grunting more subtly then speaking to arrange the ambush. That Darwinian type of thinking had turned out to be wrong once, and it now proved wrong once more.

For one thing, the development of the brain was too complex to be explained by the need for simple tools and strategies. Anthropologists started instead to look at the demands of social and cultural behaviour. The communal sharing of the food rather than the hunting of it began to seem the key to human development—focussing attention on communication and linguistic talents, co-ordination of lifestyle, division of labour, complementary skills. To many it also suggested the influence of 'woman the gatherer' rather than man the hunter. 'The force that seems to have accelerated our brain's growth,' wrote one, 'is a new kind of stimulant: language, signs, collective memories—all elements of culture. As our cultures evolved in complexities, so did our brains, which then drove our cultures to new complexities....' He aptly described the process as 'the runaway brain'.*

Thus the acquisition of large brains would also have been another biological new beginning, distinctive to the social development of *Homo*—not an extension of apedom. For the need to have such a large head even at birth meant that the well-known 'premature' emergence of the human child became a necessity. The child then stays small for a long time, learning and developing skills, language and thinking-power— and puts on that famous spurt to physical maturity at adolescence. A chimpanzee, by contrast, is already much more fully developed when born and after that simply grows steadily until it reaches full size. Childhood is a uniquely human pattern of development, and it requires a completely different pattern of growth, dentition, sexual maturation etc. while the child is learning all the complexities of the cultural milieu into which it has been born. Once again, it is the biological discontinuity between man and the apes, not their similarity, which now strikes modern researchers as significant. Indeed, it is the human's spiritual and cultural dimension of life that has determined the biological patterns that make us unique.

Moreover, the increased evidence derived especially from discoveries in Africa, had begun by the 1990s to lead to a picture of human

---

* Christopher Wills, *The Runaway Brain* (New York 1993) ; and see the discussion in Leakey, op. cit., p.105.

diffusion quite different to the gradualist, 'multi-regional' hypothesis of the transition to advanced humans all over the 'Old World'. It began to be clear that Anatomically Modern Humans had instead arrived in a stream of migration out of Africa. There had been a much earlier wave of migration out of Africa, too, which resulted in the diffusion of 'hominins' across Eurasia; but now there was compelling evidence that humans of thoroughly modern type (*Homo sapiens*) had followed the same basic trajectory at a later date—arriving in Europe around the time the Neanderthals were in occupation, and apparently bringing about the surge to modernity which we subsequently witness. Even as scientists were concluding that gradual improvement of a primitive brain could not explain the extraordinary new concatenation of growth-patterns, intelligence and cultural sophistication we then see, there came also the evidence for 'Out of Africa II' that it happened very differently: through the arrival of modernity in anatomical human form.

But if the Neanderthals were not half-baked Modern Humans as once supposed, they seemed now once more marooned on the margins. For a long time they had actually been of a certain value in the 'multi-regional' view: they had overlapped with the Modern Humans in Europe and Asia, but they had then died out quite rapidly, as if exemplifying the theory that those who did not advance in evolution got side-lined and lost out in the 'struggle for existence'. Some fossils from the Middle East, Neanderthals found in caves at Tabūn and Kebara, more Modern-looking Humans at Skhūl and Qafzeh in Israel, actually appeared to demonstrate the transition. The Neanderthals at Tabūn could plausibly be dated by animal fossils also uncovered there as rather older than the Anatomical Moderns at Skhūl. Analysis of their stone tools supported the relative dating. 'Thus the Tabūn people might represent the ancestors for Skhūl's inhabitants; so the latter would have continued to develop their stone tool-making skills in a regular way as they lived on Mount Carmel,' summarise Chris Stringer and Clive Gamble.* But new dating techniques in the 1980s again overturned the entire applecart. Closer sifting of the animal remains indicated several layers of deposits, not all the same age; and the use of thermoluminescence to track minute changes in the flints and other remains showed that the evolutionary sequence that sounded so plausible was quite out of the question. The resulting shock-wave was huge.

---

* Chris Stringer and Clive Gamble, *In Search of the Neanderthals* p.102.

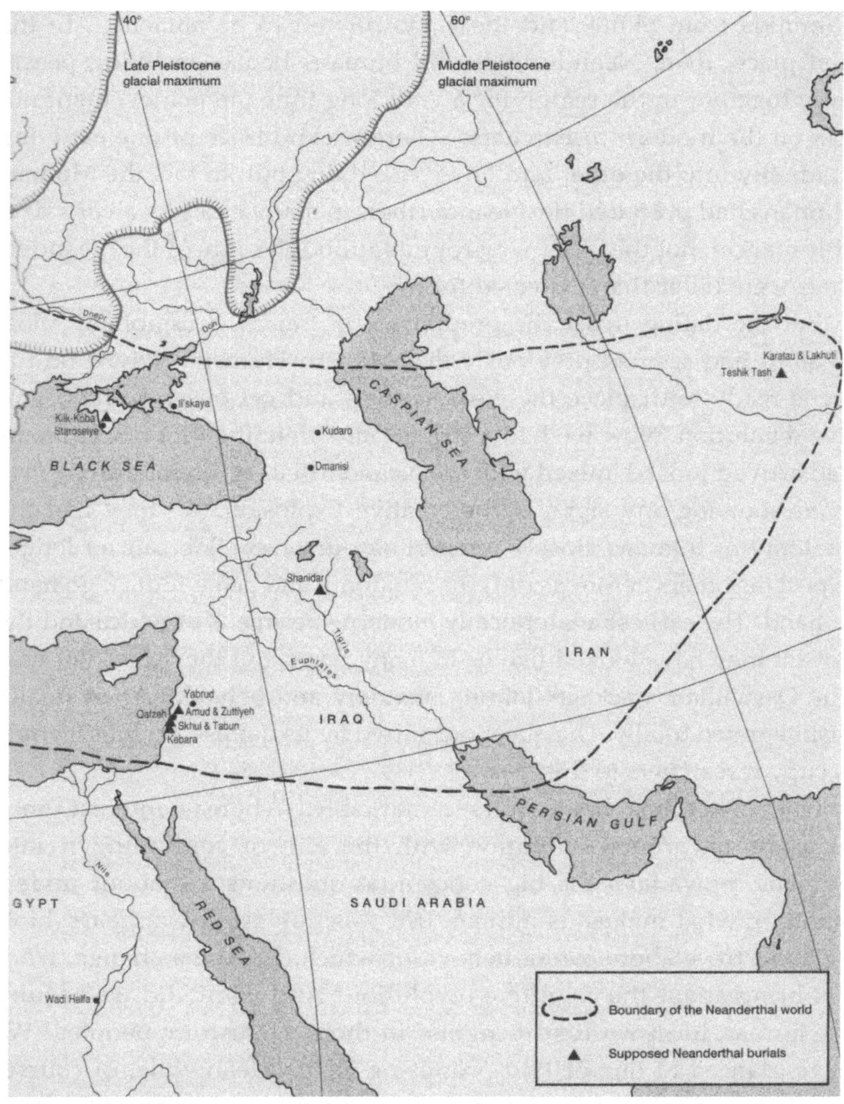

The presence of Neanderthals in the Middle East was long interpreted as showing a step in the gradual emergence of Modern Humans. The modern re-evaluation of human evolution was part-triggered by the overturning the chronology of sites such as Qafzeh and Skhul, showing that the Moderns long preceded the Neanderthals there—and long preceded the cultural achievements supposed to have been associated with their developed anatomy.

Firstly, the remains came from a much wider spread of times than had been anticipated, and in quite a different order—'implying that the Qafzeh early Moderns were not only about three times as old as the Cro-Magnons of Europe, but ... older than the Neanderthal burial at Kebara! ... The next logical step was to test the similar-looking hominids from Skhūl, and these too proved to be ancient.'* In the first place, then, Neanderthals and humans had been living peacefully together in the region for a very long time (an achievement not lost on the modern researchers). There was no trace of one evolving gradually into the other and otherwise dying out. In fact the Modern Humans had preceded the Neanderthals in many cases by a very sizeable margin, not the other way round (although some of the Neanderthals were older than expected too).

But the dating upset the proponents of 'sapiens triumphant' most of all. It had seemed previously that Modern Humans arrived on the scene ready to displace the Neanderthals and get on with the cognitive revolution. Now we had a strange situation: the Modern Humans had arrived indeed, mixed with the Neanderthals (even interbred), *long before* showing any signs of the creative Explosion. Yet they had the anatomy of humans closely resembling ourselves. 'We can no longer expect (as others before us did) that anatomy and behaviour ... go hand in hand. The earliest anatomically modern specimens of Africa and the Levant may have looked like us in many ways, but they were not us.'† The Darwinian 'package' joining anatomy and behaviour had finally disintegrated totally. The questions raised by its demise are those which occupy researchers to this day.

One effect has been quite remarkable. Whereas interest once focussed on animal evolution with the aim of continuing it into humans, nowadays the big conceptual questions are about understanding what makes us human. If it was not, as now appears, biology and the elaboration of behaviour which drove the change, what did bring about the cognitive revolution? And, How did we acquire the human form we first recognise in those prehistoric peoples? We have glanced at one or two promising leads. Meanwhile, of course, the Neanderthals have refused to stay marginal and stepped forward in their own right. We know that what we are is partly their gift. The story of human origins has only grown more interesting and more challenging as time goes on.

Supposing that humanity is just one line has been part of the assumptions that have held us back. Rudolf Steiner, whose ideas on the evolu-

---

* Stringer and Gamble, p.103. Cf. Leakey, *Origins of Humankind* pp.108-111: 'If these results are correct,' he summarises, 'Neanderthals cannot be the ancestors of modern humans' (p.108).

† Stringer and Gamble, p.131.

tion of consciousness have already helped orientate our thoughts, gave an account of human evolution which seems to speak further to the situation. He drew upon and elaborated accounts which described several forms of early man, forms whose inner life was already rich and complex though different from ours; and his underlying question was precisely that of modern research: What is it to be human? Most modern researchers would immediately shy away, of course, from his terminology of 'Atlantis' or the like. Yet in reality modern palaeogeography knows of a time when a great continent occupied what is now the Atlantic Basin. But could it have a link to human evolution? First of all, we should be clear that Steiner's ideas on the subject are not derived from Plato's Atlantis-fantasy, as is no doubt generally supposed: to understand their real significance, we must take something of a detour and recreate as far as possible his exact understanding of the archaic world, even restoring a strangely significant map. We must enter into the kind of investigations he made, which go back before the systematic studies of fossil remains such as we take for granted today, though now potentially coming together with the results of those studies in a fascinating way.* Much in his research, in fact, finds a counterpart in recent developments in science—and it is in those developments, conversely, that may prove the clearest indication of how to understand and apply his highly original ideas.

---

* We can confront the fossil evidence fair-mindedly now. It is true that in years gone by those who followed Steiner's approach were so horrified by the stooping and degenerate individuals restored from study of the fossils that they refused to relate them to his descriptions. However, we now know that the stooping degenerates never existed outside the researchers' own prejudiced minds. The remains themselves tell a different story.

Intermission

MISSING LINKS: INTELLECTUAL REARMAMENT

St Paul tells us to put on the full armour of God, to gird our loins with the Truth, to don the breastplate of Righteousness, the helmet of Salvation and the sword of the Spirit (and a few more things). Our struggle may not be exactly his, but it likewise is not against flesh and blood but is a spiritual one: an attempt to battle our way to a deeper meaning. Blake spoke about 'Intellectual Wars' which are health-giving and forward looking, as opposed to 'corporeal' ones, which are not. We need in short to arm ourselves with some important ideas for the coming effort, and having once found our way into them they will stand us in good stead for the rest of this book, at least. We will need to boil things down to very simple forms—not just for our sake as novices, but because otherwise the more elaborate concepts based upon them will remain unclear.

Evolution

The term 'evolution' is bandied about so much nowadays that we often hardly pause to think what it means. It usually stands for a complex of ideas in science, presented as so inseparable so that we often feel challenged to say on the spot to it, Yes or No. Yet in itself it is not so much a specific theory, to stand or fall, as it is a way of thinking. It is thinking in fluid terms, recognising that living things (especially) do not stay always the same. If we must hazard a definition, we may say that: Evolution is the emergence of one form out of another.* In Nature this normally happens quite slowly, though there are also instances of sudden change. Thinking about changes has for most of human history been considered problematic, even frightening, and only in quite modern science has it come to be seen as providing a solution to understanding things. There is an obvious paradox. As something turns into something else, the category we used to define it ceases to be valid, so that it is quite difficult to pinpoint what it now is and how it is related to what was there before.

---

* We may recall that there are non-scientific uses, as when an army on the parade-ground or in battle practises its evolutions, i.e. shifting from one military formation to another.

How can we pin down what something is whilst saying that it is becoming something else? The very idea of evolution can thus seem irrational or baffling.*

One way of making sense of change is to see *development*: this is where a particular state of things requires some previous stages to be gone through, and the previous stages are in some way incorporated into the later one. Many living things, including people, develop—both physically and mentally. A human adult cannot come into being per se, but develops out of the infant and the child (and biologically, we may go back much further to the embryo, etc.). All human beings follow a basically similar path of development, though in very various circumstances. The term evolution is used where the changes do not just follow a recurring path, but result in a new kind of form, different from those which have been involved in the process before.

Rudolf Steiner was a thorough-going evolutionist. The disturbing aspects of evolution, he pointed out, need not lead to irrationalism but could intimate a higher level of understanding. He built upon the earlier, imaginative but cogent approach of Johann Wolfgang von Goethe to the unity of organic phenomena amid their endless variety, e.g. in studying plant growth (*Metamorphosis of Plants* 1790). Evolution, Steiner believed, provided the fulfilment and justification of Goethe's scientific approach, which had protested against the overly systematic and classificatory science of the eighteenth century. This had merely placed similar things side-by-side, but failed to clarify their actual connection. Goethe himself did not take the step into evolutionism. That was left to Charles Darwin and Alfred Russel Wallace in the middle of the nineteenth century. The tendencies Goethe sought to deflect had by that time led to an unproductive conflict: Establishment Christianity had held fast to the old classificatory, mechanical sort of science in the belief that it alone left room for religion; evolution was therefore opposed by the full arsenal of Church and State, with ultimately disastrous consequences (one may

---

* Such is the frequent accusation against the French philosopher Henri Bergson's enthusiastic *L'évolution creatrice* (1907), which has been characterised as 'the metaphysics and epistemology of change and indetermination'. In it real things are 'centres of indetermination' and vanish as actual events or even memories. How to make use of these ideas is not clear. It is worth mentioning that Steiner could see the value of many approaches, but is unusually combative whenever he mentions Bergson! He considered that his thought repeated the mistake of the late Romantic German philosophers, who rejected thinking for the irrational flux of will, and merely renamed it *élan vital*. Cf. Welburn, *Philosophy of Rudolf Steiner* (Edinburgh 2004) pp.94ff, 195ff.

think) all round. It is important to remember that it was the Church which made of evolution an assault on religion. There is no intrinsic reason why it need be so. In Wallace's case it was decidedly not so, nor in Rudolf Steiner's. (A philosopher I knew used to remark: Things change shape, therefore there is no God—not much of an argument!*)

Yet though it certainly has led to distortions on all sides, there is something more here, in this 'Intellectual Warfare', than the fight against vested interests. Accepting evolution required more than just arguments and ideas. Resistance to it had an element of panic, because it was a new way of seeing as well as thinking; it required a leap for which the validating ideas and explanations have only gradually emerged and indeed are still far from complete. It was a hurdle of consciousness, and this is the real background of Rudolf Steiner's insistence that in order to fully develop its potential, the new science must be based upon transforming consciousness. Only then will it be able to open up the new vistas it promised. After all, looking at a plant *in evolution* meant seeing more than the present physical reality of the plant, meant seeing what it had been and could be, and in a real sense its presence in other plant-types and species. Perhaps here we touch on the hurdle in thought we met in the transition from animal to human: we cannot find it just by cataloguing the external changes in an animal form. We may remember those descriptions of Darwin's intense, one might even say meditative concentration on plant forms which sometimes quite upset people around him (and was the uncomprehending scandal of the servants!). He seemed reaching for a new way of seeing and knowing. Not grasping the profundity of Darwin's gaze, a gardener at Down House where Darwin lived recalled: 'I have seen him stand doing nothing before a flower for ten minutes at a time'—adding with the greatest benevolence that such 'mooning' was not healthy and 'If only he had something to do I really believe he would be better.'† Other brilliant scientists likewise struggled at the threshold

---

* Recalled from conversations with Jonathan Westphal at Cambridge in the late 1970s.

† Quoted in A.N. Wilson, *Charles Darwin: Victorian Mythmaker* (London 2017) p.184; Wilson's book is a robust critical look at Darwin's brilliance and his Victorian blinkers. There has been recently a resurgence of interest in his co-evolutionist Russel Wallace, who linked his ideas with a spiritualistic world-view: see the discussions collected in *Natural Selection and Beyond. The Intellectual Legacy of Charles Russel Wallace* ed. C. H. Smith and G. Beccaloni (Oxford 2008) especially the later contributions pp.253-423. For wider aspects of evolutionism, challenging reductive versions, cf. David Holbrook, *Evolution and the Humanities* (Aldershot

of a new inner standpoint. T.H. Huxley ruefully exclaimed after reading Darwin on the way new species originate: 'How extremely stupid not to have thought of that!'* Once a consciousness-shift is made, it seems so obvious.

Darwin's breakthrough to evolutionary thinking does not mean of course that he got everything about evolution right. He was limited by his own assumptions too. Indeed we have already seen that his 'package' of ideas about man and the great apes has recently had to be dismantled. His external mechanism did not provide the answers; it was a matter of discerning that 'biological new beginning' or, as one might say, a new 'biological imperative'. However, the broader assemblage of ideas based on awareness of change and development has rightly placed Darwin in the role of facilitator to the modern world, where structural and developmental understanding has become key. It was this new, dynamic and transformational consciousness which similarly formed the point of departure for Steiner's basic ideas: he was far ahead in his developmental approach to education, for example, based on teaching what is appropriate to the special organisation of the child's body and soul at each stage.† He was a bold pioneer—and his spiritual emphasis should not distance him artificially from the way that developmental and structuralist approaches have come to transform many domains of thought, from anthropology, sociology, and psychology to biology to mention just the chief. Goethe had paved the way even earlier with his botanical and morphological studies, giving Steiner a launch-pad for his own direction of thought; yet it would

---

1987); assessment of Darwin in the light of Rudolf Steiner's approach in Norman Macbeth, *Darwin Retried* (London 1974). For a positive assessment of just how close Goethe came to evolution, see the discussion by Robert J. Richards, 'Did Goethe and Schelling endorse Species Evolution' at home.uchicago.edu/~rir6/articles/Did. The basic riddle of fixing the concept of species in Darwin, Wallace and modern biology is discussed by J. Mallet in Smith and Beccaloni, pp.105-113. Steiner briefly relates his own early evolutionism and the way it brought him to Goethe's ideas in Steiner, *A Theory of Knowledge based on Goethe's World-Conception* (New York 1968) pp. xiv–xx.

* Quoted in Erik Trinkaus and Pat Shipman, *Neandertal* (London 1993) p.10.

† Not that Steiner considers that education should merely reinforce the developmental stage at which the child stands—education is an art of leading and encouraging, not just of confirming stasis. Steiner made a bold sketch of his approach in the early *Education of the Child* (London 1965) and later contributed many books and ideas furthering the work of the Waldorf Schools—nowadays found all over the world. See especially Steiner, *The Foundations of Human Experience* (New York 1996).

be wrong to look for parallels or convergences with Steiner only when they have been based directly on Goethean scientific method. Recent exciting developments such as the cross-fertilisation of cosmological and biological ideas, for instance, have come about in scientific endeavour to understand the basic mysteries of life, and Steiner's views can be both an affirmation and a further guide to such new ways of thinking. Not that all of them are congenial to all scientists—but it is a healthy sign that scientists no longer pretend that all the answers have been finalised in an orthodoxy. The acknowledgement of significant lacunae and the need to seek answers is the best situation in which Steiner can find a concrete role. Such opportunities were not always so forthcoming in the official science of his own day …. But we should turn now to some of the other new ideas that may give us encouragement.

'Anthropic Science'

In recent years, scientific thinking about our place in the universe—or perhaps the universe's place around us—has been significantly affected by the use of the so-called Anthropic Principle. Even those who criticise it acknowledge that it must be wrong to dismiss it 'glibly, as it has been a central idea for some of the most thoughtful astronomers and physicists of the last decades'.[*] It takes many forms, and not all of them are relevant to our considerations here. All of them overlap significantly with evolutionary thinking, however, and suggest a deeper backdrop to its conclusions. Anthropic thinking begins from the recognition that an astonishing multiplicity of different factors enables life (and ultimately human activities) to happen—basically the balance of Nature which provides us with a temperate environmental niche, a water-cycle on which vegetation and animal life can depend, a solar light- and heat-source in a tolerable range, and many factors more. In fact, once we start to follow the idea through, we find that the existence of these conditions makes it necessary that stellar and planetary evolution happens in such a way as to furnish a planet in the size-range and approximating the mineral, watery and aery make-up we know—which in turn means that the laws of physics must themselves be so set up as to bring about a multiplicity of

---

[*] Lee Smolin, *The Life of the Cosmos* (London 1997) p.203. It may be not too much to say Smolin wants not so much to reject the Anthropic approach so much as to replace it with a version of his own: we shall find some of his ideas of central importance to our presentation below.

quite improbable, or at the least extremely 'fine-tuned', conditions. Necessary time must also be allowed for these cosmic events to work themselves out, for life to emerge, etc. etc. We find that we are here observing and thinking about the universe, and therefore our very existence as human observers turns out to provide a key to almost everything that the universe must be, its structure, time frame, sequencing and stage of development, etc.

The fine-tuning has impressed some thinkers as so staggeringly complex and interlocking that it points to a God, who alone could contrive a world so closely fitted to our needs. But it is not clear that this is a good argument for God (though it accords with the notion in a general manner), and it is certainly not the most original aspect of the Anthropic idea. Much more significant is what it says about us and our potential as an instrument of scientific discovery. From our own mode of existing and knowing, we can directly uncover many aspects of the universe which has formed and sustains us. In this way, indeed, the Anthropic Principle has been extraordinarily productive in modern physics and cosmology.

The approach is not unrelated to the evolutionary principle, which traces the current disposition of things to prior real events. To put it less obtusely, we conclude from the bull's horns that it is a kind of animal which has developed them through butting; the older notion that it has horns so as to be able to butt with them long closed the door on scientific understanding, since it does not lead any further (further than God—how indeed could it?). But the real butting-activity of the bull dovetails into a picture of the lifestyle of large herbivores, their defensive strategies and internal rivalries, so that we may also presume the existence of extensive grazing lands, patterns of seasonal migration, climatic rhythms and in the end a whole food-chain and wider ecology. The butting-gesture similarly evokes the characteristic movement and anatomy of the large bovid type, which tells us a great deal about the environment in which such an animal operates. The 'design' of the animal effectively encapsulates the history of its evolutionary development and builds on previous animal types which must really have formed its remoter ancestors, subsequently metamorphosed to meet changing conditions. 'Seeing' those affinities is a very different sort of scientific insight—whose potential Goethe seized upon—as compared with the cataloguing and classifying approach which otherwise prevailed in the sciences of his time. That had resulted in abstract categories rather than real ancestors. Goethe used this strictly empirical 'seeing' of affinities to grasp the connection between plant morphologies, and in animal anatomy to arrive at his discovery of the intermaxillary bone—an actually existing feature which could then be externally confirmed and documented.* It is Goethe's mode of perception which is the immediate model for Steiner's deepened perception, often called his 'clairvoyance': we should not be distracted by terminology from appreciating its scientific relevance.

The clairvoyant in Steiner's scientific sense can heighten that awareness of the implicit background which reaches back into past stages of development. The Anthropic Principle is based on the idea that, in our own case, the implicit background to our development is essentially the whole cosmic order so far as we know it. We are able to grasp it consciously, in thought, because it is the universe that shaped our faculties both of sense and the heightened inner response we call consciousness through our whole evolution. (That does not mean that other orders are ruled out and cannot exist: indeed some scientists posit 'alternative universes' where laws and logic are different.) However, we can only (obviously) find ourselves inhabiting the one which fits us; a different school of thought proposes that in other parts of the universe, or in other times, physical laws themselves may be different, and gradually change. But one thing is clear. It is no accident that we encounter an organised world-order of enormous complexity; no other could have produced us. Moreover we know it to be certain and necessary that it is constituted as it is. The world is from the outset the world which will enable us to evolve, so that the human constitution thus stands written at the beginning, so to speak, of the world we inhabit; it simply requires all the aeons of cosmic and biological evolution to see us attain actuality in our present form. All the combinations and 'fine-tunings' which have produced us *must* occur, not for metaphysical reasons, but because we know we are here to comprehend them. How they happen remains a matter of evolutionary contingency, but we are the sort of being who will evolve because of the way the world is. And evolution is likewise written into us.

The proponents of Anthropic science are aware that it is not new. We find an extensive application of this idea in several strands of traditional thought.† But in Rudolf Steiner's work it had already taken on an

---

* Goethe's conception of science is not that we make an intellectual leap to a general pattern or Idea—but that the organising form is there *in* what we see, if we learn to see more deeply. We see the 'Idea' in the thing, as he said in opposition to the abstract classificatory scientism of his time; cf. Steiner, *Goethe's Theory of Knowledge*, pp.84ff.

† J.D. Barrow and F.J. Tipler, *The Anthropic Cosmological Principle* (Oxford 1986) pp.27ff.

essentially scientific form, and though his 'anthroposophical' sweep is wider than the field of physical science, it is important to be aware that he had considered acutely the methodological issues (as they are termed) and that he intended all his results to be *convergent* with natural science—not in any sense taking off from it into ideal constructs as in nineteenth-century *Naturphilosophie*.[*] He had earlier tried to elaborate the Anthropic character of his starting-point in an abortive work now called *Anthroposophy, A Fragment*, a richly productive statement which still lacks, however, some of the crucial ideas he later formulated. In one of his last writings he characterised his method as 'a path of knowledge from the spiritual in the human being to the spiritual in the universe'.[†] However, long before this he was thinking in the way we might call Anthropic when it connects with evolution. Looking at the human ability to speak, for instance, he sees in it the highest achievement in articulating thoughts, so that sounds become *words*. But it does not reveal our own state alone, but gives access to a cosmic reality. For thought of in another way—an Anthropic way—it indicates the original reality which must be of a kind to be communicated, somehow analogous to the patternings of sound that convey distinctions and discriminations, and which can thus come to expression when evolution presents suitable organs. Only a meaningful world could produce language-uttering creatures. And so in a profound sense such a creature is written into the world from the start:

> What appears last in the human being existed in the world in the very earliest times. We fancy that the human being in his present form did not exist in the earlier conditions of the earth. But in an imperfect, mute form he was there and little by little he evolved into a being endowed

---

[*] For his methodological perspective here, see especially Steiner (ed. Owen Barfield), *The Case for Anthroposophy* (London 1970) pp.26-45 ('Anthroposophy and Anthropology'). 'Rightly pursued, therefore, the two approaches, anthroposophical and anthropological, converge and meet in one point. Anthroposophy contributes the image of the living human spirit.... Anthropology contributes the image of sensuous man, apprehending in the moment of consciousness his selfhood but towering into a subsistence in the spirit' (p.44). He likens their relationship to that of a picture to its photographic negative.

[†] Steiner, *Anthroposophical Leading Thoughts* (London 1973), No. 1 (p.13). The details of Steiner's cosmological thought were all arrived at in this way. Thus for instance, when he speaks of stages of the development of the Earth, he starts by indicating how they can be reached through understanding the structural elements of human existence—Steiner, *Outline of Esoteric Science* (New York 1997) p.130, and cf. generally pp.118ff.

> with the Logos or the Word. This became possible through the fact that what appears within him later as the creative principle was there from the very beginning, in a higher reality.... What finally appears in space and time was already there in spirit from the beginning.*

The human constitution written in the very foundations of the world is not just a theory, but is directly perceptible to the extended scientific consciousness which is what Steiner calls clairvoyance. It is perceived as actively infiltrating itself into evolution, an evolutionary pressure, and Steiner can refer it as 'spiritual guidance'. It is, however, not an imported 'world-purpose' which evolution is trying to fulfil, but simply the gradual manifestation of the conditions under which evolution happens, and so comes closer to realisation at every opportunity.† It is still hidden away in the beginnings, waiting (as one might say) for everything else to happen which will lead to the expression of that deepest potentiality. It will be actually present only when every single thing that has happened in evolution has happened.

Elsewhere Steiner can describe it as follows. At the outset of evolution is found 'the archetype of the present human form'; but it has no actual existence, and is present only as a 'point' with only spiritual reality—or rather, since multiplicity and individualisation is part of what human existence means, it is a swarm of points beyond number: 'At that time the human form slumbered spiritually in the etheric point, and the whole earth-evolution was necessary in order that what rested there might develop into present-day man. Many, many things were necessary....' In fact, everything that did happen was necessary to make us what we now are. Only in these evolutionary terms can we define our humanness, and so 'gradually understand how these points became men if we make clear to ourselves all that has happened in the meanwhile'.‡

---

* Steiner, *The Gospel of St. John* (London 1962) p.27; see further the philosophical extension of Steiner's approach in the fine study by G. Kühlewind, *The Logos Structure of the World* (New York 1986).

† Perhaps such expressions shade into the figurative: but we should not look to anything except their scientific validity.

‡ Steiner, *Egyptian Myths and Mysteries* (New York 1971) p.15. The starting-point for Steiner's conception was thus in no sense an 'Idea' of the human, ready defined as in an Idealist system, any more than there is an irrational *élan vital*: The presence of the Logos-bearing human is such as we cannot express it abstractly, but only from what did happen. Humanness is not anything other than the evolutionary process that has made us, the cosmic order (Logos). In its implications this turns out to be a surprisingly rich starting-point for an 'Anthroposophy'.

To clairvoyant perception, this separating out of the totality of things and emerging life-forms from the human is a real process. It cannot ultimately be grasped in some abstraction of dialectics but in the actual evolution of humanity. Humanness is no more or less than actual evolution. When we come to consider further the question of early humans such as the Neanderthals and their Anatomically Modern contemporaries (and successors), we will need to look at the meaning of evolution itself as the comprehensive background to the question. It may seem a contradiction to say that the evolution of *Homo* was a 'new beginning', yet to derive human existence from cosmic origins. Here Steiner can help. He and his followers such as the biologist Wolfgang Schad, as well as palaeontologists such as Martin Lockley,[*] have shown that the animal affinities of humanity may then be understood in a manner quite different from the failed 'package' put forward by Darwin, and actually more integral to the thinking of modern science.

Biological Imperative, Panpsychism, Complexity and What Next?

We spoke a little defensively of the infiltration of human evolution into the events and processes of earthly conditions. Perhaps the expression is a little figurative. But its meaning is Goethean and empirical.[†] It ultimately means

---

[*] Wolfgang Schad, *Man and Mammals: Towards a Biology of Form* (New York 1977); Martin Lockley (with Ryo Morimoto), *How Humanity Came into Being* (Edinburgh 2010).

[†] Goethe was sceptical of the notion that we can avoid anthropomorphism (seeing things in human terms); instead he embraced it, though critically, as the basis of his scientific method. Perhaps one should consider this too as a stage in Anthropic thinking. Steiner explains: 'Goethe considered the sense-organs of man as the highest physical apparatuses.... Goethe declared it "the greatest misfortune of modern physics" ... that the experiments have been separated, as it were, from man. ... We do not obtain insight into the nature of a thing by disregarding the effects we observe, but this nature is given to us through the mind's exact observation of the regularity of these effects. The effects that the eye perceives, taken in their totality and represented according to the law of their connection *are* the essence of the phenomena of light and color.' Rudolf Steiner, *Riddles of Philosophy* (New York 1973) pp.195-6. Science long followed the other path, however, according to which there must be an elusive somewhat of which we predicate our knowledge derived from appearances—a tangled notion indeed. What it is in itself, we can never know. Thus living experience is an outsider, alien to the nature of reality. Rejecting this notion, George Adams extended Steiner and Goethe's insights in an Anthropic direction by arguing that the same

that we know the world is such as to produce humans, and if circumstances arise which favour some step in that direction it will naturally happen. The language is not *merely* figurative, however, and scientists talk nowadays about a 'biological imperative' which implies broadly that life will come about if it can. They speak also about the 'logic' of the situation in which life develops, considering both aspects as real determining factors.

There is no reversion here to the so-called vitalism of older times—another false turning rejected by Rudolf Steiner—which suggests that unique forces govern life: by such devices anything can (at least nominally) be 'explained'! What is implied is a new coherent understanding of Nature's workings which takes seriously the place of life in it, an enrichment of our science with its logic and laws. And in recent decades we have at last begun to get closer to it. Fifty years ago, by contrast, most scientists thought that life would turn out to be a complicated physico-chemical accident, derived from the mingling of complex chemicals (often termed a 'soup' of amino-acids) jolted into activity, probably by a freak event involving electrical energy (such as a lightning flash). If only we knew exactly which chemicals and under what conditions, we would no doubt be able to manufacture life mechanically ourselves.* 'Today, however,' writes an eminent theorist of life, Paul Davies, 'the picture of the cell as nothing but a very complicated mechanism seems rather naïve'. The properties we recognise as essential to life are not derivative from a mixing and sparking experiment, but seem rather to determine themselves. 'Autonomy, or self-determination, seems to touch on the most enigmatic aspect that distinguishes living from non-living things, but it is hard to know where it comes from,' Davies continues. Least helpful of all is the old physico-chemical approach. 'What physical properties of non-living things confer autonomy upon them?', he asks. After half-a-century doing science of this sort, Davies concludes, 'Nobody knows.'[†] Science needs to move beyond the merely physical.

Davies is contemptuous, too, of the popular misconception that if only we wait long enough, or peer far enough into interplanetary space,

---

cosmic ('ethereal') forces, working peripherally, 'create the world by the same laws of radiant perspective by which the human eye beholds it' (*George Adams, Interpreter of Rudolf Steiner* ed. O. Whicher (East Grinstead 1977) p.72). In this perspective, life is integral to the universe and a key to its structure.

* The author is old enough to remember from student days the almost canonical status of the physico-chemical theory in the form given it by Erwin Schrödinger's *What is Life?* (1943; repr. with an Introduction by Roger Penrose, London 2000).

† Paul Davies, *The Origin of Life* (London 2003) pp.8-9.

even something as improbable as life will be found to 'just happen'. He is struck rather by the immensity of the gap between non-living matter and the simplest microbial cell: 'The living cell is the most complex system of its size known to mankind. Its host of specialized molecules, many found nowhere else but in living material, are themselves already enormously complex. They execute a dance of exquisite fidelity, orchestrated with breathtaking precision ... spontaneous, self-sustaining and self-creating.'* To suppose that after long ages of lifelessness, part of a chemical 'soup' just clicked into this order and that it happened often enough for biology to take off and sustain the development of life, really does not make any sense at all. Thus in sheer scientific honesty Davies and many fellow-scientists have come to the conclusion that the material content is not all that is at work here. 'There *is* a non-material "something" inside living organisms, something unique and, literally, vital to their operation.'†

He speaks of actual forces which carry the biological information. We shall see later on that, as conceived by Davies, this bearer of the 'information' which effectively runs the organism converges in many respects with Rudolf Steiner's conception of the 'etheric' or non-material body which is the inner aspect of the living form—especially in understanding it as working peripherally, globally, rather than as contained in the material particles of the entity involved.

With variations of emphasis, a great many scientists nowadays agree that life cannot have just happened, and so they speak of a 'biological imperative', a something which makes life happen. Like Goethe and Rudolf Steiner, they do not mean a metaphysical necessity, or mere determinism, but the insistent presence of the information which brings about life and evolution. The forms of the basic animal types, in Steiner's felicitous phrase, 'come about whenever they have the opportunity'.‡ That short phrase points us, not only to living things, but to the nature of the universe they live in.

---

* Davies, op. cit., p.5.

† Davies, op. cit., p.17.

‡ One may characterise these animal types on the biological level (lion, eagle, bull) as the content which has to be externalised before the human form can appear. In them, life accommodates itself strongly to a particular niche in the environment. To the extent that the human does not overcome these animal tendencies, the typical forms 'remain, so to speak, as something which assumes its form again as soon as it has the opportunity' (Steiner, *Apocalypse of St. John* (London 1958) p.163). See below, pp. 122ff.

All the thinkers we have mentioned, including Steiner, share the determination to develop their ideas on a scientific, not on a religious basis. Nevertheless it is hard to conceive of life, if it indeed has more than a physico-chemical reality, as utterly divorced from the value of life—and so ultimately from a transcendence which is at root a religious acknowledgement. For Steiner, Christian religion made a disastrous wrong turning when it challenged evolution—and then failed to overturn it, so that it allowed itself to become divorced from cosmic knowledge altogether, becoming merely personal belief. The notion that religion could remain meaningful alongside a world-order that did not acknowledge any spiritual truth or fought it has indeed proved treacherous. Recent developments coming directly out of front-line science of the kind we have mentioned show, however, that Steiner was perfectly consistent and justified in considering spiritual ideas as independently valid in a cosmic context—as 'Anthroposophy' or human wisdom in dialogue with religious truth. They are not religion, but they may be necessary to save religion from failing in its own mission.

It is an exciting fact, at any rate, that the new theories of life have begun to open up new perspectives on the nature of the cosmos. Biology in this new mode may well be called astrobiology. For if life is clearly not a chemical property but an aspect of morphology and form, rooted in a wider structural setting in the environing cosmos, then in turn all those 'fine-tunings' of natural laws, sequences of unlikely events, harmonisations and improbable symmetries that may be called 'complexity'—are all a part of life's manifestation. So that if we ask, naturally enough, after the content of that 'information' which is the non-material organising principle of life—what can it be but that very history of evolution, the conditions which have made life possible? The astrophysicist Lee Smolin ascribes life to the whole series of 'nested hierarchies of form', from organic life and its components, up through environmental cycles to planetary and even interstellar patternings. The immediacy of life in microbes, plants and animals is an element within this larger 'life of the cosmos' and could not subsist without it.[*]

Now of course, Rudolf Steiner described the mode in which such information could be non-materially present in a living thing, as 'spiritual': as yet very few scientists would risk the potential misunderstanding that the term might cause. Davies and others do, however, argue that we must infer the existence of non-material 'information forces',

---

[*] Lee Smolin, *The Life of the Cosmos* (London 1997) pp.141ff; further discussion in Paul Davies, op. cit., pp.40ff.

operative globally rather than point-by-point, which can act directly upon organic material substance. For Steiner, the hierarchical structures are present primarily as rhythms, and the whole development of life is both a reflection and an intensification of cosmic life-rhythms. Just consider: every specific element in a rhythm is 'placed' where it is by the demands of the whole, and is part of it, although materially there are only the separate incidents which make it up.[*] Davies too, we recall, when characterising the organisation of a living cell, suggested the analogy of a musical 'score'. To many, that still seems too close to an 'idea'—a mental rather than a physical reality. Yet the possibility of consciousness in all of Nature (so-called Panpsychism, lit. 'all-ensouledness') has itself emerged as the best solution to many problems in physics as well as neuro-psychology in recent years.[†]

It is a major aspect of Steiner's cosmology, certainly, that he considered even the primordial differentiation of the world to have been simultaneously the stirring of a dim consciousness. That consciousness in the basic substance of the world is at the root of our own, which is (so to speak) distilled out of it: the primordial layer of consciousness is now overlaid by more intense developments which normally blank it out. We are still linked to it, however, and in certain states can merge back into it.

If such ideas have any validity, the long-held notion of many scientists that we are alienated onlookers, passively observing an independent and uncaring world, may finally be breaking down. Perhaps the thinker who has most extensively, and persuasively, worked out the

---

[*] 'Basically everything in the etheric body is rhythm,' explains Steiner, 'a cyclical movement of rhythm or activity, and it has a spatial character only in so far as it inhabits the physical body…. What matters with regard to the human etheric body in reality … is mobility, movement, formative activity in rhythmic or musical sequence, in fact the quality of time. Of course, this is a difficult thing for the human mind to conceive, accustomed as it is to relating everything to space; but in order to gain a clear concept of the etheric body we must try much harder to allow musical ideas rather than spatial ideas to come to our aid.': Steiner, *Art in the Light of Mystery Wisdom* (London 1970) pp.16-17.

[†] I happily exempt myself here from the need to enter here into details of the discussion. If it is permitted to conjure with names, one may mention (as well as recent neuroscientists), the great theoretical physicist John Archibald Wheeler (died 2008), and others; there is also the philosopher Philip Goff, and the pioneer psychologist William James is sometimes claimed to have held the view. A clear introduction is Keith Frankish, 'Panpsychism and the De-Psychologization of Consciousness', available at https://philpapers.org › rec › FRAPAT-28.

implications of the new biology of self-determination and complexity is Stuart Kauffman. He shares the recent rejection of the idea that life just happened: he stresses that a living thing must be whole from the outset and cannot ever result from chance combinations.* But that means that the complexity which has in fact produced life must be a feature of the cosmos much more widely, so that what produced it was an inherent tendency or a complex of natural laws that shapes and fosters complex structures. We ourselves are the product of that tendency, so that from yet another perspective we turn out to be a key to the nature of the universe which the study of inert matter cannot provide. Kauffman brilliantly shows that as they develop, things are increasingly governed by laws which bring about self-determination and self-consciousness. We—and life generally—are not the exception, or freak, but (as he movingly asserts) are at home in the universe. Kauffman is at pains to protest that his 'unrepentant holism' is 'born not of mysticism but of mathematical necessity'.† Steiner's clear-headed parallelism, not confusion, of science and religion might, if he wished, reassure him. But the presence of an overarching cosmic tendency which fosters and furthers complexity, bedding it ever more thoroughly into a meaningful universe, might obviously elicit the supposition from many that his thought implies not just that we are at home but also that we are not alone in the universe. That would be, in Steiner's sense, not a religious but a spiritual-scientific idea.

Evolution, complexity, Anthropic science, universal consciousness—these are sometimes mind-boggling ideas which we can hardly expect to exhaust without a lifetime's study. For present purposes we need to be aware of them and the perspectives they intimate. It is remarkable, to say the least, that Rudolf Steiner was able to think in many of these ways while so many among his scientific contemporaries supposed that a stark and simplistic materialism was their sole defence against Establishment religion which would deny evolution. Many scientists would nowadays admit that materialism has exhausted itself and we are probing its limitations (even if other voices only grow yet still more

---

* Stuart Kauffman, *At Home in the Universe. The Search for the Laws of Self-Organization and Complexity* (New York and Oxford 1995) pp.43-5. Kauffman also stands among those recent scientists who have achieved the enviable goal of writing lucidly and with almost a storyteller's art, as this book in particular shows.

† Kauffman, op. cit., p.69.

shrill).* If Steiner had been able to reach out to find supporting thinkers like those just mentioned, he would certainly have felt less of a lonely voice.

As it is, humanity's place in the universe has come to a head in a particularly direct manner in the upheavals of thought about human emergence—the Neanderthal enigma, the creative Explosion and the paradox of the Anatomically Modern disjoined from modern behaviour. Steiner has extraordinarily apt things to say here, whose nature means, however, that we must understand a little more about his method as well as his conclusions. That must be our focus in the chapters which follow. We need as part of that process to work out what exactly he found of value in a celebrated map—but above all, we need to understand how he believed we could recover in consciousness the remotest past, which underpins his account of evolution, leading to ourselves. There is no time to be lost.

----

* It may be appropriate to mention here the often polemical presentation of evolution by Richard Dawkins. He is such a successful publicist that many people imagine, no doubt, that all scientists agree with his sweeping conclusions drawn from biology. They do not. There is clearly much good science in Dawkins, but many of his wider views stretch the evidence—sometimes to breaking-point. His work should be read in tandem with the patient evaluation of his arguments by Anthony McGrath in a series of accompanying books. For a good many scientists, 'materialism' has outlived its usefulness as science has grappled with activity in nature. 'We have to learn that matter and its activity are not conceptually separable,' wrote J.Z. Young; 'It is those who wish to understand the activity of matter who have learned not to be materialists.' See his *An Introduction to the Study of Man* (Oxford and New York 1971) p.130.

Chapter 2

RUDOLF STEINER AND THE MAP OF ATLANTIS

A World within the World

Whether from the accounts of astronauts or satellite images of the Earth from space, from computer modelling or from sumptuous atlas-illustrations of the world, we are extraordinarily familiar nowadays with the overall configuration of our planet's surface—the face of the Earth. We recognise instantly the shape and distribution of the continents: North and South America, Africa, Asia and its appendage Europe, Australia, Antarctica. No corner of the world is nowadays labelled terra incognita nor shown with cameleopards or men with heads between their shoulders. Nor are those old projections like the *Mercator* which gave the map its 'corners' and exaggerated the land masses nowadays much employed, being replaced in home and school books by more curvaceous frameworks—that reflect the roundness of the terraqueous globe. Even the seabeds and ocean floors can be pored over with their deep trenches and ridges named and measured. All is well known, and more than a few of us modern jet-setters have likely set foot on several continents and swum in a plethora of seas.

It is also no longer a novelty to be told that the face of the planet across whose features we trace our long-haul visits to far-flung family, etc., has not always been as it is now. We know at least vaguely and in abstract terms that the rock-strata show how land has sometimes been sea, indeed that the very continents are moving beneath our feet—a startling thought still, though, to many. Drastically fewer among us would be able to recognise the face of the world even in the geological period immediately before the continents attained their current place. We still shudder, perhaps, at the thought of the Earth under us *moving*! The idea that the continents 'drift' was only accepted by Earth-scientists themselves after a desperate battle of resistance. Through much of the twentieth century Alfred Wegener, though widely admired for his meteorological work, was ridiculed mercilessly until finally in the 1940s and '50s his arguments proving the reality of the 'moving continents' triumphed, and led to the idea becoming rapidly integral to science. Now it is unquestioned orthodoxy for geophysicists, palaeontologists and climatologists. And more and more is known about the Earth's changing configurations. Increasingly we can look through the surface reality to the worlds within our world that have determined its present form.

The basic configuration of the rocks today, and arrangement of the lands and seas, stretches back into the geological period long denominated in scientific parlance the 'Tertiary'. More recently the convention of dividing this fundamental epoch into two, the Paleogene and Neogene phases, has been adopted, though familiar older ways of speaking often persist. The terms are at once names for the material strata, laid down one above another, and for the order in which they are everywhere laid down, chronicling distinctive ages of geological activity. The limestones and sandstones, clays and alluvial deposits which belong to the 'Tertiary' phase have an essential similarity to those we still see being scraped, smashed together or scattered about by wind, water or earthquakes today. The period which followed on those formations, leading up to our own present day—beginning from the last great series of Ice Ages—is often known as the 'Quaternary'. The world is here still in the process of formation. But the Earth's physiognomy does not by any means consist only of Tertiary rock-formations and modern alluvial deposits. For instance: exposed at the surface in many places, or found some way below the surface strata, we come upon a massive chalky limestone layer extending over huge portions of the Earth, nowadays mostly covered over by the Tertiary rocks which have since been heaped upon it, still showing through where they are partly eroded or have been displaced. This buried layer of enormous extent and thickness must have been laid down long prior to the phase of the Tertiary rocks. On the other hand, parts of the chalky material have actually been pushed up above as high mountain chains (such as the Alps)—one of the several reasons hinting at vast moving forces which have altered the original position of the rocks and changed the face of the Earth.

The chalky layer is of huge extent and thickness—in places it is up to 15,000 kilometres in depth as in Ontario, Canada. This is all the more remarkable because the whitish chalky rock is of organic origin, consisting of the remains of innumerable tiny marine creatures. Quite staggering amounts of material have been laid down from an evidently abundant life-process which once filled the seas. Moreover it is deposited in many cases in regular layers alternating with 'marl', reminiscent of the shell of the oyster, so that term *Muschelkalk* has been coined to describe it. The structured formation of the distinct alternating layers has led some researchers to speak of a vast eco-system or living organism, similar in principle to a coral reef which likewise forms an organic unity. Within this eco-system the tiny marine animals formed the smallest or outermost constituent part. 'When one realises that the Muschelkalk area in Europe originally stretched from Heligoland to North Africa and from western Spain to the Caspian Sea, one wonders how it was possible that this rhythmic stratification could take place over so large a surface-area!'* At any rate one must acknowledge that processes of formation were at work here which we no longer find in the modern living world, and on a scale which likewise differentiates this phase from what we find afterwards. The last great period of this geological phase, in which the process reached its height, is known as the Cretaceous Period, a term which again alludes to the chalky nature of the rocks.

The huge rhythmic living eco-systems of the seas were of course not the only representatives of life at that time. Fossil evidence furnishes us with an extensive record about the species which inhabited the land regions as well as the seas. Once again we find that a gulf separates our own plants and animals from those we discover there. To go back in imagination into the Cretaceous world is to go from the Caenozoic ('recent life') of the Tertiary and modern periods to the Mesozoic era or middle stage of life within the overall history of the Earth. We are then entering the age of the giant saurians, pterosaurs, and marine dinosaurs, which would be followed in due course by other huge animals and birds. The gulf between them and our own animals, fishes and birds is startling—though not by any means total.

And the divide is marked in the geological record by unmistakable indications of a massive catastrophe that is often termed the 'K-T extinction boundary event' (K-T meaning: Cretaceous-Tertiary)—or nowadays more correctly K-P (for Cretaceous-Paleogene). It was an overwhelming event which marks the end of the Mesozoic, with the extinction of more than 75% of the known species of animals and plants then living on the Earth. Speculation has ranged widely in the quest for what happened at this 'boundary' moment, marked by a narrow dark band in the rocks. There is certainly evidence of an enormous *tsunami*, and of thick clouds of dust in the high atmosphere obliterating the sunlight—now the dark layer which persists in the rocks all over the world. The currently favoured idea is the impact of a comet or even of an asteroid upon the Earth.

---

* Walther Cloos, *The Living Earth. The Organic Origin of Rocks and Minerals* (Forest Row 1977) p.63.

TABLE OF RECENT GEOLOGICAL PERIODS AND
PLANT/ ANIMAL EVOLUTION

Anthropozoic or Recent:

(12) Quaternary	Alluvium – marine and freshwater Diluvium – glacial deposits	(3) Large mammals including mammoth, bear, man	

Caenozoic:

(11) Neogene	Freshwater deposits, Mediterranean marine deposits	(2) Large animals including dinotherium, apes, giant lizards; palm, fig	A T L
(10) Palaeogene	Limestones, sandstones, clays	(1) Large animals including palaeotherium, birds	A
Cretaceous-Palaeogene Extinction Event			N

Mesozoic:

(9) Cretaceous	Limestones, sandstones, clays	Giant Saurians; first deciduous trees ; sponges	T I S
			L
(8) Jurassic	White Jura, Lias	Saurians, marsupials	E M U R I A

Geological and biological epochs (recent), adapted from Rudolf Steiner's table in his Frühe Schriften (Dornach 1941), along with Steiner's terminology based on the rhythm of the continent-sequence. For more of Steiner's table and his numeration of the geological phases, see Walther Cloos, The Living Earth *pp.149-151. Also included is the Cretaceous-Palaeogene Extinction event, most plausibly considered to be an asteroid collision, which figures in the mythologies as the 'destruction of Atlantis'. Despite Plato's allusion to a 'single night', the reality of change was more long-drawn-out. Steiner consistently refers to the latest Ice Age as the very 'end of Atlantis' and finds in the myths of climatic change and flooding an indication of its final stages.*

Table of geological epochs after Rudolf Steiner/Walther Cloos

The impact-event, if such it was, brought to an end a whole era of the Earth's existence. But we know that even beforehand it was not a fixed but a dynamic world-order. The long-term forces of change, however, were forces within the Earth, which have continued after the K-T extinction event too. The continents with their shell-like, piled-up deposits of chalk and other rocks were subject to more constant forces beneath them in the Earth's mantle. In the Cretaceous Period the continents we recognise so familiarly as we plan our flits about the globe were positioned, some of them, far away from their current locations. The land masses of the continents themselves are only the surface portion of a larger 'tectonic plate', partly submerged in and borne about by the heat-currents in the molten rock which lies beneath them. Parts are also hidden from our sight by the oceans. The deep heat-currents which evidently rise from far inside the Earth, are best known from the phenomena of vulcanism. Volcanic eruptions and earthquakes are now known to be linked with the movements of the tectonic plates.[*]

In this respect, our planet is far from typical of the smaller worlds which circle in our solar system, none of which exhibits this kind of inner dynamism which one may liken to a life-process of the Earth itself. They may not be wholly devoid of volcanic activity, but none of the other inner planets evidence the kind of dynamic process of change which is characteristic of the Earth, which is still uniquely active in this regard.[†]

Once the concept of the movement of the continental plates, propelled by the heat-currents in the underlying magma, was grasped and accepted by scientists, it opened the way for research to investigate the changing locations of whole land masses which we know in their present-day identities, and so map the world of previous epochs. The movement caused periodic partial submersions of some regions through 'subduction', and the volcanic activity at the margins of the plates pushed up masses of rock in others. The chalky rocks of the Mesozoic

---

[*] One of the best introductions to the subject is: J.S. Monroe and R. Wicander, *The Changing Earth. Exploring Geology and Evolution* (Belmont California 2009).

[†] 'Thus Venus lacks a system of plate tectonics like the Earth, but is instead dominated by mantle plumes. Mars, Venus and Earth convect to form mantle plumes, but only Earth recycles its lithosphere through subduction zones. We are forced to conclude that it is Earth that is unique …': so conclude E.H. Christiansen and W.K. Hamblin, *Exploring the Planets* (New York 1995) pp.244-5. 'Subduction' or forcing down is one of the fundamental processes in the movement of the plates bearing the continents.

era with their rich fossil life bear ready witness to the alternating conditions—sometimes evincing marine fauna whose fossils now appear on mountain-tops, or trees and quadruped dinosaurs where the sea now flows. Mapping the evidence of shorelines is a crucial instrument of palaeogeographical research. But much more sophisticated methods have additionally been employed: the most accurate being based on the discovery of tiny magnetic particles within the rocks which have become charged by the Earth's magnetic field. Because their positive and negative poles were originally aligned with those of the Earth's field, we can tell in which direction they were initially pointing—with which, however, their present location no longer agrees. It is therefore only necessary to reorientate them theoretically in their original position, to establish where they were when those rocks were laid down, aligned once more with the Earth's magnetic field. With enough correlations, land masses can be reconstructed and realigned to create once more the original position of the rocks and the former outlines of the continents. Correlating shoreline-evidence with the results has brought astounding conclusions, achieved by the careful amalgamation of information from different fields. Sophisticated visual manipulation in computer projections means that whole histories of movement and changing shapes are available for all the continents, going back to the earliest known times.*

The configuration of lands and seas which is thereby revealed, even just for the Cretaceous Period in the later phase of the pre-modern world, is strikingly unfamiliar. Whole continents have apparently been moved about like counters before arriving at their present locations. Even in the Cretaceous they were quite differently arranged.

Many aspects of geography gain an exciting novel sense. Once we examine them with the idea of 'continental drift' in mind, the present outlines of Africa and South America, to take just one example, tell of their historical connection in their mutually compatible shapes (an observation that had long been made, going back even to von Humboldt, who toyed with speculative ideas to explain it). Now they have moved far apart. It was only quite late, in the Neogene, it turns out, that South America went over to join North America in the position we find it now.

---

* One of the most reliable and well-presented is the online 'Paleomap' project of Christopher Scotese, which offers detailed maps of the world in all the major geological epochs: go to www.scotese.com. I am indebted to Professor Scotese for permission to reproduce his maps.

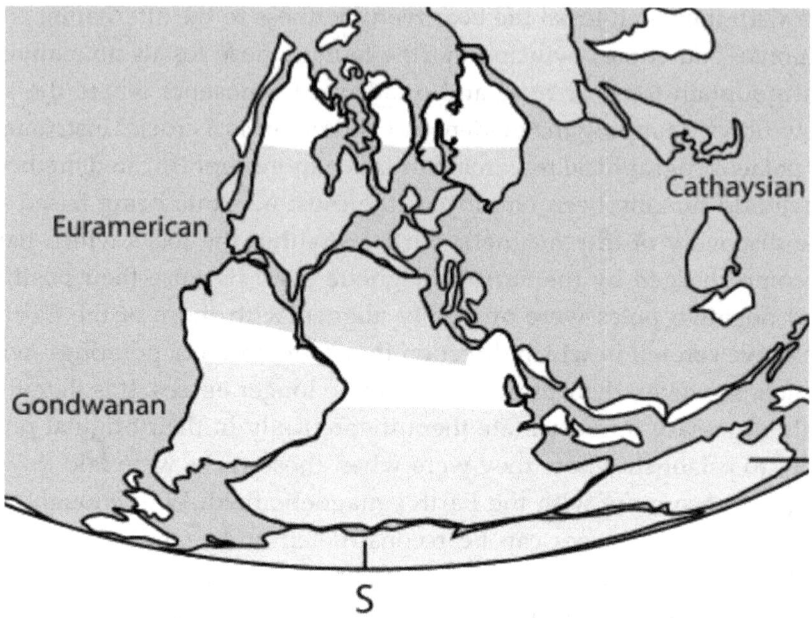

It has long been a subject of curious discussion that the continents are so shaped that they appear to fit together like an immense jigsaw. South America would dovetail with the Bight of Africa, etc. Modern-day research into the dynamics of 'continental drift' has revealed that the continents are indeed formed by the breaking-up (and eventual re-forming) of larger 'supercontinents'. This palaeomap shows the point far back at the end of the Permian epoch when the continents had not yet broken apart.

In this way we can see, as it were, a world within our world. If we turn to a map of the Earth in the later Cretaceous (see below), we notice immediately that South America is still located far down to the south, as if nestling close to the African coastline which runs parallel to it. The Eurasian land mass meanwhile is separated from the other continents by a vast swathe of water, a major sea which no longer exists, known as the Tethys Ocean.

The Atlantic Ocean, on the other hand, presents us with the most remarkable picture of all—instead of an ocean we find a huge continent, separated at one point by only a short distance from Scandinavia and bordering on the fragmentary lands which will later be Britain and Ireland. Far to the west of the continent occupying the Atlantic Basin, however, we observe that, some 300-500 miles out across a broad seaway, a great chain of mountains rears out of the water, running north to south. One day it will be the backbone of America—the Rocky Mountains. They have been pushed up out of the ocean where the Pacific tectonic plate has collided with that bearing the Atlantic continental mass. 'We know,' says one helpful summary of the situation, 'that the Earth is never at rest; we know that it is in a state of perpetual inner change and movement.... In this primeval epoch, the part of the Earth's surface which today is covered by the Atlantic Ocean was a mighty continent, while where Europe, Asia and Africa are situated scarcely any continents were as yet formed. Thus the solid matter, the substance, of the Earth has been transformed by its inner motion. The planet Earth is in a continual state of inner motion.'[*]

It may give us pause when we register, however, the fact that these words of Rudolf Steiner were spoken in 1912—when Alfred Wegener's groundbreaking (or rather earth-moving) speculations were as yet unpublished, and when the whole scientific establishment was on the verge of a thirty-years war, uniting in scepticism against the idea that anyone could show them 'a force sufficient to move a continent'! Still less were all those advanced techniques for plotting the movement of the continents which we have briefly sketched, and which would eventually vindicate him, even dreamed of. Before all this, Rudolf Steiner had already developed an approach to the changing configuration of the continents as a large-scale rhythmic process, based on the 'inner motion' of the Earth and also on cosmic relationships.

The reference to rhythmic activity was also prophetic. Modern palaeogeography—the mapping of the prehistoric world—has likewise come to the conclusion too that the movements it charts are rhythmic and cyclical: the heat from the thermal currents beneath a large continent slowly builds up, and leads to a break-up of the land mass, pushing the smaller resulting parts (which may still be whole continents as we know them!) in different directions. The build-up of heat is thereby dissipated. But the continuing dynamism from below means that the lands which have been pushed apart, floating on the underlying magma, are eventually driven together once more, forming periodically a new giant land-unit or 'supercontinent'. Then the entire process begins again—the whole extending (it goes without saying) over immense periods of time.[†] The rhythmic nature of the whole process is central to Steiner's approach too, and will form a key to his evolutionary thought.

---

[*] Rudolf Steiner, *Spiritual Beings in the Heavenly Bodies and in the Kingdoms of Nature* (New York 1992) pp. 93-4. This and other relevant material is collected in Steiner, *Atlantis* (Forest Row 2001).

[†] Prior to the emergence of the modern view, geologists had been locked in a long-running dispute between those who appealed only to gradualistic forces (erosion, displacement, etc.) and those who, finding that view insufficient to account for what we find in nature, believed that there must almost have been periodic episodes of catastrophically violent, especially volcanic intervention. The K-T extinction event proves that there have been catastrophes: but the fundamental reality is neither purely gradual, nor violent—but rhythmic.

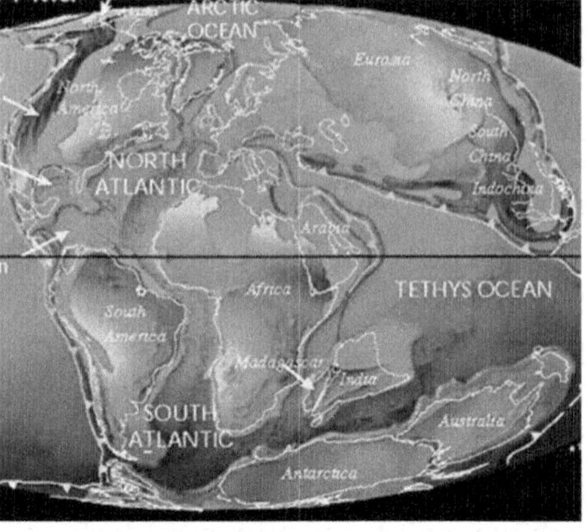

Modern palaeomagnetic techniques of mapping the ancient rocks have shown that in the late Cretaceous the continents as we now know them were quite differently disposed. Most remarkable, however, is the fact that Rudolf Steiner could accurately state in 1912: 'In this primeval epoch, the part of the earth's surface which today is covered by the Atlantic Ocean was a mighty continent, while where Europe, Asia and Africa are situated scarcely any continents were as yet formed. Thus the solid matter, the substance, of the Earth has been transformed by its inner motion.' Since that time the Atlantic Ocean has opened up through that constant movement of the tectonic plates which, as Steiner's ideas presaged, are a distinctive feature of the life of the Earth. The tearing apart of the ocean-floor in the process is the cause of the volcanic activity which threw up the Atlantic Ridge. This recent reconstruction of the 'Late Cretaceous' world (detail) is reproduced courtesy of Christopher Scotese.

From the fact that a huge continent then occupied the Atlantic Basin, Steiner referred to the epoch preceding our modern disposition of the continents, in this passage and elsewhere, as the 'Atlantean era'. He accepted the much-touted name for a continent occupying a location in the present-day Atlantic, and referred to 'Atlantis'. But we must be careful of making assumptions. He certainly knew Plato's version of that legendary 'sunken land' and its dramatic story. Yet he makes virtually no reference to it. Almost always, in fact, his use of the term 'ancient Atlantis' connotes that extensive era in the whole of the Earth's history or the entire pattern of seas and lands which characterised it, or the life which proceeded there. And hence too, as we are sometimes startled to find at first, he insistently characterises our own more recent phase of development as the 'post-Atlantean' age or world.

It is worth noting that that would not have sounded out of accord with some of his scientific contemporaries. At that time most of them were struggling, with very little success, to explain the dissemination of related species and genera found scattered right across the world on the assumption of fixed continents. Elaborate theories were devised, based upon (completely invented) 'land-bridges' which were supposed to have enabled animals or seeds to have made their way between far-separated localities. But sometimes the distances were so large that the biologists were driven to the hypothesis of more extensive land masses that must once have existed, e.g. in the Atlantic Ocean. In particular they were tempted to such speculations by the deep-sea soundings which had been recorded in the last part of the nineteenth century when exploring the Atlantic. These revealed the presence of towering mountains under the ocean, whose peaks protruded as the Azores and Bermudas. Surely, many argued, under slightly different conditions these sea-mounts could have been exposed as dry land much more extensively, until with rising sea levels or volcanic eruptions they would dramatically have 'sunk beneath the waves'? Surely the real Atlantis had been found? (The term Lemuria also derives from mainstream scientific discourse of the day.)

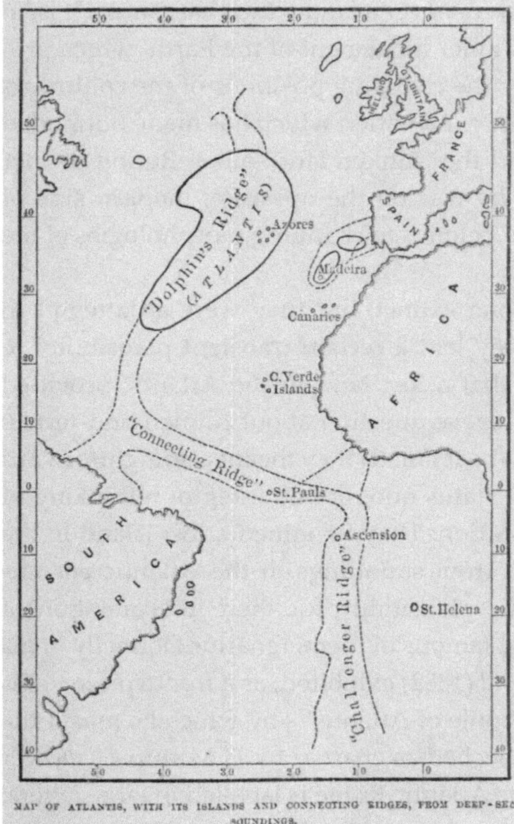

MAP OF ATLANTIS, WITH ITS ISLANDS AND CONNECTING RIDGES, FROM DEEP-SEA SOUNDINGS.

The soundings made in the Atlantic Ocean by the surveyor-ships Challenger *and* Dolphin *in the mid-nineteenth century fuelled efforts to understand the palaeogeography of the region—albeit the results were premature and misguided. The Atlantic Ridge was conjectured to be the remnant of a once protruding island in mid-ocean, and the signs of volcanic activity on the sea-floor in its vicinity were taken as indications of a catastrophe which destroyed it. This picture is now known to be entirely false. But enthusiasts for Plato's Atlantis seized upon the idea, among them the Theosophist Scott-Elliot—though the map which he published is completely at odds with reconstructions based upon the theory, like that of Ignatius Donnelly, shown here. From Donnelly,* Atlantis: the Antediluvian World *(1882).*

The evidence was never quite convincing. Nowadays we know that the notion of a 'sunken land', and linking land masses, is based upon a complete misunderstanding of the Atlantic Ridge. The ridge as we find it is part of that world-wide system of tectonics which extends all round the globe on the ocean bed, and which has even been described as the most distinctive structural feature of the Earth. It is not confined specifically to the region of the Atlantic, and even there is not the remains of any former land mass. Quite the reverse. It is generated where the deep ocean bed is expanding and separating—tearing itself apart, with resulting vulcanism in the depths of the sea. The movement which has caused the opening-out of the Atlantic Basin, thus dividing Europe from the lands on the opposite shore, creates a 'seam' and an irruption of magma from under the sea-floor. This hot fluid rock cools rapidly when it is exposed to the seawater and solidifies into the ridge that is pushed up in mid-ocean: the very same which intrigued a generation of scientists seeking to understand in static terms the changing Earth. The towering sea-mounts are made where the breadth of water is actually increasing—they were never a land which has 'sunk'. It is, in reality, a basic feature of the 'inner movement of the Earth' which provides the real explanation for the changing positions of the continents and the distribution of life, the explanation which has made outmoded the fictional land-bridges and the 'sunken land' alike. Rudolf Steiner used the term Atlantis, but he was on the opposite, modern side of the question. His explanation belongs to changing morphologies of the Earth, not to sunken realms.

Those theories are now long extinct; but they were an attempt to solve a real problem, and they lent a certain transient plausibility to those thinkers who argued that a lost land in the Atlantic provided the answer to—well, nearly every question about human and terrestrial history. By *their* Atlantis (be it noted) they meant a straightforward addition to the geographical status quo, not an integral rethinking of the continents and their evolution. They imagined a lost island in the ocean. The new information from soundings in the Atlantic encouraged them to claim scientific plausibility for their interpretation of things, so that one of the most famous of them, Ignatius Donnelly in his *Atlantis. The Antediluvian World* (1882) exhibited, as a frontispiece to his now celebrated book, 'The Profile of Atlantis'—by which he meant the cross-section of the Atlantic seabed as charted by H.M ship *Challenger* and the U.S. ship *Dolphin*. The Atlantic Ridge is labelled in large letters: ATLANTIS. The two greatest nations on Earth had conspired to prove it true. Donnelly gives full oceanographic detail (Ch. V), summarises the case from the distribution of flora and fauna (Ch. VI)—all in Part I, before going on to assemble the unanimous testimony of flood-myths everywhere, and to rewrite conventional ancient history from Egypt to South America as a series of colonial adventures of the mighty Atlantean Empire.

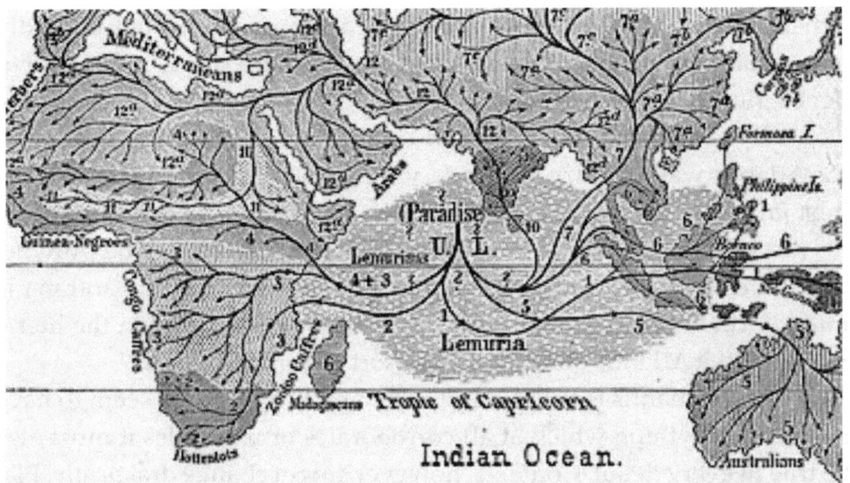

Atlantis and Lemuria land-bridges: Static scientific speculation about the dispersion of flora and fauna across supposed land-links prompted many a misguided attempt to reconstruct lost continents called not only 'Atlantis' but, as here, 'Lemuria'—terms that were taken up in a geodynamic perspective by Rudolf Steiner long before the idea of the 'continental drift' was propounded by Alfred Wegener in the 1940s. Only when interpreted as stages in the dynamic evolution of the continents would they gain real validity. From American Cyclopaedia *(1879): 'A hypothetical sketch of the monophyletic origin ... of the human race, from Lemuria over the earth.'*

In his sadly unexciting 'Map of Atlantis, with its islands and connecting ridges, from deep-sea soundings', early in the book, the lost land appears anew: it is a long bulbous island in the very middle of the ocean, about the length of France and Spain put together.[*]

And of course he refers back to Plato, the first Atlanteologist. Because some of Plato has since become scrambled in certain of these sources, it may be as well to remind ourselves of his actual claims. In two of his Dialogues, *Timaeus* and *Critias* (although the latter is tantalisingly incomplete) the ancient Athenian philosopher had recounted a 'true' story he is supposed to have learned in Egypt, which is what lends all later speculation its real significance. The story goes that long before

---

[*] Ignatius Donnelly, *Atlantis. The Antediluvian World* (repr. New York 1990); map, p.51.

Plato's time, Atlantis had been the seat of a mighty empire, and that it sank beneath the sea in a single night of destruction, leaving only a swirl of mud behind—which still apparently made the sea unnavigable beyond the Pillars of Heracles (Straits of Gibraltar) in Plato's own day. Less often recalled is the circumstance that the resistance to the Atlantean Empire was led by Athens: the Athenians are clearly supposed to find this a novel idea, since they themselves have no conception that Athens even existed in those unrecorded times. This detail is hardly ever quoted by Atlanteologists.* The mud is Plato's only physical evidence. But by some curious irrational logic it appears that among the Donnelly's and their ilk, if it could once be shown that any portion of the Atlantic was ever dry land, everything that Plato said in those writings must be circumstantially true. Most people in Plato's time, from his disciple Aristotle onward, thought the whole matter to be pure fiction. Later on, however, when Plato's thought was taken up and made into a sort of mystical religion by the Neoplatonists, belief in the literal reality of the Atlantis story became a sort of *point d'honneur*.†

If Plato's Atlantis is true, so Donnelly and his followers seem to have held, then anything which at all corroborates or resembles it must also be true in every detail. Contexts, however, might change drastically. Plato's *Timaeus* and *Critias* were now being read in the context of post-Darwinian diffusion theories (Plato does seem to mention elephants and coconuts); there was also nineteenth-century colonialism, not to mention theories about religion (naturally this must have started in Atlantis).

---

* The point is well made, however, by John Onians' chapter 'Athens and Atlantis: A Myth of Style and Civilization' in his *Art and Thought in the Hellenistic Age* (London 1979) pp.7-16.

† Proclus (AD 410-485) insists that 'we ought not to be sceptical about it'— though the kind of reality he attributes to it (a sort of archetypal instance of the rise and fall of empires, a universal truth, perhaps?) is difficult to assess. Porphyry (c. AD 235-305) leans more to the symbolic view, though symbols are intensely real for Neoplatonists. But for Plato to symbolise an important truth by something literally untrue did not seem right to Iamblichus (c. AD 242-327), who strongly condemned anyone 'setting aside the surface meaning'. With the revival of Neoplatonism in Renaissance Florence, Marsilio Ficino reasserted the literal truth of the Atlantis story, and it appears in the work of those such as Athanasius Kircher (1602-80), the great Jesuit polymath, as a large island in mid-Atlantic between Europe and America. Even amongst these learned thinkers, no evidence except Plato's say-so is adduced. The Platonic tradition of Atlantis, though manfully upheld over so many centuries, never produced anything of scientific value.

Later on, the relatively benign Atlantean imperialism described by Donnelly would be sucked into the flood-waters of nineteenth- and twentieth-century racist ideologies too.* Amid all this bizarre nonsense it was easy to forget the obvious meaning behind Plato's work.

The Athenians had a proud history of resisting despotic empires, as their war against the Persians had shown. Already looming over the horizon in Plato's time, however, were the flagrant ambitions of the Macedonian kingdom to the north which, under Alexander the Great, would push to unite the Greek and Oriental worlds under one almighty rule. Plato's reminder to the Athenians of their past glories could hardly be more pointed, even if it is frankly mythical in its storyline. His attempts to interest them in organising a better, more deeply democratic system of government (cf. the discussions in his *Republic*) and of inculcating free-ranging thought as the antidote to despotism, all have an obvious relevance. For take note—the story suggests—even the mightiest empires come to a bad end, sooner or later, just when they least expect it.

Plato's narrative has little or no connection with traditional Greek myths—including their flood-myths. It is not Egyptian either. Yet although it is a fiction as it stands, it is unlikely that Plato just made it up. The *Timaeus* which introduces it is basically an account of the nature of the cosmos, whose impressive design leads Plato to introduce the idea of a divine Creator or World-Craftsman (a distinctly un-Greek idea). In the narrative part we are also regaled with the myth of Phaethon ('Bright Shiner'), who wanted to be like a god but who fell to Earth from the chariot of the Sun. This story did become part of Greek mythology but just as certainly did not originate there; we shall see that it arrived there only quite late on. In fact it derives alongside the other materials adopted by Plato from a milieu closer to the Bible: a Maker of the world, a Lucifer ('Light Bearer') who falls from heaven after aspiring to godhead, a myth of the Flood which drowned wicked humanity, all feature in the Book of Genesis. And Plato in another work, the *Symposium*, tells of originally bisexual humanity being divided into separate sexes as a punishment by

---

* Those who wish to know may see the work of Joscelyn Godwin, *Atlantis and the Cycles of Time* (Rochester 2011); a general history of ideas about Atlantis, not always very illuminating, is furnished by R.P. Kershaw, *A Brief History of Atlantis* (London 2017). For example the German ethnologist Leo Frobenius claimed in 1910-1912 that his Inner Africa Research Expedition had uncovered the Atlantean (i.e. non-native African) origins of the Yoruba culture in Nigeria—as recorded in his books which in certain circles had considerable influence. Needless to say, Rudolf Steiner never engaged with such ideas.

the gods, much like the biblical 'male and female' man in Genesis who is divided into Adam and Eve. The *Timaeus* and its fragmentary companion *Critias* in particular offer extraordinary parallels to the mythology of the Bible. The emphatic single night of destruction is similarly typical of biblical vengeances, as with Sodom and Gomorrah, or the night of the Angel of Death passing over Egypt. There is definitely some truth in the old quip that Plato was 'a Moses speaking the language of Athens'.* Where exactly he acquired his mythology, and whether it sheds any light on paleogeography are both questions we must still defer for the present. But it will be the mythology reflected in Plato's warning tale, not his fiction itself, that will concern us.

At the close of the nineteenth century, Donnelly fostered a widespread eagerness (one might say a fad) for information about Atlantis, of a type which temptingly appeared to accord with developments in the evolutionary understanding of the Earth's geology and geography. Plato was at hand to bring the lost land intriguingly and vividly alive. But the trend also encouraged another line of research, altogether more interesting, to put forward its conclusions which, though tarred by the same mistaken obsession with the Atlantic Ridge, fundamentally drew on quite other sources of knowledge.

An Emerging Story and a Startling Map

W. Scott-Elliot's *The Story of Atlantis* was published in 1896, and from the same author came in 1904 *The Lost Lemuria*. From 1925 onwards the two short works have been repeatedly re-issued as a single volume. The author was a Theosophist, and his account of events has come down in effect as the definitive version of an Atlantis story which is not based on Plato.† One of the founding fathers of the Theosophical

---

* The myth of Phaethon (*Timaeus* 22c) is discussed in connection with Near Eastern Lucifer mythology in Neil Forsyth, *The Old Enemy. Satan and the Combat Myth* (Princeton 1987) pp.67ff, 130fff (to which we shall return); a full-blooded approach on these lines to the *Timaeus*, though to be treated with caution in details, is Margaret Barker, *The Great High Priest* (Edinburgh 2003) pp.262-93. Plato recounts a myth close to the biblical Adam-and-Eve story, *Symposium* 189d. Biblical vengeances in a single night famously include Sodom and Gomorrha (Genesis 19:23: the smiting of the Egyptian firstborn (Exodus 12:29); the fall of Babylon, Daniel 5, 30-31; etc.

† One of the few places where there is an apparent reference to Plato in Scott-Elliot's book is actually a misunderstanding (shall we call it). Its

movement, A.P. Sinnett, furnished the book with a Preface explaining that it was based on something which was also attracting a good deal of interest at the time. 'For anyone who will have the patience to study the published results of psychic investigation during the last fifty years,' he wrote, 'the reality of clairvoyance as an occasional phenomenon of human intelligence must establish itself on an immovable foundation.'* Less widely known were the results of its systematic use to investigate the happenings of past history, even that of the remotest times. But Sinnett pointed to the phenomenon of what we would now

---

author had evidently read A.P. Sinnett's *Esoteric Buddhism* (1883), which summarises Theosophical cosmology and teaching about successive continental cycles. It tries to uphold a tenuous link with Plato's story, while basically acknowledging that the real Atlantis was older by far. The land, says Sinnett, 'the destruction of which is spoken of by Plato, was really but the last remnant of the continent ... its last island, that, translating its vernacular name, we may call with propriety Poseidonis' (*Esoteric Buddhism* repr. London and Adyar 1972) p.49. Thus it is clear that Sinnett invented the name Poseidonis (whatever was its vernacular original) for the land which Plato called Atlantis in his Dialogues. The name Atlantis had been appropriated to the Theosophical idea of an archaic continent/age. Sinnett's invention of a left-over island conveniently linked the Theosophical cycles with the Greek account and the name appealed to those with a basic knowledge of the *Timaeus* to help them out, since Plato does of course repeatedly mention the Greek sea-god Poseidon, the patron of Atlantis and begetter of its first dynasty. Sinnett's notion that Plato's Atlantis was actually Poseidonis unfortunately led Scott-Elliott to suppose, wrongly, that Plato had used the term. H.P. Blavatsky perhaps did not help with some slapdash remarks, when she said about certain Atlantis traditions: 'whether it was the "Poseidonis" mentioned in "Esoteric Buddhism," or the Continent of Atlantis, does not much matter' (Blavatsky, *The Secret Doctrine* (1888; repr. Los Angeles 1974) vol. II p.265). Apparently Blavatsky was not much taken with Sinnett's clever idea, though she also speaks of 'the submersion of Plato's Atlantis, or Poseidonis, known to the Egyptians only because it happened in such relatively recent times' (id. p.314). Was it this phrasing which confused Scott-Elliot? See his *The Story of Atlantis and The Lost Lemuria* (repr. London and Adyar 1968) p.2. Whatever else, all this goes to show how little Scott-Elliot's account was based upon Plato!

* A.P. Sinnett in: W. Scott-Elliot, *The Story of Atlantis and The Lost Lemuria*, Preface pp.v-vi. The extent to which acceptance of psychic and spiritualistic phenomena had indeed permeated the thinking of the period in the way Sinnett suggests can now be appreciated thanks to the very extensive documentation by Janet Oppenheimer, *The Other World. Spiritualism and Psychical Research in England 1850-1914* (Cambridge and New York 1985). For the important role of Theosophy, pp.159ff.

call regression, where under hypnosis a suitable subject can be induced to relive and describe events of his or her early life, or even to recite as if by an eyewitness happenings that apparently relate to a former life, or to places and times far removed from the present. 'And from this thought we may arrive by an easy transition at the idea that in truth the records of Nature are not separate collections of individual property, but constitute an all-embracing memory of Nature herself, on which different people are in a position to make drafts according to their several capacities.'*

It is not initially clear, however, quite how Scott-Elliot really worked. He himself was a Scottish laird, though involved with the London Lodge of the Theosophical Society as was his wife Maude Boyle-Travers, referred to sometimes as 'Mary'. She was a psychic, and her spirituality so impressed A.P. Sinnett that he regarded her as in touch with one at least of the Masters of Wisdom, who were the true leaders of the Theosophical movement. They were the highest adepts of occult learning, and they had communicated with Sinnett himself as well as with Madame Blavatsky in a celebrated series of letters which seemed to arrive, miraculously, from far away places. The Theosophical movement had no higher authority, so his trust in 'Mary' was praise indeed. Over a period of more than ten years she acted for Sinnett as a 'scryer', going into trance states and reporting her visions—including some of Sinnett's past lives going back to ancient Egypt. Following a mesmeric 'experiment' soon after their first acquaintance, intensive sessions began in 1886, continuing whenever she was in London.† It is natural to assume that it is to her that Sinnett refers in his Preface to *The Story of Atlantis*. The part of her husband William would have been to guide the research and relate the results to the findings of modern science, meaning of course the discovery of the Atlantic Ridge and the usual range of other issues concerning the distribution of flora and fauna, etc. There is a strong though not wholly clear tradition within the Theosophical Society, however, that the real source of Scott-Elliot's knowledge was not uniquely his wife's clairvoyance. It is reported to have been

---

* Id. p.vii.

† A.P. Sinnett, *Autobiography* published as a supplement to *Theosophical History* (London 1986) from a typescript copy in Adyar, the movement's Indian Headquarters, pp.33-4, 41-2, and p.45.

none other than Charles W. Leadbeater, then certainly a rising star in the movement—with some truth, perhaps, but that statement is not without its own difficulties.*

Significant as we may find Sinnett's claim, therefore, the background and presentation of the work hardly inspire serious confidence.

---

* The tradition rests on a letter of C. Jinarājadāsa, a later President of the Society, written in 1947. His comments come therefore from long after the appearance of the book—or rather books: for his remarks basically relate to a work called *The Great Law* (1899). This gives an interpretation of the history of religions strikingly in accord with the ideas of Annie Besant and Charles Leadbeater. He claims that they in fact wrote it, but that Scott-Elliot lent his pseudonym (!) W. Williamson to the work to give it greater circulation. (Scott-Elliot used the name again subsequently for a book about Jesus Christ.) In *The Great Law* there is a chapter on prehistory which is a rehash pure and simple of *The Story of Atlantis*, which material, however, Jinarājadāsa specifically says was the work of Leadbeater, only aided a little by Annie Besant. Leadbeater is thus in effect credited with the work of 1896 as well, and especially the maps. It must be said that this is not the way that Scott-Elliot himself speaks of his sources (e.g. *Story* pp.77-8: see our Appendix). And it is a somewhat surprising claim, since Leadbeater did not himself publish anything 'more' on this subject for the next seventeen years, whilst his study *Clairvoyance* of 1902 spoke of his ultramicroscopic researches into 'occult chemistry' but treated clairvoyance in time only as a way of seeing the future, not as a way of exploring the past. It was Sinnett who had written about Atlantis in *Esoteric Buddhism*, and it was with Sinnett that Scott-Elliot and his wife were intimate. 'Mary' had spent innumerable hours in mesmeric states, whenever she was in London, responding to his queries. This recalls Sinnett's statement in the Preface (p.x) that 'Every fact stated in the present volume has been picked up bit by bit with watchful and attentive care, in the course of an investigation on which more than one qualified person has been engaged, in the intervals of other activity, for some years past.' He knew well about putting together, 'bit by bit', the results of such research, to establish his own sequence of lives. There, on the face of it, lies the obvious background to the book. Would the educated clergyman Leadbeater have made the gaff about Poseidonis? And would he have submitted to have his clairvoyant experiences picked over 'bit by bit by committee'?—surely not his style. However, it would be unwise to rule Leadbeater out (cf. Sinnett's 'more than one qualified person'). The maps will need further investigation when we try to understand the transmission which lies behind them. The future President Jinarājadāsa was a protegé of Leadbeater's. When Leadbeater was disgraced, but subsequently allowed to rejoin the Society, an outraged Maude Scott-Elliot (as 'Mary' had now become) left the movement in disgust. Did the President's memory quietly minimise her, and exalt his patron, in loyalty? Jinarājadāsa's letter is now in the Theosophical Society in America's archives. The relevant extract can be found quoted at https://theosophy.wiki/en/William_Scott-Elliot.

Nonetheless, the content which emerged from these researches, most especially the celebrated 'Map of Atlantis', seems to me of extraordinary interest—far more so than could be expected from the incompetences of those involved in its publication—and is not without a connection to Rudolf Steiner's views on the subject. Certainly he found there an approach and a body of ideas that helped him to elaborate his own profounder thought.

In the early years of the twentieth century Rudolf Steiner wrote a series of articles in which he gave first expression to his views on prehistory. They were collected into book form in 1903.[*] Steiner at that time was well known as a spokesman for modern evolutionary theory and its intellectual ramifications in science and philosophy, personally linked with the most prominent German evolutionist Ernst Haeckel, about whose importance he wrote a book. Steiner at that stage had won considerable recognition as a philosopher—though he had been educated in the Technical University, which in those days meant that it was still hard to gain recognition or even qualifications in 'academic' circles. He came from an essentially scientific background and even in his pure philosophical writings he was concerned essentially with the issues raised by modern science, such as freedom and determinism. Thinkers such as Haeckel (something like the Richard Dawkins of his day) were committed to a blusteringly deterministic approach. But this, countered Steiner, was not really consistent with their own evolutionism. His *Philosophy of Freedom* argued against the notion that man could be understood as if fixed at a single stage of development, meaning he could not possibly be 'free'. His own work was, Steiner wrote, an attempt 'to do for philosophy what Darwin had done for the natural sciences' (Ch. 12). Philosophy, for him, did not prescribe 'answers' to the puzzles about human knowledge, but revealed them as 'riddles' which turned attention back on the mind and its ever-changing perspectives. Hence his philosophy took the form of a history or evolution of thought,[†] and the great thinkers were

---

[*] Translated as Steiner, *Cosmic Memory* (New York 1971).

[†] Hence his stated aim of 'developing through the account of the history of philosophy, philosophy itself'—Steiner, *Riddles of Philosophy* (New York 1973) p.xxiv. As an example of his method, one might take his short treatment of Einstein, whose way of thinking is treated not as an answer to questions of physical reality, but as drawing our attention to our own situation as thinkers which must be overcome through a further development or spiritual science: 'That the theory of relativity forces us to think in this way constitutes its

seen not as building a timeless structure of truth but as vitally connected with the wider struggles of their time. He wished to bring evolutionism into thought in a very radical way. (We recall how discovering evolution was itself a matter largely of finding such a new perspective.) Man could only grasp his freedom by learning to think in a free way, in what would be in reality a further evolution of his being ('the spiritualised theory of evolution carried over into moral life').*

From the beginning of the twentieth century, Steiner sought to elaborate these ideas in the approach he termed 'anthroposophy' (sketched out in principle in the 'Conclusion' to *The Riddles of Philosophy*). Science for its own consistency needed to be open to less closed kinds of thought if it was to carry through the programme of evolutionary understanding. Nowadays there is increasing recognition that evolution at the outset had been interpreted, e.g. by its co-discoverer Alfred Russel Wallace, as implying a further dimension of life.† Taking 'development' seriously implies that there is more to things than is manifest at any one time; the implicit presence of the whole environment in the individual entity is a clue to the nature of consciousness. But evolutionary biology cannot make full sense, as Steiner saw, if we continue to think only materialistically even in the physical sciences. It required that extension of consciousness to which we have referred. As Janet Oppenheimer has already remarked, he was by no means the only significant figure to find in the esotericism of Theosophy some positive signs of a new world-view. It is certainly wrong to think of Steiner's decision to work with it a change of direction, or a conversion. Esotericism appealed because it offered what he needed to advance in his own scientific thought. One way of unifying Steiner's own development right up to the completion of his *Outline of Esoteric Science* (1910) is to trace his continued aspiration to apply evolutionary understanding to the cosmic world-picture as a whole.

At any rate, Steiner acknowledged in his own *Preface* of 1903 that he was impressed by Scott-Elliot's results, and accepts them fundamentally into his own approach. We shall shortly see why. But he declines to

---

value within the development of world-conception' (op. cit., p.444, and generally pp.442ff).

* Steiner, *The Philosophy of Freedom* (London 1970) p.169.

† Cf. C.H. Smith and G. Beccaloni (eds.), *Natural Selection and Beyond. The Intellectual Legacy of Alfred Russel Wallace* (Oxford 2008).

review them point-by-point.* Specifically, he now wishes to *supplement* the 'external' information given by Scott-Elliot (which I take to mean more or less physical-geographical information) since his own interests lie fundamentally elsewhere. Certainly he has not taken anything specifically from Scott-Elliot, still less attempted in that amateurish fashion to prove it from the discovery of the Atlantic Ridge. It is not easy to know whether he draws on any direct acquaintance with Scott-Elliot's sources as such when he mentions behind them 'the knowledge hidden within the theosophical movement'. More likely, he means simply on his own supplementary research and his own esotericism on which it was based. His obligations to the conditions under which he received this knowledge are mentioned occasionally in the course of the book. Later he ceased to work in this way, and simply presents his own researches whose methodology we shall shortly attempt to clarify.† What we can say is that *The Story of Atlantis* provided him 'in externals' (i.e. physically and geographically) with conclusions acceptable as a starting-point, which must mean fundamentally convincing and compatible with his own.

So after all our unpicking of the story behind the *Story of Atlantis*, we may state some basic conclusions: from clairvoyant 'regression' techniques, and from contributions of other kinds, Scott-Elliot believed he could amplify in detail the Theosophical teaching about an archaic continent which he is confusedly content to identify with Plato's Atlantis. Virtually nothing that we learn in *The Story* actually comes from Plato, however. On the other hand, it presents us with some startling information which, whatever its ultimate source, breaks through the limitations of its chequered presentation. Though we cannot know exactly what Rudolf Steiner thought valuable in Scott-Elliot, those 'external' factors of physical geography are undoubtedly worth looking into further. The Atlantis-Map is crucial, in view of Steiner's account of the changing Earth in 1912, and especially nowadays, when we have available the palaeogeographical reconstructions already referred to. The Map is likely to be the esoteric core of the clairvoyant experiment, transcending the 'explanatory' framework (see further Appendix).

---

* He did point out some mistaken aspects of the research behind it, however: see *Theosphy of the Rosicrucian* (London 1981) p.41.

† He is palpably fretful of the situation already—i.e. being 'still obliged to remain silent about the sources of the information given here…. But events may occur which will make a breaking of this silence possible very soon'—*Cosmic Memory* p.41.

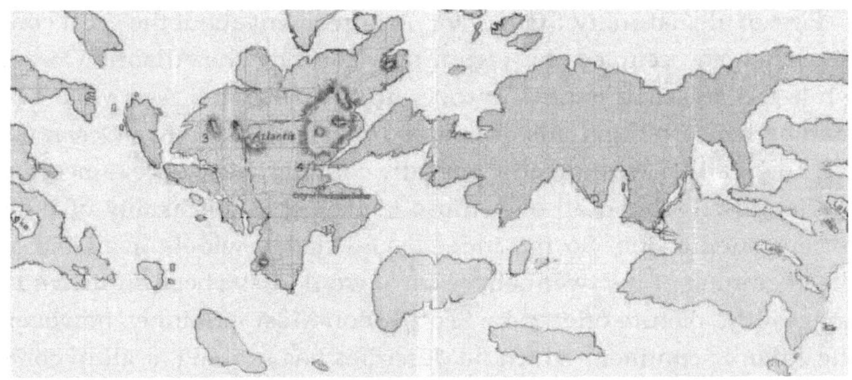

Primary Map of Atlantis: Scientific efforts to reconstruct the pattern of the globe in former geological ages were held back by the resistance to the 'moving continents', and confused by notions about the Atlantic Ridge. Meanwhile a map was published by the Theosophist William Scott-Elliot in 1896 which formed the basis of his account concerning the evolution of the Earth and humanity, collected in The Story of Atlantis and The Lost Lemuria *(1925). The map was developed through techniques of spiritual regression and 'cosmic consciousness', though its original provenance remains obscure. Scott-Elliot's work was approved 'as regards externals' (= geophysical features?) by Rudolf Steiner who extended such techniques and sought to make them genuinely scientific. Scott-Elliot relates the map to the era when 'the amphibian and reptile forms which then abounded had about run their course, and were ready to assume the more advanced type of bird or mammal' (p.43) and in harmony with this suggestion the map invites close comparison with modern reconstructions of the Late Cretaceous (cf. p. 70 above). Many features seen in this primary map turn out to be startlingly accurate. The present-day Atlantic Basin is occupied by a vast continent, connecting at its tip with the land now forming Scotland and Ireland—altogether in a different league from the protrusions based on speculation about the Atlantic Ridge. To the west a series of presumably mountainous islands may be compared to the modern reconstruction of Laramidia (the later Rockies), then separated from the large land mass and the basis of later North America. South America is probably to be recognised much farther south, still differently orientated, and not yet joined to the Americas complex—a genuine feature which could be interpreted at all without an understanding of continental dynamics. Other continents appear in a southern grouping indicating their origin in a large land mass including Antarctica, Australia, etc. In modern constructs, large parts of Africa and Arabia are submerged by sea (Tethys Ocean), and this feature is still more exaggerated in Scott-Elliot's map. Some 'alternative Cretaceous reconstructions' likewise suggest a vast open tract of water. (The map is reproduced here freed from the heavy grid and the doubtful correlations with modern geography of the Theosophical original.)*

First of all, naturally, Steiner was in agreement about the great continent which occupied the region now forming the Atlantic Ocean, while the area that is now Europe, north Africa and Asia was submerged beneath hundreds of metres of water. The Tethys Ocean (as we now call it) was already partially familiar to science, since the marine fossils found all over those lands tell unmistakably of their former submersion. So far, indeed, knowledge widely available at the beginning of the twentieth century would have been sufficient to suggest the picture offered by Scott-Elliot. Most strikingly, however, the Atlantic continent which he describes has nothing at all in common with the supposed exposure of lands round the Atlantic Ridge, as some scientists of the time were willing to hypothesize; Donnelly himself could only make the evidence stretch to a modest island halfway across to America, and there was no way that the 'land' there could previously have been more extensive. (We must remember that there was at this stage no accepted idea of the continents moving.) Scott-Elliot's Atlantis is vast:

> The continent of Atlantis itself ... extended from a point a few degrees east of Iceland to about the site now occupied by Rio de Janeiro, in South America. Embracing Texas and the Gulf of Mexico, the Southern and Eastern States of America, up to and including Labrador, it stretched across the ocean to our own islands—Scotland and Ireland, and a small portion of the north of England forming one of its promontories—while its equatorial lands embraced Brazil and the whole stretch of ocean to the African Gold Coast.[*]

In the face of this, Scott-Elliot's inevitable reference to the soundings taken by the good ships *Challenger* and *Dolphin* seems even comically inept. What we have here is outside the Donnellian framework altogether. It is comparable instead to the results of that much more recent reconstruction of the Cretaceous world which show the Atlantic Basin ('the whole stretch of ocean to the African Gold Coast') occupied by the huge continental block which in scientific parlance goes by the name of Laurentia. Included on the Laurentian tectonic plate are lands that will be parts of the present-day United States, but also Greenland and—again just as the 'Atlantis' of the 1896 map indicates—connecting to portions of Britain (Ireland, the Hebrides). The basic similarities of contour with *The Story*'s map are astounding.

---

[*] *Story* p.18.

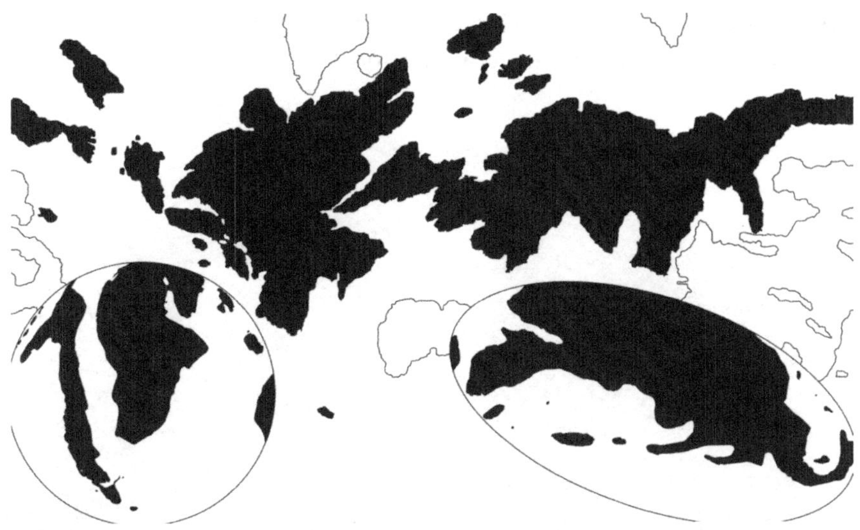

Comparison of the modern geomagnetic and the 1896 maps reveals an astonishing agreement in the size and contours especially of the Atlantic continent and Eurasia. This concurrence of results could hardly be guesswork based on contemporary ideas, and indeed could not even be interpreted at the time when the notion of moving continents was unknown to science: the secondary maps added in Scott-Elliot's book (see our Appendix) purport to show subsequent geographical changes still based on a static model (and the Atlantic Ridge!) with little scientific plausibility. The 1896-map continents are here shown for comparison against geomagnetic outlines, inset, after Scotese.

Scott-Elliot's palaeomap, far outstripping his flimsy science, gives us a continent which is a worthy model for a continental evolution that will lead to the modern disposition of the world, as it is constituted nowadays on the basis of 'continental drift'. The positioning of the Atlantic continent, as Scott-Elliot says 'stretched across' toward Africa and Britain, simply means that the Atlantic Ocean did not yet exist. Scott-Elliot of course could not have drawn upon plate-tectonics, which explain how it subsequently opened up. Other details are likewise curiously suggestive in the context of modern science. It is interesting to note for example the several long, in effect virtually continuous islands off the western coast of the continent, extending even farther to the north and almost meeting the far tip of Eurasia. Even though the published map is sadly almost completely devoid of any relief-indications, these certainly invite comparison with the orogenous formations nowadays termed Laramidia—the chain of the 'Rocky Mountains' embedded in present-day America. (The Western Interior Seaway which then stretched between them varied considerably in width by a few hundred miles during the Cretaceous Period.)

The independent chain of peaks once rising out of the ocean from across the Western Seaway, known in palaeogeography as Laramidia, has turned into the backbone of the subsequently developing continent of North America—forming the Rocky Mountains.

Other features of the map may be surveyed briefly here—before we come to tackle some of the problems in understanding what it tells us.

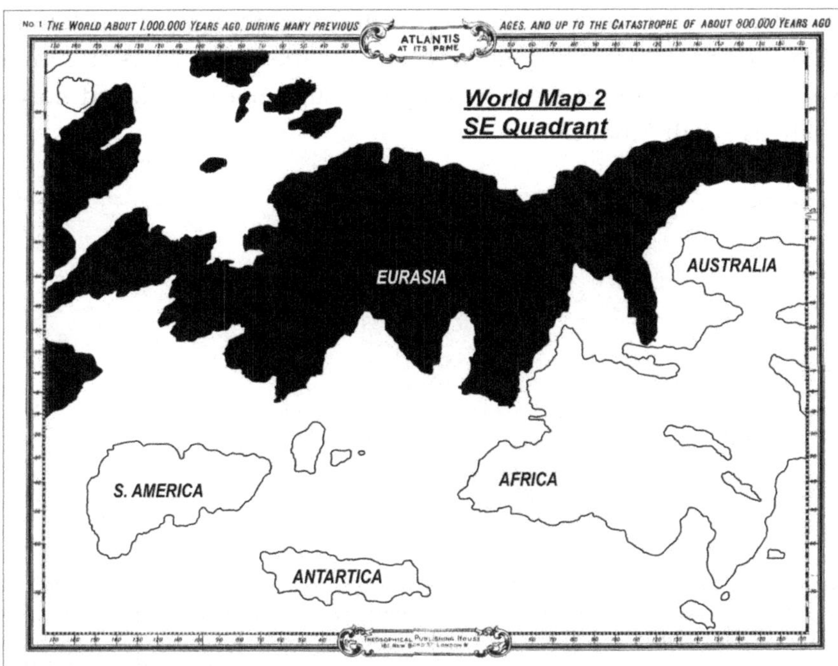

The map published by Scott-Elliot is here shown in outline with suggested identifications of the land masses as they have evolved into our modern continents through movement of the tectonic plates. Those movements have since been traced by geomagnetic methods—see next map.

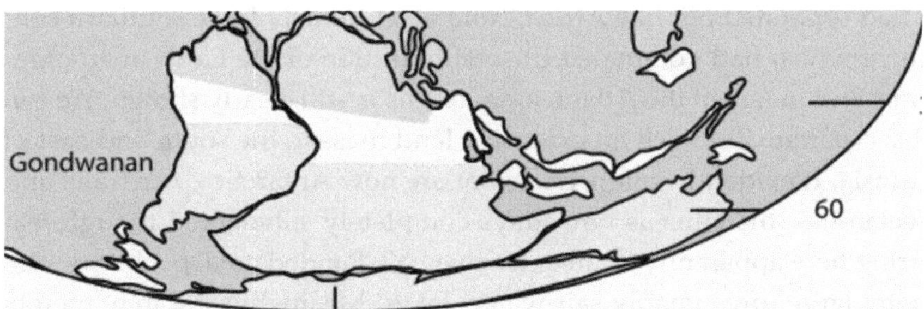

Scott-Elliot's map of the southern hemisphere suggests that a large continent, which he termed Lemuria, broke up into smaller land masses and formed new combinations. In modern research such a continent is designated by the name Gondwana, from which Antarctica, Australia, Africa and South America originated.

Africa was indeed extensively submerged in the later Cretaceous, and the wide ocean dividing it from Eurasia was already surmised from fossil evidence. In the map, there appears a substantial land-formation, shown partly coinciding with the present Horn of Africa, which it is undoubtedly tempting to consider part of the African continent. But its placement so far south seems exaggerated, and in view of our knowledge of plate-tectonics and the movement of the land masses, it is more plausible to identify it from its shape (though not yet aligned north-south) as South America.

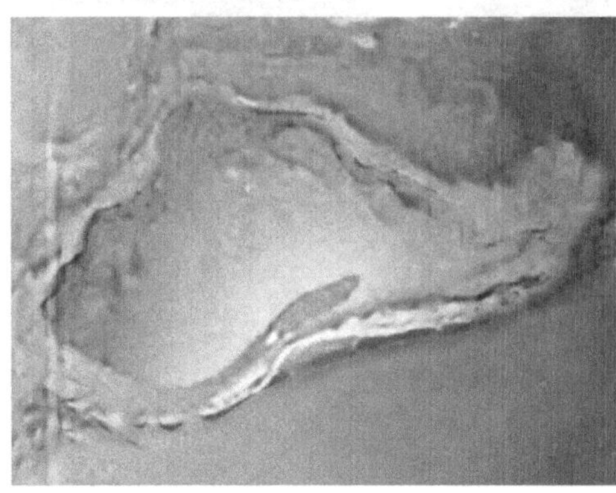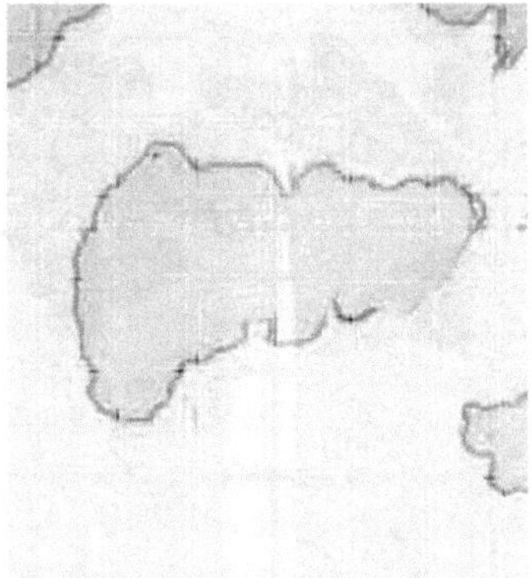

The map published by Scott-Elliot shows a land mass in the southern ocean which by its shape may well be identified with a still unattached South America, and by its location indicates the break-up of the large supercontinent which is now called Gondwana, equivalent to Lemuria in Steiner's understanding. From Scott-Elliot (1896) and outline after Scotese.

South America at the period under consideration was still far removed from its later position joined to still formative North America. It had separated off, like Africa, from an extremely large southern continent which had dominated the configuration of the Earth in an older time: and in fact in the Atlantean map this is still clearly shown. We can observe there just such an extensive land mass to the south and east of Eurasia. It evidently comprises what are now Antarctica, Australia and Zeelandia—the latter is nowadays completely submerged though featuring here apparently as an extensive half-flooded area, pocketed with many large (presumably salt-water) lakes. Meanwhile we may merely comment on the striking general agreement with modern research as to the shape of Eurasia, with its European 'promontory' and protracted Asiatic extension, although there is a sense from the Atlantean map that overall it has been positioned there rather too far south. (It is of course nowhere said how the correlation of the map with the modern world in the Mercator projection in the published version was determined.)

I would urge that the several general observations we have just made concerning the two world-maps, for the Cretaceous and 'Atlantis', arrived at by utterly contrasting methods, cannot conceivably be the result of chance similarity. If once we forget that notion of trying to prove Plato, then (1) the representation of a very large continent where the present-day Atlantic Basin lies, together with (2) a Eurasian land mass strikingly concordant with that reconstructed by modern means, (3) a South America still far removed from its current location and in relative proximity still to the supercontinent from which it broke away, also (4) shown and including Australia along with much more in its very extensive southern territory—all these things are too similar to geomagnetic and geological reconstructions of the late Cretaceous world to be accidental. And since those modern techniques were as yet unheard of, and even the notion of attempting them was dismissed, and would be dismissed for thirty or forty years afterward, it is indeed hard to see how the 1896 map could have been the result of happy guesswork based on current knowledge. Indeed contemporary knowledge of the historical geology of the continents and oceans was so much in its infancy that it led to quite ineffective rationales and inadequate foundations (Atlantic Ridge, land-bridges etc.). By whatever means we are to account for it, where Scott-Elliot could not, and whatever sceptical caveats we may want to keep in place, the Atlantean map was (and is) a stunning achievement.

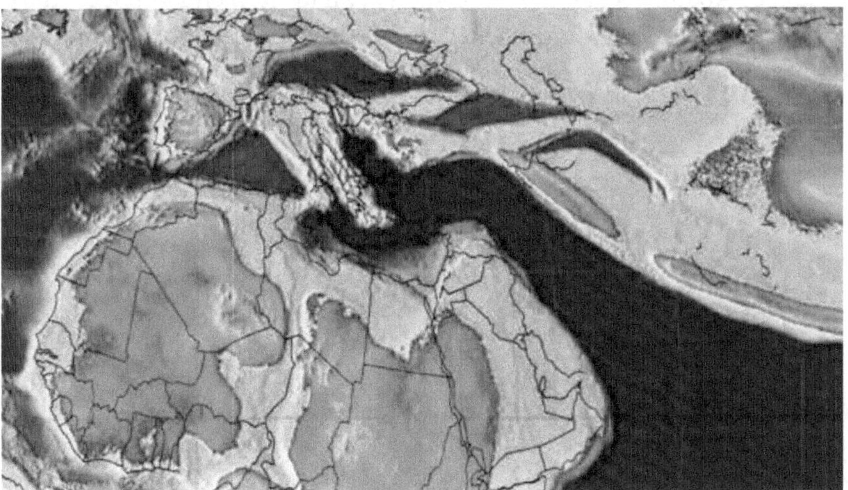

A striking feature of the 1896 map is the indication of a vast tract of open ocean where one would expect to find signs of formations related to modern Africa. Modern palaeo-mapping of the Cretaceous indicates high sea levels and the submersion of large parts of North Africa and Arabia. At least one alternative reconstruction shows even more dramatically an ocean-expanse that bears comparison.

Difficulties there certainly are too. But there are also many more details which open up significant connections between the results, as one might say, of ultra-modern science and clairvoyant vision. Many features which were impossible to comprehend, and which may have complicated the presentation in 1896, can be more clearly interpreted (like the moving South America) in our own day. And again, many of them go far beyond the possibility of accidental resemblance. We may take one final example: the late division of the ancient continent into two parts described by Scott-Elliot and shown in one of his additional maps.

Scott-Elliot reaches out for explanations not just to the work of *Dolphin* and *Challenger*, but to a farrago of causes great and small. Not all the catastrophes which destroyed Atlantis, he explains, were like Plato's single day and night of devastation: many were 'comparatively unimportant landslips such as occur on our own coasts today. When the destruction was once inaugurated by the first great catastrophe there was no intermission of the minor landslips which continued slowly and steadily to eat away the continent.' This is his way of reaching out to the popular understanding of how Atlantis 'sank beneath the waves', supposedly reflected in all those folklore versions of sunken lands from Lyonesse to the Dvāraka where once Krishna reigned. But piecemeal submersion of large areas (one way or another), until everything has finally gone, makes little geological sense in relation to a continent almost as large as Africa. In fact, the occult version of Atlantis bears no real relation at all to 'sunken land' legendary. At best, Sinnett's Poseidonis tacked it on, but no explanation beyond Plato's fictional one is mooted.

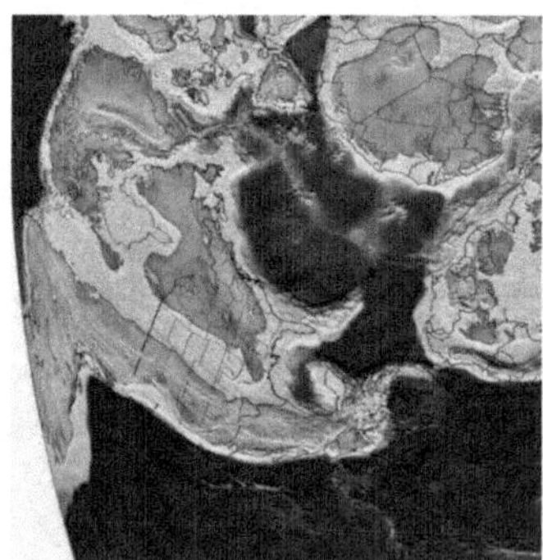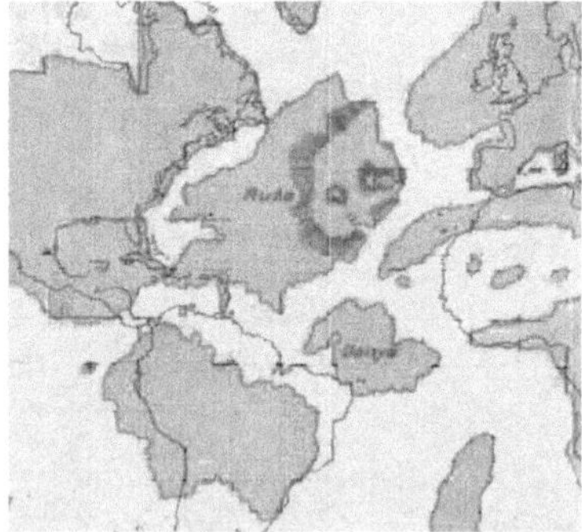

Map showing division into 'Ruta' and 'Daitya'. Although the supplementary maps in Scott-Elliot's book are of lesser scientific interest, one of them does reflect important ideas concerning the development of the continents that were known prior to his published account—ideas which once more could not have been inferred from scientific knowledge of the day. It shows the result of the still higher sea levels in the latest phases of the Cretaceous, when the continent occupying the Atlantic Basin broke up into two halves, said to have the names Ruta and Daitya. Confirmation of the truth behind this development was obtainable only with the rise of geomagnetic techniques. Here Scott-Elliot's subsidiary map is compared with a particular modern version of the final configuration of the Cretaceous. From Scott-Elliot (1896) and Nicolas Scotese, Late Cretaceous Atlas paleomap.

However, what Scott-Elliot relates of the subsequent shape of Atlantis makes extremely interesting sense when once we think in terms of *rising and falling sea levels*—never once considered in the book. But in that context paleogeography has once again come to remarkably concordant results.

Sea levels rose dramatically in the last phases of the extensive period which, as we have seen, is basically named from the Cretaceous marine deposits. Smaller time-divisions are distinguished by scientists on the basis of local strata which are exposed by circumstance at various sites, lending their name to a subdivision which denotes a time-period within the large geological phase ('chronostratigraphic stages'): Maastrichtian, for example, or Campanian. In the late Cretaceous stages, rising sea levels across the whole globe did indeed split the Laurentian land mass into two main parts—a larger northerly, and a somewhat smaller southerly, pair of islands.* By the time known as Campanian (a late division within the Cretaceous) the northern one has become partly fragmented by the extremely high level of the ocean; the Western Seaway is very broad and large areas are flooded in what has now become the Hudson Bay region. Meanwhile nearly all of later Europe is under water, with the exception of some small parts of Ireland and Scotland, the Iberian plateau and the Norwegian mountain chain; large areas of Africa and South America are likewise submerged.†

It is not massive vulcanism which has produced this situation—but more, if one may be forgiven the expression, a Universal Flood.

Scott-Elliot quite remarkably describes this late division of the large continent into two lesser land masses, northern and southern. He characterises this phase as the result of gradual processes, rather than outright catastrophic agencies—only he fails to comprehend the underlying reason. He transmits the terms 'Ruta' (for the northerly), and 'Daitya' (for the southerly), as the names of the two lands. The names show that the interpretation of the Atlantis story has been affected at this point by the lore of the Hindu Purānas, where they refer to semi-divine races. The races are interpreted in connection with Atlantis by Blavatsky in the *Secret Doctrine*, and it was doubtless there that Scott-Elliot found a reference to 'the famous island of *Ruta*, and the smaller one of *Daitya*' which must be distinguished from 'the main continent of Atlantis' and

---

* See Nicholas Scotese's Map 21 in the *Atlas of the Late Cretaceous* (Paleomap Project) and our detail from it with comparative map.

† *Atlas* Map 19.

its large-scale disaster. There is nothing in Indian folklore to suggest, however, that the two islands were a stage in the evolution of a large continent at all, which is the way the story is developed by Blavatsky and Scott-Elliot. But Blavatsky's reference does attest that the interpretation already existed before *The Story of Atlantis* was written, even though it is a story she does not exactly tell.[*] She also knew that, whatever exceptions might be claimed, 'it was several million years ago that the main Atlantis perished'.[†] Attempts to pull it forward into historical times, and link it to the vagaries of Plato's storyline, we too must strenuously resist. To seek its human reference in Plato's mythic world is to overshoot the mark the other way, actually making it too late to chime with the time frame of human evolution; the Cretaceous Period is, on the contrary, far too early. Yet Atlantis will reveal its meaning for humanity's place in the world as we come to understand Rudolf Steiner's thought more exactly.

---

[*] Naturally one must bear in mind the esoteric, i.e. protected nature of the materials concerned, but alongside the interesting aspects of the primary map there are many disturbing features in the Atlantis-presentation by Scott-Elliot which could only be finally clarified if we were to know just what came from where. Does all the material come from a unified tradition, or is it compounded from several as often seems the case? Did either 'Mary' or the 'committee' sift out the valid components satisfactorily? These questions have a certain fascination in themselves, which I attempt to elucidate in the Appendix. I conclude that the primary map of 'Atlantis at its Prime' (the one we have discussed) is an original source, and goes back considerably prior to Scott-Elliot. I offer reasons to suppose it may well stem from the great magus of Elizabethan times, John Dee. The Renaissance voyages of discovery and rapid advance of the sciences of geography and cartography, in which Dee was centrally concerned, provide the context for an interest in mapping the disposition and history of the continents. Other features of *The Story of Atlantis* cannot be much earlier than Scott-Elliot's own time, particularly as regards the secondary maps.

[†] Scott-Elliot, *Story* p.17; Blavatsky, *Secret Doctrine* vol. II pp.222, 314.

Chapter 3

STEINER'S 'CLAIRVOYANCE' AND EVOLUTION

Suppressed Faculties

We have seen that there are then many reasons for thinking that Rudolf Steiner accepted, in the 'externals' of geography at least, the paleomap of the world which formed the basis of *The Story of Atlantis*. It fits exactly the descriptive paragraph which we began by quoting, in which Steiner explained the inner dynamism of the Earth and the ever-altered face of the world.* It accords too with a scattering of significant details mentioned by Rudolf Steiner, e.g. the fact that the present-day island of Ireland is not actually a part of Europe's geological configuration, but a remnant from the land mass which once occupied the Atlantic Basin— Atlantis as Steiner called that older formation. The opening-up of the Atlantic Ocean left it stranded, so to speak, and now widely separated from its original geological context.

Now it would not have been possible to ascertain that fact from the conventional methods and acquired scientific knowledge of Steiner's time. Only through the advances in understanding that came with the acknowledgement of 'continental drift', and detailed modern geomagnetic reconstructions, can it be confirmed today. Yet it is already indicated in the map which stands at the head of the series in Scott-Elliot's first pioneering book at the close of the nineteenth century.

It is worth reiterating that his primary map contains so much that is astonishing, so much which simply could not have been arrived at through the ordinary science of its day, that its chief points will bear repetition briefly once more. If it contained nothing further, it would be remarkable for the vast extent of the Atlantic continent, out of which the North American land mass will develop while the Atlantic Ocean will open up to divide it from Europe; remarkable too for the general size and disposition of Eurasia, and the extensive ocean which floods the space now occupied by the southern continents; more remarkable even for the indication of a huge southern land mass which appears to include the several continents of Australia, Africa and Antarctic: but it

---

* Steiner, *Spiritual Beings in the Kingdoms of Nature and in the Heavenly Bodies*, quoted above p. 66.

is breaking up and has apparently spawned a South America which has still far to travel before it will form part of the configuration we know as the modern Americas. (Some of Scott-Elliot's 'Lemurian' material would also fit well with modern understandings of the older history and break-up of the supercontinent Gondwana. But we might mention also that certain factors in the evolution of the Earth's surface are lacking—the extraordinary course of development of the Indian subcontinent, for instance, as now understood.)

Rudolf Steiner evidently concurred as regards the external framework. However, he rightly criticised other aspects of Scott-Elliot's work; only in conjunction with Rudolf Steiner's spiritual-scientific approach can we really make use of it. More crucial in that context than the acknowledgement of any specific geographical results is, of course, the readiness to accept that those results can be reached by 'clairvoyant' means. Scott-Elliot understood, in general terms, the reason for having a basis such as the maps and 'records': as a step to seeing. (He meant a spiritual apprehension of a world that no longer existed, outwardly; but of course any map, still more than a picture, always has its reality only in the unrepresented, the world to which it refers.)* Steiner had early adopted the principle of always confirming everything he said by his own researches. The notion of meditating upon traditional symbols or formulae was, he believed, a way that had been right in the past. But science had brought a different kind of objectivity. He sketched out the methodological implications in certain of his philosophical writings, where he spoke of a parallelism but even more importantly of a *convergence* with the results of natural science.† Hence he would, I think, have been better pleased that it is nowadays possible to bring out the significance of his work, e.g. in conjunction with the geomagnetic mapping of the continents, and the

---

* The evidence is clear that maps did not originate from mundane necessities, but from a desire to locate earthly theatres of action within a larger design. Note also that a map does not represent, but retains an abstract relationship to areas which it covered in which the concrete content is not shown: see the fuller discussion of this background sketched out in Appendix I.

† See for instance Rudolf Steiner, *The Case for Anthroposophy* trans. and ed. by Owen Barfield (London pp.25-45).—see especially the chapter called 'Anthropology and Anthroposophy' (pp. 25-45). It is worth noting here that his approach has little in common with the so-called German *Naturphilosophie* of the nineteenth century, which proposed a 'higher level' interpretation to be placed on the results of natural science giving it an overall philosophical 'meaning'. Anthroposophy contrastingly aims to capture the dynamic *in* the process of discovery and interpretation.

recognition of the driving forces within the Earth. Rudolf Steiner arrived at his understanding from the 'other side', one might say. But it would be wrong to suppose that science has ever been without the dimension to which he thereby drew attention. Here we will have to correct some of the distorting lenses that have frequently misrepresented the scientific venture in the interest of polemical concerns. Of course empirical data are crucial. But science has emerged only with the development of large and powerful organising ideas, and a dynamic of further discovery. Its ideas are not like those formerly familiarly in theology or philosophy, profound though they were. Science has to generate ideas that reach out, interact, modify, and cohere in a dynamic way, and eventuate in a unity with the given facts. Such inwardly transformational thinking is now well-known to have come from the Hermetic and 'magical' thinking revived at the Renaissance. Its inner dynamism also has a special connection, as one might say, with evolution. The notion of a kind of active thinking that is intimately aligned with the workings of Nature may also be not unrelated to 'clairvoyance'.

'Convergence' does not mean that Steiner's object was merely to second-guess the future direction of empirical scientific research. 'Anthroposophy' is the aspect of knowledge, which he rightly saw as needing to be reinstated in modern science, which concerns our own place in the cognitive process and in the world where we hold the unique position of 'knowers'. It is well-known that the issue of our own role as observers has cropped up in the context of modern physics, and in a still more comprehensive way in 'anthropic' approaches to scientific knowledge (touched on earlier). The concerns which motivated Steiner were in fact the issues modern science has needed, often rather controversially, to sort out. Steiner seems to go beyond their accepted parameters, however, with his terminology of 'clairvoyant' and 'occult'. But he wanted to develop on his own account a coherent model of research that started from our *inner* connection to the world at the foundation of our knowledge, and that could therefore be investigated only by an extension of consciousness. The basis of it was evolutionary. It may be reduced to two maxims: firstly, that we are not lookers-on from nowhere, but an integral part of the world we know; and secondly, that our knowledge changes us in response to our ever-changing relationship with the rest of the world. Our knowledge and our consciousness have grown out of the world, and that decisive moment when we know ourselves as grasping our surroundings objectively is the end-product of a less conscious but just-as-important deeper-level engagement. An understanding of what goes on at such deeper levels would actually clarify the basis on which science/knowledge rests—the prehistory of our consciousness, so to speak. Such a 'science' would have a validity and inner certainty of its own.

It is all the more important to emphasise the human context of Steiner's 'extension of consciousness', because it is otherwise too easy to suppose that clairvoyant knowledge is somehow alien to normal life, or that by laying claim to it Rudolf Steiner somehow meant to claim a superior vantage-point which we mere mortals must simply accept (or not). Perhaps some do take his results as a series of anecdotal glimpses into a domain that is otherwise impenetrable and unknown to us. Yet that goes against everything that he said on the matter, and it goes against his fundamental aim: the deepening and extending of human life and creative thought, not a leaving it behind. We may start to understand what he really meant if we turn, for example, to his account of extended consciousness in an early essay. In it he touches first on a familiar philosophical point, with a well-known and slightly mediaeval slant. (Aristotle was then and is still a potent influence on the life-sciences after all.) It is about how we categorise things as the basis of our knowledge.

There is a distinction to be made between this wolf before us in its paddock at the zoo (a 'particular' wolf), and the idea of wolf in general (the 'universal' wolf) which covers all such animals, of whatever size, sex, age, condition of health etc. Without the universal concept, we would not know that this was a wolf we see. Yet from another point-of-view, the idea 'wolf' appears to be just a compilation of what we have observed of actual wolves put into a general term. We rather assume that such an idea is just a something in our heads. But the clairvoyant, he says, engages in a mental process which makes the picture-idea wolf into a real experience; he *sees* the universal wolf.* He does not mean by this, of course, simply to say that the mediaeval philosophers were right about universals: for their philosophy on this subject, whether 'realist' or 'nominalist' in tendency, certainly had no room for clairvoyant perception! Rather, he indicates, what the mediaeval logician grasped abstractly and in a rather stereotyped way, can be explored concretely, as a morphological reality, through the extended consciousness he speaks of.

----

* *Philosophy and Anthroposophy* (London and New York 1929) pp.37-9. The renewed Aristotelianism of Steiner's teacher Franz Brentano at the University of Vienna was one of the major founts of developments in twentieth-century philosophy—such as Phenomenology and psychology. For Steiner's relation to Brentano's ideas, cf. *The Case for Anthroposophy* pp. 69ff.

We have remained, both in science and religion, Steiner argues, rather mesmerised by this moment of the particular illumined by the universal in our knowledge of things. Much effort has been spent trying to reduce the one to the other or vice versa. Materialism has sought to hold firmly to the reality of the particular, deeming the idea a mere abstraction—but even in its more sophisticated forms, the result was an impasse in which it seemed that ideas must fail to provide real knowledge altogether (the so-called Kantian view that 'things-in-themselves' are beyond our scope, a view firmly espoused by a surprising number of scientists in Steiner's time, against which he endlessly contended). At the other extreme, ideas lack teeth (so to speak) except as they are applied concretely, and if stretched beyond their range became vacuous.

Steiner is determined to inject a new practical spirit into the debate. Firstly, it is obvious that we always bring wider experience to our knowledge of the particular. We see, for example, a gate which has been left open. The perception is informed throughout by our knowledge of the context, of the kind of activities that go on in the farmyard, the carelessness of the farm-hands in tying it back, the distractions caused by outside events in the lane, etc. But is it not extraordinary to paint the picture as it often has been, as a process of building-up from particulars ever higher level abstract concepts spreading out farther and farther away from the immediate scene? In reality we do not just generalise more and more. Our concepts enable us to *see more in* the particular situation through bringing to it that wider knowledge: surely we are more fully and deeply engaged with the scene before us, recognising the way the event has happened, as we enrich our concept through broader understanding of the background? Steiner's intellectual hero, Goethe, in his scientific writings, spoke of the 'perceptive power of thought' and realised that it was exactly this kind of insight, or a seeing-into-how-things-happen, rather than endless classifying particulars under general laws, which was crucial to the kind of understanding we call scientific. However, science long remained subject to the ideal of classifying things under general headings—and the recognition of the true process of scientific insight is still partially held back.*

---

* Cf. Steiner, *Philosophy of Freedom* (London 1964) pp. 41ff. Wittgenstein more recently unpicked the kind of wrong-thinking that long influenced the official description of science. He showed that it is never possible fully to isolate the 'particulars' of knowledge—'just what we see'. Mary Warnock in response drew the interesting, and rather Goethean conclusion, that therefore 'we must think of perception as containing a thought-element, and, perhaps, we must think of

Now if we reach real insight out of what comes to us initially as a thought-component, and not only by adding-in more sensory data, and if we grasp it in its coherence and immediacy going beyond the bare perceived object, we have already in principle opened the way to that deepening of the perception of reality which Steiner was accustomed to term clairvoyant. In essence, clairvoyance is a consciousness which is able to see farther into the web of factors out of which we normally see only the open gate. Everybody can sense abstractly what lies in the background; and there is much evidence that some exceptional people are able to reach a perception of what actually did happen. For if ideas are not basically mirror-images made in our heads but stem from our larger but unfocussed participation in reality, it seems that we actually know more than the sharply defined content of our consciousness.

Suppose we could become more fully conscious of what we always discern semi-consciously as playing into the larger situation, as the implicit history behind what we see just now before us? For most of us the 'meaning' of what we see is formulated as an idea, or thought. But if thought is not just made up of abstractions-from-reality, but takes us somehow more deeply into the reality of things, it could become more than we normally mean by just a thought.

What really happens when we immerse the immediate scene we perceive before us in thought and 'understand' it? Treating either the thought or the thing as ultimate has always led to philosophical dead-ends. Steiner's response is breathtakingly simple. He asks instead how the polarisation of thought and thing arises, and suggests that it is a way *we* carve up the world so as to know it. Knowledge is our affair, as he said. It is quite wrong to suppose we have to presuppose ideas and particulars as a starting-point, only to discover that it is difficult to see how we can know the world at all in terms of them. Nor are they (as other thinkers propose) a trap out of which we can never escape. They stand for the current state of our mind. They represent the state of our knowledge of the thing in its context. More can be found out, in which case the situation will be made fluid again until we reach some new conclusions,

---

thinking as containing a perception-element': Warnock, *Imagination* (London and Boston 1976) p.192. Goethe used this perceptual-thinking approach to arrive at the discovery, for example, of the intermaxillary bone in the human skeleton. Similarly, in his botanical researches he showed the actual basis of the unity of all plants in contrast to the external divisions catalogued by Linné. The history of science can furnish other examples (e.g. the circulation of the blood), though they do not fit the prevailing materialistic textbook.

deepening our perception of the object, and fix upon the same specific things in a new aspect, and changed a little in the process ourselves.

The fluid intermediary condition when we are re-assessing and redefining our ideas is one of which we are not normally very conscious. But that is what interests Steiner, since out of it issue the clear but polarised opposites (the thing, and the idea we have which explains it). If we want to get to the reality of them, we must become conscious of that phase of engagement in which we are usually not very conscious. Such an approach, says Steiner, 'embodies the search for a cognitional method in response to which the real world will reveal itself …. Anthroposophy presses forward to the perception that a new consciousness must be developed, issuing from ordinary consciousness as, for instance, waking consciousness from the dull dream consciousness. Thus the cognitional process becomes for Anthroposophy a real inner occurrence extending beyond ordinary consciousness'—indeed, requiring it to be deepened and expanded.*

Take for example our awareness of space. Through the ideas of geometry etc. we are able to understand the familiar things around us as objects in space, not because we impose those ideas upon them, but because our own nature belongs to the same spatial universe. Our consciousness does not just appear from nowhere. We have evolved our detached consciousness, looking on at things spread out around us in space, by accommodating within our mode of life that uprightness which we early on struggle to maintain. Though as yet unconsciously, we learn to balance, to orientate ourselves in a complex, highly unstable and adaptable way to the space we inhabit, constantly re-equilibrating our position; and it is that unconscious complex adjusting-process which reappears on a higher level, becoming our structured, reflective awareness of space—and, through a still further distillation, in the structuring activities of language and thought. With these we 'describe' space. We do not invent them ('posit them'); one might rather say that we discover our deeper participation, in which the wider factors 'ingredient in'† a situation involving us can come to expression, heightened, made conscious, as thoughts—though thereby they also lose a certain force. We find the reality in them again when we uncover the activity which belongs to us as we define our place in the total scene: knowledge is a continuation, Steiner insists, of the real drama of evolution which has brought about our emergence on the scene.

---

* Cf. Steiner, *Philosophy and Anthroposophy.* p.12.

† A.N. Whitehead's felicitous characterisation of the way a concrete situation incorporates wider factors.

Vastly more is present, therefore, when we relate ourselves to something in space, than just the final picture of our distance from the object. A history of unconscious activity underlies it.* And if once we become aware of it, we find that this complex of activity is not subjective, or abstract like a governing 'law', but is essentially the reality of our own growing and evolving spatial identity. We should envisage an overall phenomenon of spatial perception which contains ourselves and the object which is the immediate focus of our awareness—rather than trying to get outside our own standpoint to categorise what is there, purely objectively. This is both impossible and unnecessary. We ourselves best reveal the phenomenon of the world, because we are rooted in it and part of it. Our organs are formed by it. And the realisation that the 'thought-component' in our experience represents, on further examination, something with a history and an active role in our development, points to something wider, joining us to the world we inhabit. This can well come together with the scientific approaches which are nowadays coming to ascribe consciousness in some degree to all of reality, not just to ourselves as onlookers (so-called Panpsychism).†

The suppression of the active component in our engagement with the world, so as to think about it, is of course a necessary one—for our consciousness. But then, purely in order to understand that situation, we need in turn to adopt a viewpoint which goes beyond that of our ordinary consciousness. Otherwise we shall come back again and again to that alienation from which so much contemporary thought cannot escape. 'Man's isolation from the world can be overcome,' argues Steiner, 'by seeing that he has to give a provisional definite form to his ego so as to suppress from his consciousness the forces that unite him with the world

---

* The process is repeated with every growing-into-consciousness. Developmental psychology can examine aspects of it through studying children's thought. Two interesting points which emerge are 1) that adults have comprehensively suppressed awareness of how they organised their experience previously, and are astonished when they do see into it; and 2) the researchers nevertheless find themselves tending to adopt child-perspectives, slipping back in to them as they do their studies. Thus they remain vital. Steiner's own method brought him an unprecedented ability to comprehend the child's developing consciousness, as utilised in Waldorf education.

† See above pp. 59f. The way that Steiner instead points to our integral value as a key to the manifestation of its truth, is also akin to those trends mentioned briefly there which are termed 'Anthropic', and to the study of 'laws of complexity', through which (e.g. through our organisation as conscious beings) a higher level of reality is able to make itself known.

.... The ego owes its self-awareness to the fact that it spreads a veil over the knowledge of the world. It follows ... that it conceals at the same time the deeper foundations in which this ego has its roots.'* Becoming aware of the activity behind our thought releases us again from our subjectivity, set over against the object, and integrates us once more with the world.† We achieved a moment of clarity by narrowing our view—but we need not remain trapped in that moment forever. Active self-knowledge will also uncover those 'deeper roots' that join us to the world, and raise them from the dark penumbra which our own self-development has cast across them. When Steiner calls this activity 'clairvoyance' we must realise that it is indeed a development beyond the consciousness which grasps itself firstly in isolation, but it is a development rooted in that very consciousness. By clairvoyance he never meant something which is arbitrarily tacked on to our normal human faculties. It is something which broadens and deepens our objective mode of thought, showing how we ourselves in our subjectivity are part of a greater consciousness.

Other thinkers of Steiner's time were also feeling the need to challenge the notion of thoughts which denied them their inner depth and potential for further unfolding. The psychology of Freud, for example (whose intellectual background overlapped in fascinating ways with Steiner's!) observed that in some of his patients an idea could appear and reappear, and have a powerful effect even when it was not consciously present in the mind. Indeed it seemed that precisely because it was not present consciously, it expressed itself in disturbing ways, and that a 'cure' consisted precisely in bringing it to 'normal' conscious recognition. Freud was thus led to his famous theory of the Unconscious.‡ But Steiner did not believe in a special aspect of the mind which was defined by its un-consciousness (and nor after a time did Freud!). For Steiner the forces behind the idea-in-consciousness are an expanded awareness of our connection to the world; Freud's follower Jung already recognised

---

* Steiner, *The Riddles of Philosophy* (New York 1973) pp.450-1.

† Cf. Steiner, *Human and Cosmic Thought* (London n.d.) pp.9-16.

‡ The best introduction to Freud's thought is still Richard Wollheim, *Freud* (London 1971); see particularly pp.159-60. Freud in his later work abandoned the simplistic contrast between conscious Ego and 'the Unconscious', concluding that the Ego comprised more than just conscious activity. Steiner was connected in manifold ways to those individuals who started the development of psychoanalysis, notably through Josef Breuer (Freud's original collaborator) and Friedrich Eckstein, among others. Steiner gave some extremely perceptive lectures on *Psychoanalysis and Spiritual Psychology* (New York 1990).

their 'archetypal' content. The disturbed working of the trauma, or event that cannot be assimilated consciously, might also find its rightful place in such an account. There too the environment becomes present to us in a way that the ordinary consciousness cannot cope with. Instead of pointing to an unconscious mind, his approach moved rather toward the recognition of a wider consciousness in Nature.

Consciousness and Evolution

Our consciousness is thus not something arbitrarily set up against Nature, mirroring it from without, but an intensification of that dimmer conscious presence, distilled from the sort of awareness slumbering in Nature. Its complex structures, including psychological constructions which express our 'self', incorporate and overlay the simpler structures on which they build, and like those of childhood mentality they are never actually lost. Now, reversion may, however, not be simple slipping back. For example an element of infantile play may be incorporated on a higher level in brilliant imaginative or artistic activities. Rudolf Steiner considers that it is possible to delve into the primordial levels of the consciousness in ourselves and in Nature through certain meditative techniques—that scientific extension of consciousness of which he wrote. He does admit that breakdowns of the higher-level consciousness do occur, and experiences belonging to the duller 'Nature-consciousness' may resurface as spontaneous 'clairvoyant' perception, which he thus terms 'atavistic' clairvoyance, using a technical term for the way that features of a distant ancestral form (atavisms) may reoccur in organic evolution. But he held that it is also possible to recover them in a more sophisticated way by enhancing in parallel, rather than dulling down, our ordinary awareness.[*]

By means of suitable training, it is possible therefore in Steiner's view to go back to the most basic consciousness of nature, and thereby to become aware of the outer condition of perception corresponding to it. We must envisage the formation, for example, of the relationship of the Sun to the Earth which enabled life. Far back, before even the differentiation of the solar system, the substantiality from which the Sun and

---

[*] The 'atavistic' i.e. backward evolutionary character of spontaneous clairvoyant faculties was recognised by those scientists like F. W. H. Myers who likewise attempted a synthesis of psychical research with evolutionary theory—cf. *The Other World* pp. 266-325. Steiner offers a still more comprehensive grounding.

planetary organisation will emerge may also be experienced clairvoyantly as a dark consciousness, still dimmer than the deep-sleep consciousness we fall into at night. At this stage, the future-substantiality of the solar system occupies the space of the orbit of the present-day planet Saturn—volumetrically the future solar system. Such outwardness as is associated with it is still long prior to the differentiations of the elements found today, but Steiner calls it physical in the 'anthropic' sense that we still have it as the germ of our body as the kernel of our becoming, associated with fundamental workings far below the level of consciousness.[*] He is working from what is within us to what underlies it as a cosmic formative force/consciousness. This is the primal state, the furthermost that can be inwardly revealed to us, reachable through our having derived thence our physical body. In the 'Old Saturn' phase even the dim consciousness of Nature, Steiner says, 'was just evolving. It was dull, but far-reaching.' Yet one can even now enter into that consciousness, 'and in such a state the individual is able to move far, far over the Earth and to give expression to cosmic facts.... This consciousness, though dull, embraces vast regions.'[†] The differentiation of the solar system from this primal state will also be a process of the refinement of that primal consciousness, which contains the original 'information'—which can be regained. (Organic and human evolution will still further individualise and heighten consciousness, which always rests, however, upon these basic structures.)

Steiner also gives the instance of a case of 'atavistic' reversion, where a girl who suffered (probably) an allergic reaction fell into a deeply unconscious state, in which she nonetheless 'made all kinds of drawings to which she added names'—drawings of cosmological import, of which her conscious mind understood nothing.[‡] But Steiner maintains

---

[*] The anthropic sense of these primordial cosmic descriptions in Steiner, i.e. conditions we know about because they are observed shaping our own existence, is clearly brought out e.g. in his *Outline of Esoteric Science* pp.129-30 .

[†] Steiner, *Rosicrucian Esotericism* (New York 1978) pp.65-6.

[‡] Loc. cit. A very similar (if not identical?) case was discussed by C.G. Jung in 1902 in his study of 'The Psychology of so-called Occult Phenomena', translated in Jung, *The Psychology of Occult Phenomena* (Princeton and London 1977); for her diagram and discussion of her 'mystic science' cf. pp.42ff. To understand it Jung says he 'waded through occult literature ... and discovered a wealth of parallels with our gnostic system, dating from different centuries, but scattered about in all kinds of works, most of them quite inaccessible to the patient'—op. cit., p.vii. But Jung's language belies his own long-term engagement with occult ideas. W. McGuire in the Preface points out that Jung initially

a parallel conscious perspective on his own cosmic perceptions. Nevertheless he recognises the cosmic import of the girl's spontaneous drawing, or cosmic schema. It reflects a real underlying consciousness in us all, in which awareness expands to cosmic proportions—and Steiner even observes that in fact such cosmic consciousness is relatively easy, while detailed observation of more particular matters clairvoyantly is much more difficult!

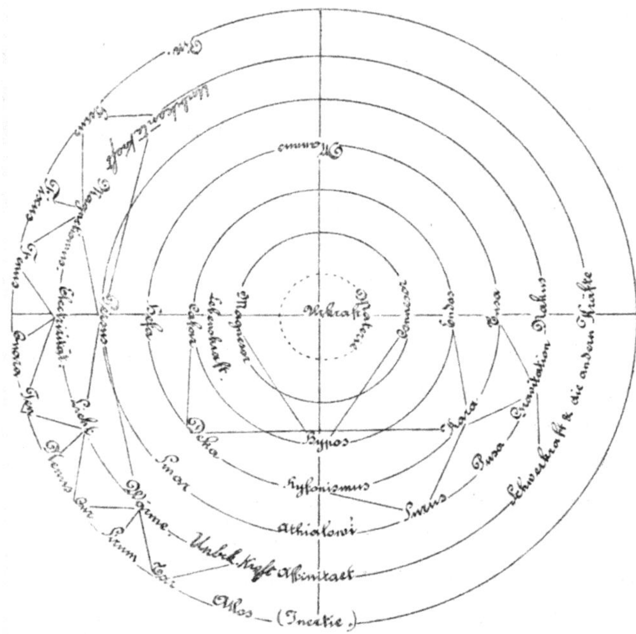

'*Cosmic Diagram' spontaneously produced by one of Jung's patients during semi-somnambulistic states, showing the interaction between the Magnesor and Connesor powers at the basis of the universe. Rudolf Steiner described an almost identical case-history. Such intuitions of complex structure not consciously thought out essentially resemble archaic or shamanic cosmologies, later rationalised into our own astronomical worldview. From C.G. Jung,* Psychiatrische Studien *(1943). Jung considered such patterns as 'discoveries' which could be compared or assimilated to the forces of physical science (see e.g. the schema in his* Synchronicity *(London 1972) p.137.*

affirmed these and similar sources of knowledge, before backing away, mainly perhaps on the advice of Freud (pp.vii-x) who insisted on the subjective nature of the Unconscious. Privately, however, it is often suggested that he continued to believe in the possibility or even accepted it outright. For Steiner on the primal flow (a sort of will-to-be) of the 'Old Saturn' state and his most determined effort to extrapolate the structures which emerged there, see his *Inner Realities of Evolution* (London 1953) pp.13ff.

When we attempt to decide whether this perspective can be accepted as valid, we should bear in mind that our basic cosmological ideas to this day derive ultimately from similar shamanistic trance-projections. The schematisation of the concentric planetary orbits (conceived as defined by spheres carrying the physical planets), with the stars beyond, is undoubtedly a transposition of ecstatic or trance-states, whose successive levels are so pictured in shamanic symbolism. The shaman is then 'travelling' among the spheres. How indeed could the idea have arisen observationally? Looking up at the sky, there is nothing to suggest that moving lights there are arranged concentrically. In an exemplary way the image arrived at inwardly fuses with them, and their inner significance is revealed, as a result of our deeper-level connection rising into consciousness.*

Now if that is so, we at least have the rudiments of an understanding for that expansion of consciousness in time and space which was necessarily involved in our case of recovering ancient Atlantis. There the reversion and the conscious control were divided between persons. Steiner never, so far as is known, employed such a division, but retained his own conscious control. Steiner claims a high degree of certainty for his methods, which we must scrutinise further. But the basis for many of his further ideas is the notion we have explored, that our consciousness has in its development of greater inner complexity, suppressed the experience of certain cosmic dimensions. Even so, those dimensions

---

* See M. Eliade, *Shamanism* (Harmondsworth 1989) pp.275ff. A version of these spheres is found already in Babylonia in the sixteenth century BC (Kassite period), though evidently it remained esoteric: cf. J. Lindsay, *Origins of Astrology* (London 1971) pp.30-31. Travelling through these spheres to the stars was important in ancient Egypt, as later in Hermetic teachings, and we now know that the modification of putting the Sun in their centre was the result of further Hermetic-spiritual ideas influencing the Renaissance: Frances Yates, *Giordano Bruno and the Hermetic Tradition* (London and Chicago 1964) pp.153ff. Nevertheless, the picture once given, one can only be dazzled by what has been aptly termed 'the amazing mathematical feat' of Plato's pupil Eudoxus, when he hypothesised that each sphere had its motion modified by two lesser spheres and that 'by assigning certain speeds and directions to the three rotations you get a compound motion' from which he exactly and correctly derived 'the observed motion of the sun and moon. Similarly the motion of each of the planets may be decomposed into four rotations': so described by D. Ross, *Aristotle* (London 1964) p.96. Both schematically and mathematically, humanity's discovery of the cosmic image may be compared to the kind of mental leap later made by Leonardo at Imola (see Appendix), rather than step-by-step empirical discovery. We need to look to a similar background too when we come to the next great leap by Copernicus and Kepler—based on the harmonics of the Platonic solids!

are still within us, and can be reawakened pathologically, or on a new level through further development consciously undertaken. Or to put it another way: the cosmic consciousness which lies far back in our origins can re-emerge in a new modality, as 'spiritual science', through deliberately directed human development.

Clairvoyance and Evolution

The idea of recovering deeper-level connections upon which our higher consciousness depends, as a key to the cosmic states which primordially corresponded to them, is perfectly intelligible in principle; but it might seem doubtful whether it is feasible to define each stage conclusively. Rudolf Steiner contends, however, that such thinking is available as the starting-point of his anthroposophical venture; it is, actually, none other than evolutionary thinking.

First of all, he reminds us that evolutionary thinking corresponds to an understanding of ourselves as developing and growing; it could not be applied to Nature until we had discovered it out of our modern sense of individual potential which strives to be fulfilled. We noticed earlier how scientists struggled at the threshold, until an inner step could be made. Seen in this way, conversely, scientific thinking can itself become a spiritual exercise as we think in terms of the dynamics of knowledge as we sketchily laid out just now. We must consider more deeply the emergence of evolutionary thinking, as opposed to the abstract kind of thinking which seizes only on the moment of separateness in knowledge—the kind of knowledge which characterised the cataloguing and comparing approach of pre-evolutionary thought. Great benefits came from this phase, and Steiner of course does not dismiss them. But they led into an impasse which still limits much of our thinking today—from which anthroposophy as the spiritual correlative of evolution offers a way out.

The great achievements of eighteenth-century science, in the phase which is classically recognised as the Enlightenment, reflected a much less individualised society, and were in essence classification (taxonomy). Everything was put in its place. One need only think of the great Carl Linnaeus (1707-1778), the 'founder of botany', who set the standard in comparing and arranging methodically the different types of plant and inventing a precise nomenclature which classifies every specimen by genus, family, and species.

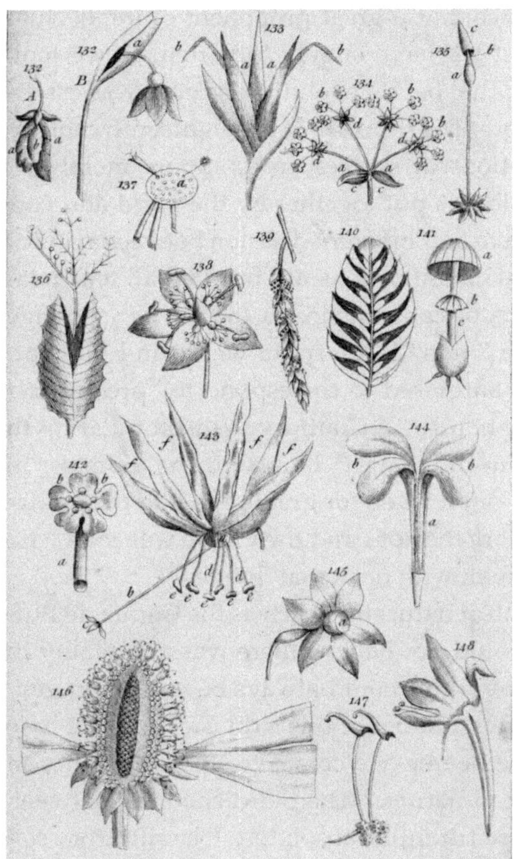

In the eighteenth century Carl Linnaeus (or Linné) gave plant taxonomy the status of a leading scientific approach. His precise observations of the features shared and not shared by different plants enabled him to integrate them into a complete systema naturae. *It did little, however, to show the real connections between the similarities, and in his time the notion of species verged on becoming an empty abstraction—and not only in botany. Only the variety of individual specimens was considered real. A famous conversation of Goethe with Schiller revealed the gulf between this and his new way of thinking: Goethe's specific attention to growth and transformation (metamorphosis) in plants led him to grasp the living unity behind the variety of forms, paving the way for modern developmentalism and evolutionary approaches. From Linnaeus,* The Elements of Botany *(Biodiversity Heritage Library).*

We still use his system of identifying plants—albeit with some necessary adjustments made since his time. We admire his immense knowledge and his ability to pick out features crucial to comparing one plant-taxonomy with another. We observe how religiously the eighteenth-century scientists took classification, articulating no doubt God's perfect creation. But classification places us, for all our admiring recognition, in a singularly external relationship to things. It lays them out in groups, placing together things with common features; things with fewer common features form the next class, which introduces in turn some new features indicating a further group where they are dominant, and so on *ad infinitum*. The species are merely a way of pointing to one shared feature here, another there, and we catalogue them methodically purely and simply for our intellectual convenience. If we ask: How does it happen that some plants are formed like this and others like that?—there is no reply. We do not enter into the process of plant development, merely note external peculiarities.

It is a *sine qua non* of science; but a great proponent of intellectual history points out to us what we have largely forgotten, as we look back on Linné's achievement: that before evolution, science was essentially classification and no more. The thinkers of the eighteenth century considered that 'our conceptions of species are therefore merely ... combinations of ideas of attributes put together by the mind and corresponding to no fixed objective and inherent division between natural things.... And thus biological classifications are but verbal, and relative to varying considerations of convenience in the use of language .... Even the nominal essence "man" is a term of vague and fluctuating import, which cannot be supposed to correspond to "precise and unmovable boundaries set by nature".'* Contrary to the popular myth that Darwin 'dethroned' man and ranged him with the animals, in pre-Darwinian science it was often taken for granted that in the gradations of animal forms there were the apes and then there was man who shared many features with them, with no actual divide.

In biology some of the greatest naturalists such as the Comte de Buffon (1707-1788) found they could not believe there was any reality in 'species', but that intermediate forms could always be found: 'In general, the more one increases the number of one's divisions, in the case of the products of nature, the nearer one comes to the truth; since in reality only individuals exist in nature.'† The old dichotomy between particulars and universals here triumphs absolutely. Classificatory science is based entirely upon that moment in the dynamic of knowledge in which we polarise specific object and general law, highlighting the particular but reducing everything that relates it to other things to a purely abstract category. We may indeed exult in our moment of clarity; but find ourselves self-defeated, cut off from reality. Even mathematics, as pure reasoning, was felt to be without grounding to real distinctions.‡

Evolution did not come just as a new way forward in science. It rescued science from becoming a barren exercise in analytics. Subsequently

---

* A.O. Lovejoy, *The Great Chain of Being* (Harvard and London 1964) p.229; the quoted phrase ('precise ... nature') is taken from John Locke.

† Buffon, *Histoire Naturelle* (1749), First Discourse, cited in Lovejoy, op. cit., p.230.

‡ Hume, who followed this style of thinking to its logical conclusion, could not think that even the clear and necessary divisions of geometry could have any real correspondence with the 'prodigious minuteness of nature': cf. David Hume, *A Treatise of Human Nature* with an Introduction by A.D. Lindsay (London 1961)—cited, vol. I p.xxii.

Buffon's own observations broke through the web of abstractions, when he showed that a species can only reproduce with its own species (cross-breeds losing the ability to do so). His ability to see a real unity behind the classificatory label is one starting-point in understanding evolution, but few thinkers of the time could follow him. The crucial problem of how to reconcile the variety of forms with the existence of distinct species would only be resolved through evolutionary thinking, revealing the way species change to generate many gradually shifting types. But resistance to making that intellectual leap was considerable. The Church fanatically favoured the classificatory approach, indeed would fight Darwin to the death—even the 'death of God', if by that we mean His demise as a significant agent in scientific debate.[*] But the deeper problem was that to go over to evolution (and evolutionary schemes were on offer long before *Charles* Darwin) required a dramatic liberation of thinking from that kind of classificatory-surface-unreality approach that left human knowledge suspended, to use a phrase from a great poet of the time, in a 'Void outside of Existence'.[†] What was at stake for science was the need for a way which can ground us as thinkers in the reality underlying living forms and the order of the cosmos.

Making a start on the necessary way of thinking was far from straightforward.

Evolution is not a concept that has come down to us with a venerable history from classical thought, but is rather an unprecedented idea.[‡] Goethe's concept of development or metamorphosis was a huge

---

[*] Margaret Jacob, *The Newtonians and the English Revolution* (Brighton 1976) points out that, especially in England, liberal-minded churchmen had adopted a version of Nature as made up of analytically separate parts, operating like clockwork, with a notion that it 'left room for God'. 'The linkage … was so indissoluble that only in the second half of the nineteenth century, when Darwin successfully challenged that explanation … did the whole edifice of science-supported religion come crashing down. Darwin's science was lethal to religion only because … a particular explanation of the natural order had gained acceptance as an indispensable support for liberal English Protestantism' (p.16). The corollary is that Christianity (or religion) need not continue tying itself to such thinking for ever. Nor need it reject evolution.

[†] Blake, *Milton, A Poem.*

[‡] Pioneering sketches there certainly already were. Erasmus Darwin, a highly regarded scientist if more patchy poet, advanced a comprehensive scheme of evolution in his *Zoönomia or, The Laws of Organic Life* (1794). Andrew Wilson exhibits commendable scepticism in the face of his grandson Charles Darwin's claim that he knew nothing of him!—Wilson, *Charles Darwin*

step forward. Though his approach was swept aside in the widespread enthusiasm for the ideas of Newton and his mechanist successors, it has recently come to be re-evaluated more positively. The young evolutionist Steiner found his views extraordinarily exciting, and wanted to use them to give a foundation to the theory, which he still found lacking in the materialistic ideas of Darwin and Haeckel.[*]

Breaking with the approach of Linné, the *Metamorphosis of Plants* was less concerned with the separate types of plant than with the growth-process which had originated them all. Goethe was not aiming to articulate an 'idea' that would cover all known varieties, but at a new way of looking at the plant. (Cf. Darwin in his garden fixated on a flower.) By scrutinising in empirical detail every stage of growth, Goethe concluded, one can identify the basic element in all plant growth, namely the leaf. All the other parts of the plant can be seen to originate through special modification of the leaf, and likewise all varieties of plant originate by means of particular metamorphoses. His observation was tremendously detailed and convincing, as was his sense of the whole. There is no question of a 'mechanical' combination, adding or subtracting this or that. The plant is at each stage a powerful living unity which generates all its parts. It could assume many different forms, but each species has a unity of its own. Though he did not derive them historically, in sequence, it is also clear that all plants are connected, concretely, through their common generation. And so Goethe arrived at the 'basic plant' (*Urpflanze*) which is developed in all of them.

In recent decades evolutionary scientists have belatedly jumped on the bandwagon of using development (metamorphosis) as a key to biological change (so-called 'evo-devo'). In that very modern context, the work of biologists who have pursued Steiner's Goethean line of evolutionary thought, such as Wolfgang Schad in his work on the morphology of mammals, has been hailed as delivering a new level in the understanding of evolution. In the opinion of one professional in the field, 'Schad's insights amount to a new and revolutionary, integral theory of evolution.'[†]

(London 2017) p. 54. However, it remains true that Erasmus Darwin could say nothing about the basis on which evolution could come about.

[*] Steiner is undoubtedly correct to point out that evolution is in origin a Christian idea: Steiner, *The Fifth Gospel* (London 1995) pp.9ff and cf. Ernst Benz, *The Mystical Sources of German Romantic Philosophy* (Pennsylvania 1983) pp.32, 36, etc..

[†] Steiner, *Theory of Knowledge based on Goethe's World-Conception* (New York

For Rudolf Steiner, the approach pioneered by Goethe constituted a new way of seeing things that was not alienated from reality—his 'perceptual power of thought' brings us to a discovery, not a theory: the modifications of the leaf into stem or root, the presence of a particular bone. Goethe observes how an actual plant-form originated, a dynamic process in which the plant as observed was one phase. We see its morphology, therefore, as the result of its changing growth and development, and find the same factors at work in other plants. Such insight is not merely a schematic 'idea' like the categories laid out by Linnaeus, which lead to no new reality but compare external features. The evolutionary idea is unprecedently close to the actuality of the plant or the organism, since such an idea is nothing but our perception of the shapes it has gone through and which have produced what we see. We see nothing but a leaf in various forms, and to 'understand' more is to see the leaf in more and different transformations; the idea is the plant seen through all its history. In a celebrated conversation in 1794 with Friedrich Schiller (an encounter often mentioned by Steiner), Goethe made this point triumphantly—though he did not entirely convert his future friend. He was to describe afterwards in an essay called *Happy Encounter* (1817) how he, the enthusiastic researcher, had finally tangled with the mind known for being an arch-classicist in poetry and formalist in philosophy. Hearing of his morphological idea, Schiller replied coolly that he was speaking of a 'general concept', an *Idee*; no, Goethe insisted, or if so, then he has arrived at the sort of idea one can see with one's own eyes, a real experience. It would grow in one's garden.

Arriving through a morphological thinking-process at the transformations which have generated an organism is the exemplary form of that 'perceptual power of thought' which Goethe described. It is startling to see modern scientists coming to the same discovery, and realising that it was the old analytic-classificatory way of thinking which actually concealed the dynamic hierarchy of forms, as an element in which each present organism can now be seen. Goethean thinking, comments Lockley, 'portrays organisms as manifestations of a higher level of organization that was previously hidden from our view simply because we were looking with analytic eyes, rather than with the synthetic, organic

---

1968), with Steiner's later Preface pp.xv-xvii; Wolfgang Schad, *Man and Mammals: Towards a Biology of Form* (New York 1977)—and for the enthusiastic reception given to Schad's approach see e.g. Martin Lockley, *How Humanity Came into Being: the Evolution of Consciousness* (Edinburgh 2010) p.177; see also pp.53, 181-94, etc..

eyes of nature'.* The kind of knowledge he means also makes us aware of something else in the equation. Does 'the innovative approach of Wolfgang Schad,' he continues by asking, 'tell us something about mammalian consciousness or indeed about our own human consciousness? As we shall see, the answer in all cases is probably yes.'[†]

Evolutionary thinking involves us to an unprecedented degree in the world we know. We can enter into it because we too are at home in the world, and that 'higher level of organisation' which is revealed to our view, no longer concealed from our analytic knowledge, is our own extension into the world as well. Our intensive consciousness is ultimately continuous with the resonances in Nature itself. Thus already in scientific cognition, as Rudolf Steiner himself said, 'the true or genuine Man makes his presence felt in dim feelings, in the more unconscious life of the soul'.[‡] In a strictly evolutionary sense, the emergence of human knowledge begins to be recognised as an active force, or 'real event', in a world which is after all the organised environment that has shaped and honed us, which is part of ourself.

This 'Anthropic' turn begins to hint at the idea of humanity's implicit presence in the deeper processes of evolutionary development.[§] And if we break out of that polarisation of knowledge which gives us momentary clarity but veils our own deeper role, we discover the consciousness of such a human being—without relinquishing the advance in self-consciousness we have gained. Such 'clairvoyance', as Rudolf Steiner insisted, is not only compatible with modern science, but can be seen growing out of it as the fuller implications of evolutionary thought are felt. He characterised it in a more homely way. When we see a person of adult years, we can ask how he is organised, how he functions, etc. But when we get to know him, we may also see

---

* Lockley, *How Humanity Came into Being* p.53.

[†] Lockley, op. cit., pp.182-3.

[‡] Steiner, *Philosophy and Anthroposophy* pp.11-12.

[§] In all the kinds of new thinking to which we referred (above, pp. 46ff), the higher laws of complexity that point to the nature of the world can only be grasped through changing ourselves—our analytic kind of consciousness to a concretely perceptive one. They can be formulated only in relation to our own role as revealing them, in which sense they are all variations of the Anthropic idea. For many researchers and scientists, however, these implications lie at the far margins of thought. Rudolf Steiner takes them methodologically as a new starting-point based on self-transformation.

expressed in the mature man the young man whose youthful character is now fulfilled and realised. What Steiner meant by clairvoyance shows us the world within the world, in which our own human reality was working its way to expression, while we were still a part of the greater whole.

Scientists have recently found their way to the apprehension of higher-level organisation from a different side—namely, from an approach to biology growing out of astrophysics. We found it notable that these thinkers nowadays have little sympathy for the once much-touted idea of life being an electro-chemical freak show, which began when amino acids were jolted accidentally into unaccustomed activity—or for any of its somewhat more sophisticated formulations. One of them points out the enormity of the gap between these uncomplicated chemicals and the stunning complexity of even the simplest organic cell. Such ideas cannot explain how life perpetuates and increases complexity, and many scientists incline rather to the view that life is not a peculiar freakish domain, but an indication of tendencies or even laws of complexity inherent in the universe.[*] Many talk of a 'biological imperative' implicit in the evolution of the cosmos.

Lee Smolin notes that the same 'ecological' patterns and structures of evolution seem to occur on all levels of the universe. It is these structural features which underpin the workings of life, which he regards as part of a 'nested hierarchy of self-organised systems that begins with our local ecologies and extends upwards at least to the galaxy'.[†] Life is not to be predicated just of a particular organism, but refers to its place in a vast network of interactions. A fundamental concept in evolution is the way that aspects of the environment are progressively internalised by living things (so that reproduction by egg-laying in a suitable pond or shallows is in higher animals incorporated into the reproductive system, for example). Living things become independent, or freer within their environment, by *incorporating* it. Human beings have taken the process to its furthest degree. But for us to do so, the universe must itself be 'fine-tuned', and able to nurture us in complex reciprocity. Its structures must dovetail exactly on every level. Ultimately, we are internalising the cosmic patterning which Smolin describes as the life of the stars.

---

[*] Schroedinger, *What is Life?* (1944). Paul Davies, *The Origin of Life* (London and New York 1999) pp.3ff.

[†] Lee Smolin, *The Life of the Cosmos* (Oxford 1997) p.159.

Paul Davies, another astrophysicist interested in the nature of life, has characterised these structures as the 'information' which enables organisms to grow and develop. And building further on ideas of complexity (like those of Stuart Kauffman) he has come to theorise that it is most probably misguided to seek a material substrate, a medium on which this information is 'written'. In his further extension of thought, biology requires nothing less than 'a major departure from the traditional reductionist view of the world, in which forces between inert particles of matter, and information is treated as a secondary, derivative concept'. He proposes that biological information is an independently existing presence, a specific reality 'that can be traded by "informational forces" in the same way that matter is pushed around by physical forces … with real causal efficacy, rather than a merely qualitative description of how complicated a system is'.[*] Otherwise we cannot account for the way 'life works its magic, not by bowing to the directionality of chemistry, but by *circumventing* what is chemically and thermodynamically natural'.[†] In these passages Davies comes extraordinarily close to Steiner's conceptualisation of the 'etheric forces' which determine organic form. And like Steiner, he does not believe that the information is lodged in particular component parts, but globally in the whole; the individual letters which constitute a proper spelling are physically no different for being part of a word which means something. From the latest scientific ideas, Davies comes to the same conclusion: 'both quantum mechanics and relativity suggest that information is a global rather than a local physical quantity. You cannot simply inspect a location in space and discern information'.[‡] Steiner himself already pointed to mathematical ways of thinking 'peripherally' rather than 'pointwise' in order to capture the essential nature of the etheric forces.[§]

The Archetypes of Life

How real are such information-bearing types or patternings in themselves?—it is oddly hard to escape that ingrained suspicion that they are

---

[*] Davies, *Origin of Life* p.242.

[†] Op. cit., p.237.

[‡] Op. cit., p.44.

[§] See George Adams and Olive Whicher, *The Plant between Sun and Earth* (Colorado 1982) pp.35ff and further discussion in our next chapter.

a reflection of the effects which have accumulated rather than an actual force in Nature. However, the contrary view has received increased support since a celebrated debate among biologists concerning a hypothetical scenario: What would happen if life had to start all over again? Cambridge Massachusetts and Cambridge (UK) took broadly opposite sides, the former suggesting a new random direction and results, the English biologists providing evidence that it would repeat itself, because the 'same basic patterning is too deeply engrained in the fabric of organic process'. The latest assessments of the question tend strongly to agree that 'there is supporting evidence for the latter view'.[*] If material life were wiped out, the same types would assert themselves.

According to Steiner, the clairvoyant can see these archetypes of life and, accompanying them, the human archetype in which all the rest are in a sense contained or combined. It was Steiner's description of the three fundamental animal types and their relationship to the human form which helped orientate and direct Wolfgang Schad's much admired contribution to biology. Sometimes he specifically connects the clairvoyant's vision with the evolutionary phase he associates with Atlantis. The animal archetypes were already present and, as he says, 'remain … as something which assumes its form again whenever it has the opportunity' but also 'the germ of the spiritual man was there, but he could not yet develop an individual form'. Elsewhere he indicates that these patterns are derived not from earthly conditions but from the stellar dimension of the universe.[†] How Steiner characterises further the emergence of the human in his spiritual science will occupy us shortly. He conceives it in the modern evolutionary perspective. But let us also note that the vision of the animal forms and man is one of the oldest, most consistent and central religious experiences of mankind. And it is always focussed on man's relationship to the animal forms *among the stars*. It is the vision of Paradise.

Pardes is already a Persian loan-word in the Bible's story of Genesis, and belongs originally to ideas about the religious significance of the sacral ruler in the ancient Middle East, and the Hebrews amongst others took over the idea. It expresses the cosmic sources of the power belonging to man in his highest dignity, and is imagined as a sort of garden with exotic

---

[*] Lockley, *How Humanity Came into Being* p.171.

[†] For the connection to Atlantean evolution, Steiner, *The Apocalypse of St. John* (London 1958) p.163; starry archetypes, Steiner, *The Spiritual Hierarchies and their Reflection In the Physical World* (New York 1970) pp.98ff.

plants and animals where the ruler appeared symbolically as a sort of Primal Human Being, a child of Earth and heaven. For the 'garden of God' is the starry sky. By his wise dispensation of power the King keeps the world connected to its divine origins, controlling the forces of Nature in imitation of God. Zoroastrian mythology describes the origins of man as having taken place in the centre of the world, where heaven and Earth meet, by a river, while over him grew the Tree-of-all-seeds and alongside him was the Bull, the archetypal sacrificial animal which is symbiotically linked to human life. The Tree shows life linking heaven and Earth, and the sacrifice of the Bull to God intensifies life into a religious act, being almost a self-sacrifice, and so renews the world ever and again from its source. In Babylon, the temples were structured in layers representing the seven heavens, and at top contained a garden with trees and rivers that mythically flowed down from heaven (Genesis still has 'waters above the firmament' to puzzle the commentators too). All these representations were likewise attempts to recapture the primal sources of life and community with Nature. And we now know that the caves of the Upper Palaeolithic, full of visionary animal-forms that surround the shaman in the spirit-world, are a still earlier statement of the same idea.

Around them animal and star-symbolism came together, as they did still more in Egyptian versions associated with animal-formed gods and with the Pharaoh, one of whose mystic forms is the polymorphous Sphinx. The King in Paradise was watched over by the guardian Cherub, which term also appears in the Bible as the name for the animal-archetypes around the divine Throne (a later version of the same symbolism).[*] Rudolf Steiner pointed out that the archaic religious vision of man in harmony among the stars with the animal-formed gods, forms a sort of radical ancient alternative to the modern materialistic attempt to place man among the animals. Indeed one may say that Evolutionism, albeit unconsciously, is trying to articulate one of

---

[*] The Zoroastrian version is presented in R.C. Zaehner, *Teachings of the Magi* (London 1975) pp.40-41. The wider diffusion of this incredibly rich symbolism looms large in religious studies: a useful first survey is F.H. Borsch, *The Son of Man in Myth and History* (London 1967)—for the Tyrian version, which is close to the Bible, cf. pp.105ff. The imagery of Paradise was especially embodied in the Temple, in Babylon and Egypt but including the Jewish Temple in Jerusalem, where the priest-king (or god-king) presided. When the Temple was destroyed and the people driven into Exile, the prophet Ezekiel (Chs. 1-2) had a vision of the abiding elements of the vision, with the four Animal-forms round the mystic Throne. Both this and the Paradisal-Creation opening chapters of Genesis were 'esoteric', i.e. only the initiated were allowed to discourse upon them.

the deepest, oldest religious visions of mankind—doubtless the religious intensity of Christian orthodoxy's struggle against Darwin is a dim recognition of a profound rival!* The traditional choice of Lion, Eagle and Bull alongside the Human represents a deep intuition of the animal types in their diverse morphological significance. Martin Lockley and Wolfgang Schad show in depth how our affinity to these types is a seminal thought for biology.

Animal-human harmony did not last. Death came into the world and all our woe. When the Primal Man died, his seed ran down upon the Earth and from it sprang the first male and female couple, the generation of birth and death. From them the vision of Paradise was hidden. The Bible presents a particularly moralistic recension of the myth, but Mircea Eliade has shown that the loss of Paradise is an archaic theme that has been extremely pervasive, going back into some of the most primitive strata of human spirituality. In India, the mystics' prehistoric god Shiva was already shown as Lord of the Animals, and most likely associated with transmutation of sexual energies, fusing them once more to regain the primal state. Signs of this symbolism are known already in the rediscovered Indian prehistoric sites.† For it is central to this mythology that the ideal state, now lost, can be regained by a seer—who thereby becomes also a leader, sacral King or founder-figure.

Rudolf Steiner, of course, saw things more scientifically; but he stands in a long line of those who saw that a larger consciousness of humanity's connection with the cosmos had been overlaid by the development of the limited ego. It can be recaptured in the nostalgic pictures of Paradise in mythology—but also through a forward-looking scientific understanding of the archetypal picture of man along with the animals that are related to him. But in that form it also poses a question which we have to face: for surely the archetypes, evoked in the imaginative picture of Paradise, are known solely in the actual forms which men and animals now have? Or what does Rudolf Steiner mean when he claims to 'see' clairvoyantly the archetypal patternings which play themselves out, for our normal faculties, only in visible Nature—to 'see' for instance the animal types? Or to 'see' man

---

* . Rudolf Steiner, *Egyptian Myths and Mysteries* (New York 1971) p.132.

† 'The Nostalgia for Paradise in the Primitive Tradition', in M. Eliade, *Myths, Dreams and Mysteries* (London 1972) pp. 57–71. For the prehistoric Shiva as *pashupati*, Lord of the animals, see Gavin Flood, *An Introduction to Hinduism* (Cambridge 1996) pp. 28–30 for the prehistoric evidence.

already emerging in the course of his development like the youthful man who can still be seen behind the form of his maturity?

The complex structuring techniques that shape our experience of coherent space reach back, as we have seen, into our earliest experiences of orientation (balance, gesture). But ultimately they are also a record of the fine-tuned cosmic structurings that enabled our emergence as upright beings in the first place. The situation may recall the use of magnetic traces in the rocks to indicate original orientations of the strata, and evoke a former world. Could we similarly discover a non-material reality whose 'information' can be mapped in a similar way—the cosmic prototype of our being? The world-order is so complexly equilibrated that it seems, to some scientists, to map out the possibility of our human evolution. Could it be possible to *see* there, through transformed perception, our archetypal forms? That would imply that they must be more than mere maps. Or how must we conceive the body-of-information that so-to-speak makes us run?

Some insight, it is often suggested, might come from information-theory—and even more from artificial intelligence. It is true that computer-scientists often incline to the belief that cognitive activity is merely a string of information (algorithm), self-consistent, which could potentially therefore be made independent of any human subject. Highly evolved organisms and their complex cognitive activities need not be much different from computers (machines capable of running a 'program'). Roger Penrose, however, who knows as much or more about artificial intelligence as anyone on the planet, pointedly reminds them that this is forgetting something. By the time we reach the stage of algorithmic expression, he points out, the act of knowing is fundamentally over. It is not just a matter of finding a closed set of formal rules: 'for how are we to decide what axioms or rules of procedure to adopt in any case when trying to set up a formal system' unless we first look 'in accordance with our intuitive feelings about "self-evidence" and "meaning"? The notion of self-consistency is certainly not adequate for this.' Rather, he says, 'we *see* the validity' first in a way that cannot be just trundled out algorithmically. 'The kind of "seeing" that is involved in seeing the *truth* of something,' he adds, 'is not the result of the purely algorithmic operations that could be coded into some mathematical formal system.'[*] (Moreover, there are deep mathematical problems with the idea of formal systems treated as if they could be free-standing entities.)

---

[*] Roger Penrose, *The Emperor's New Mind. Concerning Computers, Minds and the Laws of Physics* (Oxford 1989) pp.144-6.

Penrose takes us back to the moment of understanding which is still a 'seeing', like Goethe's, though from it we afterwards separate-out its elements in sequential order (directly mathematical or otherwise); we can then run backwards and forwards through it as much as we need to. Rudolf Steiner's theory of how we fix the opposite elements in cognition might also come to mind, for remember that this is also his account of how we make ourselves. If Penrose's description has something essential in common, so too does Owen Barfield's in the sphere of the history of consciousness. For the meanings we think of as abstract, he points out, only became so by a process of gradual dissociation from a richly concrete wholeness of perception. At the beginning of language, meaning is still part of a directly perceived whole. In ancient languages, the motive power of the wind, as a force that moves things about, could not be distinguished in thought from the inner motive power, welling up from within, which we have come to call spirit. Barfield shows that it is not possible to understand this as metaphorical (i.e. a transferring of already fixed meaning-counters), for this theory disappears up its own tail by implying that meaning itself must have originated as metaphorical, which is an absurdity.*

Barfield's original directly perceived meaning rather corresponds to Penrose's kind of 'seeing' where we recognise the right application, the immediate truth; afterwards we can disentangle its implications one by one and string them out algorithmically. But at the outset, the meaning of words is concrete, perceptual, the meaningfulness still implicit in the perception itself. And because it is perceptual such insight fuses immediately with experience. In seeing things moved about (our 'material' aspect) the meaning of the action (our 'inner' aspect) is indistinguishably present. Ancient humanity saw the spirit/meaning in things, and the truth of a meaning was an actuality, so that knowledge was unquestionable, revelatory—indeed overwhelmingly so. There was no room for issues such as 'Is this really so?', or 'Is this the way to look at it?'. For those to be conceivable, we had subsequently to limit the intensity of the experience and obtrude our self into the picture. The modern equivalent moment of what Penrose calls 'seeing', when we fuse our insight with what is there before us, is perhaps only a faint echo of the total conviction which permeated archaic meanings.

---

* One of the best short introductions to his work in this sphere is Barfield, *Speaker's Meaning* (London 1967), see especially pp.56-67. And see now the collection published as Barfield, *The Rediscovery of Meaning and Other Essays* (Connecticut 1977).

Barfield has brilliantly employed his insights in the history of consciousness. Barfield's theory indicates that our present polarised consciousness, and our usual sort of meanings, emerged out of a unified,— Goethean, 'clairvoyant' or spiritually perceptive—one. In Barfield's thought it is treated as touching upon the sources of creativity in the arts,[*] and of course in the approach to knowledge itself in pre-modern times, in a brilliantly illuminating way. The way that we learn to find a 'self', as we gradually separate-out a primary fullness of experience, is an important element in his analysis. It is important to note that we cannot start with a 'self', but with experiences that bridge over the modern divide of individuals one from another. We are talking about a body-of-information which starts in a unified intensity, a meaning-entity that separates subsequently into analytic component parts.[†] In our standpoint we have shifted a step back from the external notion of knowledge, to the kind of being which is present in knowledge, where in the immediacy of knowledge self and world are originally fused. Children's consciousness still contains much of that primary awareness, when they move with the wind, etc. Their inner intention is still carried by the formative forces of the environment.

All this is to suggest that information-entity which structures a living organism is not, after all, anything very like the setting-up of a computer with its software, designed simply to run algorithms. A complex organism does indeed embody concretely a range of articulated activities, whether physical manipulations, or speech, or thought-articulations. But the insightful life, thought and perception behind it should be a spiritual immediacy that realises itself through inhabiting and elaborating a living body. The best model is the incarnation of meaning in the word, as in speech we experience our entire living organism on a higher level, and we can come to express ourselves and

---

[*] This was Barfield's starting-point in his *Poetic Diction: A Study in Meaning* (London 1952). Valuable for our viewpoint here is his demonstration that early language is not referential, i.e. pointing to this and that, but a 'holophrase' in which everything is presented undivided, pp. 83ff.

[†] It is important for understanding Steiner's radically evolutionary approach that he does not begin his anthroposophical account of the spiritual reality underlying our development with monads or spiritual selves, but rather with prototypical *human experiences*—specifically sensory yet also spiritual experiences (perceptions). Only at a relatively late stage will these be incorporated into a 'subjectivity'.

our meaning through it.* The body-of-information which creates its embodiment in this way is not like an algorithm or a computer-program. It is a blinding insight into the possibilities of life, and meaning, and evolution at the heart of the universe, from which all the particulars and oppositions and interconnections will be afterwards sifted out. Whatever kind of existence we choose to assign to such entities, we can see why their revelatory presences were venerated as worthy to live and move in the Garden of God, or to shine out magically in the caves of the Upper Palaeolithic.

Animal archetypes Baron Rosenkrantz's version of the Seal with the archetypal animals and emergent humanity is reproduced in John Fletcher's Art Inspired by R.St. *p.83.*

Rudolf Steiner gave indications for a number of 'Seals' representing stages of metamorphosis and other aspects of animal and human evolution, which a number of artists realised in paintings or in images incorporated in the organic space of the Goetheanum buildings. Here we see the archetypal animal-forms with emergent humanity. Seal by Arild Rosenkrantz (Private Collection).

* We touched previously on Rudolf Steiner's insight that here we have an intimation of the power which was working in the cosmic beginning and in the process of our bodily origination.

Chapter 4

ANIMALS AND ATLANTIS

One of the most quizzical elements in the Atlantean researches of
Scott-Elliot and others is the combined presence of man and of prehistoric flora and fauna. Human activity is said to have played into
animal evolution. Taken literally it is sheer fantasy; but we have
seen that man's place in evolution can be understood more deeply
as a key to the laws of biological complexity, and a driving force
behind them. Rudolf Steiner was already able to characterise our
relationship to the other kingdoms of Nature from such a dynamic
viewpoint. His ideas show how it could indeed be meaningful to
associate human development with much older stages of the Earth,
and that in turn will help us to understand what happened in the
subsequent emergence of earthly humanity—manifested eventually in that remarkable moment of breakthrough to modernity
from which our explorations began.

Unplatonic Ideas

Many things are starting to come together. Rudolf Steiner's associating of evolutionary insights with a transformation of consciousness,
enabling us to perceive higher-level complexity, and biological-information forces which reveal our own deeper connection to the cosmos, has
turned out to be strangely in accord with trends in the recent unfolding
of science: whether we think of panpsychism in physics or astrobiology.
The fixed frame of mind which had dominated science before evolution
was good for categorising things, but it obscured the network of kinship,
the common origins and manifold transformations which binds things
together—including ourselves. Morphological awareness of our shared
history is already one step in the 'clairvoyant' transformation of consciousness of which Steiner spoke.

Moreover, this new kind of thinking—and it is genuinely new in the
last century or two out of all human history—shows that we potentially possess a kind of penetrative way of 'seeing', capable of revealing to us a world of autonomous complexity than the understanding
which merely lays out isolated facts, things in sequence. Rather than

acting through specific components, these formative features may indicate 'biological-information forces', actively diffused in the cosmos. And many scientists are increasingly struck by the potential of the 'Anthropic' approach which points to features in the universe, that we know has given rise to ourselves; it argues from what we are to the forces and structural features underlying our evolution. Even on a minimal assessment of the direction implied by these significant scientific trends, Steiner's idea of the human as a fundamental element in the make-up of the cosmos, which has actively realised itself through our evolution, bound up with the gradual intensifying of the initial dim consciousness in things, need not be a barrier to the scientific mind today. Indeed, the convergence is striking.

Rudolf Steiner goes considerably further in his researches—or it would be better to say, since his trajectory is that of 'anthroposophy', he brings to bear upon evolution resources including his expanded perception as a way of seeing the past. His account of the method of extending that moment of conscious 'seeing' is detailed and consistent, and if we are still inclined to doubt how it could be made to yield detailed concrete results, we may admit that Steiner himself repeatedly emphasised the difficulties!* But the convergence with natural scientific approaches may actually guide us in understanding his account, which in turn confirms and furthers the insights which have emerged from those remarkable developments in the sciences of life of which a rather piecemeal exposition has been attempted above. Moreover, if we accept the implication of the non-mechanistic origin of the organising spirit which sets going the process of life, we shall understand how Steiner's immediacy of perception leads us not just to a cosmic order with abstract laws, but to the presence of living structures with a history which includes the forerunners of ourselves.

Steiner's descriptions notably include the perception of the animal 'types': these emerge out of the fundamental trends of evolution; and equally the form of human existence, which is in some ways the realisation of the whole structure of the cosmos. These are patterns revealed in the complexity of life's unfolding and actively working itself into multiple, individual existences. His clairvoyant descriptions of them have a grandeur and beauty which we must also attempt to acknowledge. Emphatically, however, they encapsulate the beauty

---

* See for example Steiner's in *The Inner Realities of Evolution* (London 1953) pp. 7ff.

of a world-in-becoming, not a transcendent world which is supposed to be the timeless reality beyond it—though it might well be called its 'hidden side', from the standpoint of our ordinary categorising consciousness. We must never for a moment think of these living types or of man as anything like Platonic ideas, with the implication that the nature and being of man is already complete from the outset. Steiner insists on the contrary that these possibilities themselves are evolving. Modern science could be helped to understand such processes, argues respected biologist Judyth Sassoon, by Goethe's approach to the organic type. 'In his scientific studies,' she writes, 'Goethe described living organisms as generative entities, constantly changing and redefining themselves with respect to each other and their surroundings.' Hence, she adds, 'recent developments in evolutionary biology... may be of interest to Goethean scientists. In the past few years, some mainstream biologists have argued that the current evolutionary model requires updating.... The proposed "extended synthesis" includes several ideas from evolutionary developmental biology such as phenotypic plasticity, that present a more dynamic and responsive model of organisms than previously. This shift in the concept of organisms seems to converge towards a Goethean perspective and could provide a basis for valuable exchange between mainstream biologists and Goethean researchers.'[*]

What man can be is dependent upon the realisation of the possibilities at each stage, not a goal or an ideal. The pattern of 'information', the manual which will produce human beings is itself constantly being updated, and if it is guided by those presuppositions inscribed in the cosmos itself, it is yet subject to all the gradual sifting and separating, which is vividly described as a spiritual process by Rudolf Steiner in a quintessentially evolutionary way. Events in the far cosmic environment are already a part of that development, establishing some of the fundamental rhythms and regularities which so amazingly cohere to produce life.

We can follow one description of Rudolf Steiner's in a little more detail. Clairvoyant perception sees in the primal evolutionary state of things, he says, the etheric 'point', which is to say, something occupying no physical space. (It can be considered, however, as capable of expansion to an infinite periphery.) He vividly describes the primal

---

[*] Judyth Sassoon, 'Goethe's Spirit Haunts a new Dynamic Biology', in the journal *Elemente der Naturwissenschaft* 113(2020), pp.5-18 (abstract)—available at https://elementedernaturwissenschaft.org/en/node/1499.

reality of the human form, likening it also to the etheric plant-archetype that the clairvoyant sees as yet unrealised in the plant:

> a light-form, a beautiful form, which in reality is not there, but rests slumbering in the point.... It is a form that is distinct from a physical man, as different as the archetype is from the plant. It is the archetype of the present human form. At that time the human form slumbered spiritually in the etheric point, and the whole earth-evolution was necessary in order for that what rested there might develop into present-day man. Many, many things were necessary for this, just as much as is also necessary for the seed. The seed must be sunk in the earth, and the sun must send its warming rays, before it can develop itself into a plant.[*]

The etheric point, he continues, is at the same time a mass of points, indicating the mass of individual human beings who will evolve. Gradually, very gradually they will be able to become aware of their archetypal existence, and realise that everything which has since happened in cosmic evolution is necessary for their awakening to consciousness. Yet there is already,

> floating before their souls, as a picture, a thing which was already present as a picture when the sun was still united with the earth.... This picture is the meaning of earth's evolution; to bring this picture to realisation the sun had to separate itself from the earth and the moon detached itself. Through this man could become man.[†]

We can only offer the briefest attempt at characterising Steiner's extensive presentation of human evolution, as it comes closer to conditions which make possible the realisation of that original 'point'.

This journey commences, however, in ages long before the emergence of animal life. At that primal stage we may discern man's implicit being, slumbering implicit in the basis of things, hardly awakened in the primal dull consciousness which is inherent in the universe. In a previous chapter we touched upon this original relation of the human body, bodily actuality, to the stage when the solar system was still just forming. That stage of consciousness can still re-emerge under certain conditions in modern mental life. And we saw how in recent cosmological thinking like that of the astrobiologist, Lee Smolin, the structuring activities eventually actualised in organic forms are also nowadays recognised as part of a chain or hierarchy, continuous with these processes of cosmic formation.

---

[*] Steiner, *Egyptian Myths and Mysteries* p.15.

[†] Steiner, op. cit., p.36.

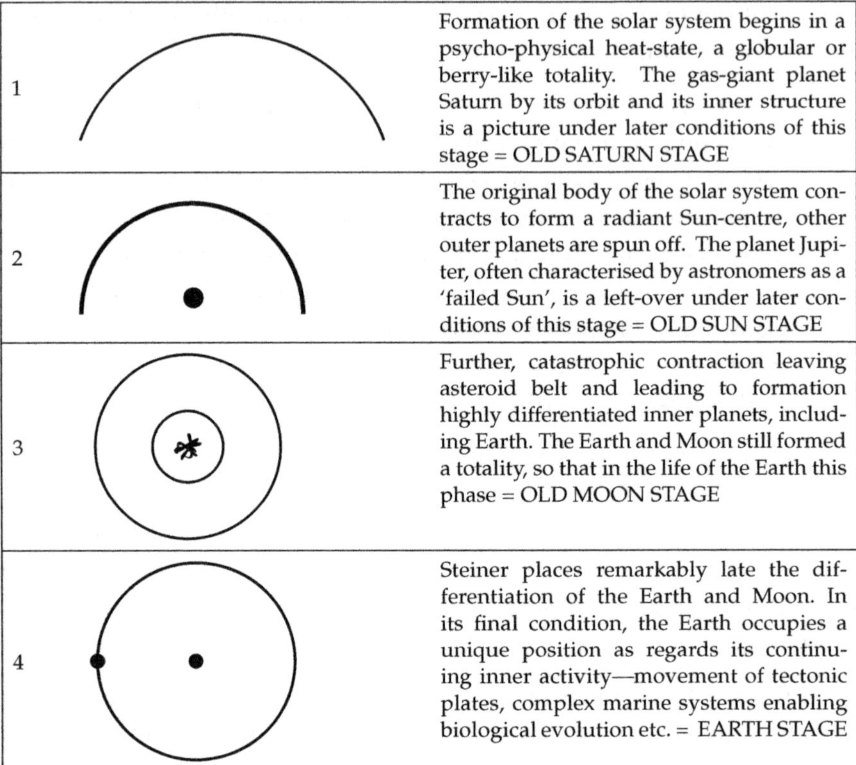

1		Formation of the solar system begins in a psycho-physical heat-state, a globular or berry-like totality. The gas-giant planet Saturn by its orbit and its inner structure is a picture under later conditions of this stage = OLD SATURN STAGE
2		The original body of the solar system contracts to form a radiant Sun-centre, other outer planets are spun off. The planet Jupiter, often characterised by astronomers as a 'failed Sun', is a left-over under later conditions of this stage = OLD SUN STAGE
3		Further, catastrophic contraction leaving asteroid belt and leading to formation highly differentiated inner planets, including Earth. The Earth and Moon still formed a totality, so that in the life of the Earth this phase = OLD MOON STAGE
4		Steiner places remarkably late the differentiation of the Earth and Moon. In its final condition, the Earth occupies a unique position as regards its continuing inner activity—movement of tectonic plates, complex marine systems enabling biological evolution etc. = EARTH STAGE

Rudolf Steiner considered these stages as constituting a fundamental rhythm underlying all development—e.g. of the human form. And it already reflects prior developments in cosmic-spiritual evolution, to which the planet-names are also archetypally applied. The life of the various stages can be described as the emergence and development of beings, hierarchically interrelated, within which our own life takes its place. Steiner illustrated the rhythms here outlined in many ways, e.g. in artistic 'Seals' showing the metamorphosis of forms. Cosmological Stages of the Solar System according to Rudolf Steiner (after his lectures especially such as Spiritual Hierarchies and their Reflection in the Physical World; Inner Realities of Evolution etc.)

For Steiner, the four main stages of the formation of our solar system are one highly important presupposition for the nature of present-day life and our own existence. We have mentioned that the starting-point for his descriptions is fundamentally that heat-charged state of the substance which filled the whole area of the future planetary system, subsequently defined by the orbit of present-day Saturn and from which the other giant outer planets were thrown off. After this 'Saturn' era, a second stage is the contraction of that nebulous mass, which intensified into the Sun as a star, around which the subsequent formation of the whole system will revolve. A third stage, and a further progress toward materialisation as it would be recognised today, nevertheless still sees the substance of the planets in a partially differentiated state, along with the solar centre and the remaining

Saturn-body; and only in the last, or fourth, do the planetary bodies (especially the inner planets) take their present form, while on our cosmic doorstep—as it were—Earth and Moon assume their separate and significant relationship. Material substance itself may have been in some ways different from its present condition during many of these stages.* But more importantly, these structures are already conceived as mapping out the kind of developments that will take place, later in our history, on the level of organic life. They are, in modern language, not only material processes but, globally, they are already accruing the 'information' that will enable the complexities of life and eventually the further refinement and concentration of (human) consciousness.

It is easy to see that in the case of plant life, for instance, the most basic structure is that of an organism membered into its earthly environment, drawing substance from it, but relating for its energy and unfolding life to the source of light, the Sun.† Thus a plant could never be understood purely out of local environmental factors. The cosmic phase of star-formation, when the Sun originates out of the primal materials of the spiritual-material entity, first produces a system where it has substance around it which is illuminated and influenced from without. This is a primary phenomenon on which a whole series of developments will be built upon the same motif. It is also the first step in a tendency which will play itself out in the establishment of planetary-gravitational rhythms:‡ the rhythms established will produce the possibility of planets with balanced environments where stable conditions may foster organisms; and those organisms may

---

* It may also be admitted that Steiner's account of the forming of the solar system differs in some aspects from current cosmological thinking, though the measure of convergence is also basically very great. For cosmological purposes, the account in the *Outline of Esoteric Science* needs to be supplemented by Steiner's descriptions e.g. in *The Spiritual Hierarchies and their Reflection in the Physical World* (New York 1970).

† In detailed scientific fashion based upon Steiner's approach, this starting-point is impressively extended into a detailed understanding of plant morphology in George Adams and Olive Whicher, *The Plant between Sun and Earth* (London 1980; Colorado 1982). For a valuable introduction to underlying ideas of point and plane (cf. Steiner's 'etheric point' which can be considered as expanded to infinity), see also Ernst Lehrs, *Man or Matter* (London 1951) pp.210-220.

‡ A study in the line of Steiner's thought is Hartmut Warm, *Signature of the Celestial Spheres. Discovering Order in the Solar System* (Forest Row 2010).

take advantage of the interplay between the changing conditions of warmth and light and the nutrient-rich environment. The present-day plant has all those structural differentiations written into it—and could not exist if there were not those vast preliminaries of exactly such planetary orbits around a radiant Sun. It presupposes likewise differentiation of the planetary environment so as to allow, on at least one world, the interaction of cosmic irradiation and available chemical materials to furnish the context for living organisms. For Steiner, the Sun with its light is therefore at no juncture a mere material agent, for even in the first ages of its development as it differentiates itself from the proto-planetary materials around it, it is already establishing a pattern that will be perpetuated in the process which we see embodied in what we term a living organism—a plant. In its higher-level complexity, the second or Sun-stage of our system can be called a sort of plant-phase, in which man is implicit in an etheric form, such as organises plants even today.

These organising forces, which reveal themselves on a higher level of complexity to transformed vision, are what Steiner indeed often terms 'sun-forces'. They obviously have features which to our ordinary type of awareness seem like 'ideas'. If we are to understand their cosmic origins, however, it is important that we do not imagine them just as stark, abstract types of the subsequent real plants which have been shaped by them in their later evolution. There is obviously no place for a house-plant or an oak tree in the cosmic environment of the 'primal Sun' stage. The cosmic structures of that time were utterly different. But they must nevertheless be considered as already intimating the differentiations we will see much later in physical-biological existence. Steiner's extended perception is able to see them as actual centres of life, far back in cosmic time and written, so to speak, across the cosmos.

These 'life-bodies' are the supra-individual aspect of living things, types as they are understood in the Goethean world-order. Evolutionary thinking, after all, always brings us to a perception of concrete ancestral forms, and real beings—in Goethean thinking these *are* the explanation.

The puzzle of how to think of an archetype, which can resemble an abstract idea in some ways, but generating endlessly different forms, came a step closer to conceptualisation with the advent in science of so-called 'Chaos theory'. It has been described as a thinking for the evolutionary age. New mathematical approaches demonstrate that the same principle will not necessarily (as used to be presumed) always produce the same outcome; rather the same starting-point will

In association with Rudolf Steiner's ideas, Theodor Schwenk's finely observed studies of vortices in water and air led to the insight that rippling and swirling movements, though regular, did not merely repeat specific patterns. One impulse produced rather a range of gradually changing modifications that could develop in a great variety of ways. His observations helped inspire a whole new science, and the new mathematics of 'chaos theory'. From Theodor Schwenk, Sensitive Chaos.

generate constantly different results. Most famously, the 'butterfly effect' exhibits the train of unpredictable consequences from the tiny movement of a wing. 'Chaos' it may seem, yet the theory makes sense of it. Thereby it has come closer to describing the actuality of living form, which never jerkily repeats itself. Rudolf Steiner and his followers, particularly Theodor Schwenk, once more pioneered aspects of these new modalities of thought.*

Science in general has by no means fully freed itself from thinking in fixities. Schwenk opens to our rapt attention a new world of flowing forms, which arise where liquids move and swirl, and is able to think in their terms. Nothing repeats itself exactly in the streaming currents and flow-forms, yet there is a deep coherence which again reveals a higher order of complex behaviour to which we were previously blind.

We may now see in the arising and transformation of flowing forms a manifestation of that same 'etheric' dimension that is active in the

---

* Cf. Theodor Schwenk, *Sensitive Chaos* (London 1965). The significance of the new mathematical approach itself is made remarkably accessible in James Gleick, *Chaos. The Amazing Science of the Unpredictable* (London 1998); for Schwenk's contribution, pp.197-200.

life of organisms. The etheric world ultimately manifests itself as life in its glory and complexity. Its archetypes are flowing 'chaos', which is yet the sure begetter of the rich morphological variety we know (and belong to). However, Schwenk's work is able to suggest how the laws of the arising and passing-away of forms were already manifested in the cosmic separations of our Sun and planets, when matter was in a state less solid and dense than present-day substance. They already constituted a life-process. At the same time, this world with its laws of streaming form, is already a partial realisation of the deeper potential in the cosmos to produce our human mode of being and consciousness. In one sense, the primal world of these cosmic manifestations can even be described as man at a certain stage of realisation.

Thus we are not to suppose that there was an inevitable process whereby the conditions for the eventual existence of human beings simply unrolled toward a determined end, like the working-out of a mechanical-abstract principle. As a matter of fact, Rudolf Steiner's account rather significantly stresses the problematic, disrupted nature of human origins, rather than depicting the process as a straightforward line of emergence. What he actually describes is much more interesting from an evolutionary viewpoint. If we now skip along in evolutionary time to a stage when those etheric plant-archetypes had emerged from the human-material matrix, and were able to be operative in an earthly environment, we have some richly illuminating indications.

Steiner beautifully describes how the Sun-stage of cosmic organisation constitutes a sort of archetypal plant-existence. But it was also a step in organisation that brings human existence from its primal potentiality into a fuller expression. The Sun-environment polarity, in which all plants still live, was then in its original cosmic form: but that structure would continue in many further adaptations. It would live on in plant life, with all its variety—occupying a diversity of environmental niches. And also, as a structure, it would continue to find a place within the emerging human, not lodging itself in a specific niche but being incorporated into further levels of complexity. It is however the same archetypal structure which continues in human reproduction in the womb, where the child grows and is fed by the blood from the sun-like heart above. Rudolf Steiner can rightly say therefore that man already existed in a certain form. 'Man' began to stir to awareness in the sunstirred vapours of that ancient planetary environment, still united with the form of the plant.

The Polish-French mathematician Benoit Mandelbrot, best known for the 'Mandelbrot set' and his studies of 'fractals', explored mathematical models that from identical starting-points produced different outcomes (often termed 'chaos'). He found ways of representing these progressions, often with myriad transformations, and the results have proved relevant to the understanding of changing, flowing forms like those of Schwenk, in 'ragged' formations like seashores, and in living organisms. His work has been considered the mathematics appropriate to the evolutionary mode of thinking.

Rudolf Steiner's description captures the beauty of the scene:

> The luminous masses surged back and forth. Within them were all the human beings of today, woven through by all the spiritual beings, which rayed forth light in manifold grandeur and beauty. Outside was the earth-cosmos in its great variety.... As though by an umbilical cord that sprang from the divine, man hung upon this totality, on the light-womb, on the cosmic womb of our Earth. It was a collective world-womb in which the light-plant man lived at that time, feeling himself one with the light-mantle of the Earth. In this refined vaporous plant form, man hung as though on the umbilical cord of the great mother, and he was cherished and nourished by the whole mother Earth. As in a cruder sense the child of today is cherished and nourished in the maternal body, so the human germ was cherished and nourished at that time. Thus did man live in the primeval age of the Earth.*

---

* *Egyptian Myths* p.54.

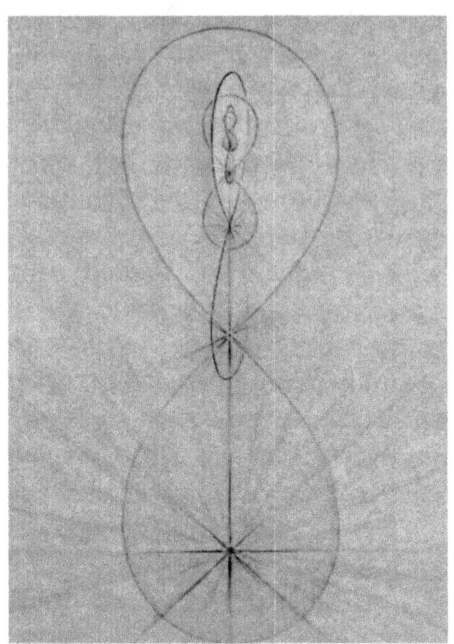

The plant between Sun and Earth. From a dynamic perspective, a plant is not so much an object-to-be-classified in Linnaean style, as an expression of structures that must be conceived in the setting of Earth's relationship to the Sun in cosmic space. George Adams and Olive Whicher exemplify this approach: 'Consider how each node evolves in a growing plant. It begins its existence … where at the growing-point of the stem the primordia of the leaf-buds are born. It is at this moment beneath the sun-centre and above the more physical and radiating realms of root and lower stem which carry it and provide substance. Ideally it is poised radiant below and the sun-centre above, like the lemniscatory crossing-point in the bipolar space.' We see this morphology in the picture of the 'Decussate Form of Phyllo-taxis'—which is to be found for instance in the Stinging Nettle. The plant is not assembled mechanically; the lemniscatory space can rather be 'potentized, renewed again and again at a higher level—from node to node up the stem'. In evolution, as Rudolf Steiner observed, the incorporation of such features at a higher level in more complex forms is a basic princi-ple. From Adams and Whicher, The Plant Between Sun and Earth.

Beautiful as this is, we must take it quite scientifically. The basic life-sit-uation of the 'plant between Sun and Earth', into which organic plant-life will subsequently pour itself, comes recognisably before us. For that life-process of the primordial plant has been internalised by man (and viviparous animals) as the foetus in the womb. The child does not yet have its own life but lives and grows out of the life of its mother, under her warm heart, as the plant receives its energy from the Sun. The Sun-Earth relationship which determines the plants is brought within the being of human-animal reproduction, takes place inside it. We see it in the growing child in the mother's womb. The structural possibility of human gestation is incorporated, we may say, at this cosmically early stage, as a kind of heightened plant-process. But at that stage, of course, none of this potential could yet be realised in organic life. The organic situation of embryonic development, however, is a perpetuation of once macrocosmic flowing/streaming structure.

Then Steiner describes a further stage of cosmic development, a condensation of the planetary substance to a denser fluid state. There was not, however, as yet, a distinct atmosphere—'but going outward, the waters changed into heavy mist, which gradually became more refined': 'It contained various materials in a state of flux, which were enveloped by mists that became ever finer until, in the highest spheres, they became extremely rarefied. Thus did our earth once appear and thus was it altered.'* In that separation of denser substances (still fluid, however) from more rarefied states, we come to a divide in the direction of evolution. In the ecology of the gradually separating waters and the land, the etheric life will be able to flow down into emergent vegetation-archetypes in all their incredible variety of forms. But the slowly intensifying consciousness which is the prototype of man is linked with an increased structural complexity that provides a 'resistance' to environmental pressures, which cannot yet be given outer expression. The 'plant-process' stays internalised, and will be the basis later of an embryology.

The plant 'idea', in evolutionary thinking, means precisely the polarisation of Sun-energy and Earth-environment and the forms that can stream in-between. Its realisation will ultimately come about under more earthly conditions as the physical life of the plants—locked in specifically to that niche in the ecology of the whole where water-Earth is shone upon from without. But the prior cosmic situation, as we have seen, will also be incorporated and internalised as a prototype of the animal-human complex of gestation.†

The effect for the first stirrings of human existence was a holding back from dependence on external conditions, as these changed in the

---

* Steiner, *Egyptian Myths* p.55. Such a planetary structure is reminiscent of modern reconstructions of the giant outer planet Jupiter. The stage of cosmic contraction at the juncture described is no longer that which fills the whole Saturn-orbit, but the smaller (but still vast) orbit of Jupiter. The present planet can be understood as a vestige of conditions which then prevailed; it is often termed by astronomers a 'failed sun'. Cf. Eric H. Christiansen and W. Kenneth Hamblin, *Exploring the Planets* (1995) pp.319ff. Recent additions to our knowledge showed the virtual absence of any solid core.

† These complexes only become physical and organic, of course, in subsequent earthly conditions. Indications of specific stages of early cosmic development, still shown in human embryology, are described very fully in the work of Rudolf Steiner's pupil Hermann Poppelbaum, *Man and Animal* (London 2014).

way Steiner indicates. The cosmic 'plant-form' in its original significance is preserved, but now on a higher level. In one sense it is held back, though on another it is rescued from the hardening into physical conditions which affected the evolution of vegetation, and becomes part of a 'higher' development. The human etheric archetype remains submerged in the physical reality of that time, keeps incorporating developments into itself and still pervades the environmental totality, resisting specific realisation. Such are the kinds of cosmic structures and processes which Steiner perceived among the prerequisites of our evolution.

It will still be long ages before it can find the right material conditions for its realisation on Earth, requiring an organic complexity far beyond that of the plant. In animal and human development, the polarity of light and darkness with its strong rhythmic aspect owing to the movement of the cosmic bodies, will also be absorbed on a still higher level into the alternation of sleeping and waking consciousness, freeing the living organism still more effectively from the plant's dependence on the outward condition of light radiating its energy into the earthly environment. Day and night become an inner life-process. Man thus becomes ever more a 'world in himself'. So much must suffice as an indication of the process, which Steiner characterises in many details, whereby the animal and vegetable kingdoms can be considered as stages 'left behind' from the evolution of the human form, and coming earlier to expression in the physical conditions on Earth and becoming bound up with them.[*]

It is only with the resource of the rich mineral and organic compounds, formed in the gradually separating fluid and gaseous conditions of the later Earth, that evolutionary developments related to the animal realisation of these structures become possible. We move on now to a stage when the Earth has a planetary existence and an oceanic organisation. Both plants and animal life become biological realities. The plant still reaches out toward the light source, but in the darker watery Earth primitive animals are able to evolve ways of conserving themselves and seeking nutrition through moving about. As a higher-level transformation far beyond the plant-like dependency stage, and even primitive animals, the fish-phase of animal evolution will present an organic realisation of free movement in the currents and streams of the ocean. In fish, reproduction still takes place when the eggs are fertilised in the watery, nurturing environment. In the higher

---

[*] Steiner, *Egyptian Myths* pp.38-9; fuller accounts include Steiner's *Outline of Esoteric Science* pp. 222ff.

animals and in human beings, the watery environment will itself be internalised and embryonic growth made independent of external accidents and attacks in the womb. The fishes, however, have settled into the niche that provides a setting for them to perpetuate fish-behaviour—as long as conditions remain similar. The colonising of an external environment, however, runs the risk of becoming a dead end. Some ninety per cent or more of all species are known to have become extinct.

In all these phases of cosmic and biological evolution the human being is not what we might term a mere theoretical possibility, but a real cosmic situation—even if it has physically ceased to exist. Though it is a potency that will appear physically again in a more complex morphology, when that becomes possible. It remains something still open to myriad possible developments which might help it come about, whereas the lower forms and fishes, contrastingly, have made their evolutionary commitment. Yet they are not simply devoid of human relevance. For what lives in the fish outwardly will need to be a part of human existence, incorporated into the greater complexity of our organisation. Evolution will eventually fashion the physical organism that can incorporate man. Steiner describes how the life and form of the fish awakens an interest—a 'longing'—on the level of still ethereal humanity. The human archetype cannot yet be realised under such conditions, but it can feel drawn toward the fish as they develop in the ancient oceans and seas, in a kind of first fascination with an animal existence. And thereby the dim consciousness which so far had risen only to sleep-like intensity, could begin through this animal engagement to alternate with a sharper awareness: a waking-state prefiguring objective perception, though still at a rudimentary stage. The directing of this fascination, as it were, upon the proliferation of lower animals accompanied man's archetypal existence from this time.*

Perspective: From Animal to Human

Rudolf Steiner contrived to investigate many significant moments in evolutionary history, cosmologically remote or subsequently organic, in his attempt to convey the impetus of evolution that extends into and informs the present-day natural order we know—the world within the world. We return now to the stage which he designates 'Atlantis', and

---

* Steiner, *Egyptian Myths* pp. 229–30; cf. *Outline of Esoteric Science*.

we may consequently begin to understand his account of human existence at that time, which is the focus of our principal interest.

Even when contemplating some of his most challenging assertions, we have also seen that modern ideas from the forefront of cosmological-biological thinking have taken huge strides in the direction of his basic ideas. The new 'Anthropic' principle in science helps us makes sense of his 'anthroposophical' approach. Also the idea that the rhythms and structures of the cosmos and our planetary system are hierarchically 'nested' in a continuous sequence with those of life, and still more the idea that cosmic structures may in a 'global' way function as 'biological information-forces' (etheric forces), need no longer stagger the mind of the progressive scientist.

At the bottom end of the hierarchy come individual species and varieties, in a specific niche. In the case of these individualised forms, if the niche is subject to too much change, the species closely geared to it will become extinct—almost every species in the fossil record eventually dies out. On another level, however, the incorporation of the otherwise external and unstable environment into the organism of the more advanced animal makes for a new independence—so that an ever greater degree of autonomy and inner elaboration become possible.

The actualisation of simpler life-forms is thus one side of a larger evolutionary process that also opens the way for more complex realisations which, when subsequently incarnated, will maintain themselves within the environment more successfully and not so easily succumb to its instabilities. Steiner *sees* this vividly and concretely as the separating-off of the simpler, more ecologically narrow life-forms, each of which, however, also indicates a further, more elaborate realisation that can emerge in subsequent evolutionary opportunities. Wolfgang Schad has demonstrated in fine detail how the three basic animal types represent divergent patternings which are all synthesised in the human.[*] Hence it is not a simple process of extrusion: there is in tandem a waking up to a sense of affinity—that driving fascination with unfolding life which is part of Steiner's version of the biological imperative. Finally, human morphology will constitute the greatest incorporation of many

---

[*] The far-reaching scientific value of his analysis is stressed by Lockley, *How Humanity Came into Being* pp.181-196 including diagram. The three types are small animals dominated by the head-sense system; carnivores, centred in rhythmic and circulatory activity; ungulates, living in metabolic and limb processes. The threefold tendencies correspond to Steiner's mature anthroposophical teaching on the threefoldness of psycho-somatic human existence.

adaptations which we see separately in their environmental niches; the human is held back from specific adaptation, and shows the fullest possibility of autonomy and versatility we find in the universe (at least *so far*, as G.B. Shaw might have added).

All this might seem on one level to be saying no more than that, as more complex life-forms come about through evolution, the possibility of human life comes fundamentally closer. Yet it would be quite wrong to see just a kaleidoscopic accumulation of developments.

We recall that there is a considerable consensus that ingrained evolutionary tracks would steer life again on the same course, if circumstances arose, recalling Steiner's comment on evolutionary developments coming about when they can. Such deeply scored tracks are structural complexes like Schad's animal types, or still more like those analysed by Smolin: of biology, temperature, climate, ecology, and ultimately cosmic relationships extending even 'up to galactic level'. What Steiner's clairvoyance sees is neither an abstract law necessitating a particular course, nor an accidental concurrence of conditions, but precisely such structural patternings through which the human potential of the universe, hidden at first in a point and 'actually nothing', is gradually, opportunistically, bringing itself into being. For what is possible in the fine-tuned ordering of the cosmos will happen when it can— rather like finding the words to formulate one's meaning as one learns a language. The pressure for it to happen is a purely spiritual reality at work, not a 'life-force' or a 'world-purpose' or an irrational *élan vital*. It is more essentially like a situation which asks for a moral answer, or the meaning which emerges in the combination of signs, or like the 'forces of perception' which give to the sensory qualities of a painting an inner life. This is the truth in the old myths, like those of Paradise, which rightly bring the moral dimension into their account of the world and human origins.* But can this possibly be called scientific? Perhaps it may help if we formulate the question in a more familiar guise.

We are accustomed to speak of 'higher' and 'lower' forms of life, or to refer to the 'increasing complexity' we observe as evolution proceeds. But there are many biologists who dispute just that. There are, they counter, just different developments as conditions change, and a

---

* The implications even of 'the Fall' are by no means basically negative: the Fall implies Paradise, and a seer, sacral king or a charismatic leader who could claim that vision could give direction and meaning to his society. (Only the Judaeo-Christian version makes the loss irremediable, and therefore makes us totally dependent on a redemptive act.)

series of evolutionary solutions to the challenges they pose (a 'kaleido-scope'). Anything else is merely subjective, like our ready preference for cuddly mammals over spiders or crocodiles. Many evolutionists insist that we must be deluded even in wanting to see an increasing scale of complexity, which they see as a notion based on outmoded assumptions that there is purpose in Nature. If there were such a straight-line course discernible in Nature's workings, it would inevitably drag in the old pre-evolutionary assumptions of a set goal to development and some sort of metaphysical 'necessity' outside Nature itself. They indicate the many far-from-straightforward developments in evolution, suggesting no such linearity exists.

But for others, their determination *not* to find a tendency is itself just as unscientific: and they point out that it would be foolish to conclude, from the fact that a path does not go in a straight line to its destination, but dodges about through ups and downs, that the path does not lead anywhere. It would indeed be failing to see the tree for the wood. In fact, it is another case of the need to see, in Penrose's sense, what cannot be demonstrated point-by-point. Christian de Duve is one among scientists who maintains that if we pierce through the luxuriant side-growths and view the evolutionary tree dispassionately, 'the structure of the trunk, with its progressive rise toward greater complexity, is clearly evident'.* His view is the more valuable in that he looks for a purely physical explanation, and believes we must work out physical laws which determine life's rise.

Recent explorations of complexity, however, suggest that there may be a different approach. There are signs of a third way through the impasse between reverting to out-dated thinking and a new reduc-tionism. Much thought has lately been given precisely to complexity in Nature—notably arising from the work of Stuart Kauffman. Arguments still rage over the best way to think about it. Is it simply basic physi-cal law that culminates in complex structures such as life, or is it that higher-level laws, a deeper meaning, can manifest themselves when a critical level of complex organisation has been reached?

The latter approach is the kind which Rudolf Steiner sometimes used to characterise moral 'necessity': he does not believe in timeless ethical abstractions, but in moral perception which is another aspect of the spiritual awareness that informs human knowledge. An eth-ical imperative makes itself felt when a being endowed with such

---

* Discussed in Davies, *Origin of Life* p.253.

conscious power of perception is present in a scene. The moral force of the situation reveals itself only then: so for instance if someone has an accident on a remote mountainside, no one can be blamed for failing to rescue them if no one has the relevant awareness; but if once seen by a passer-by, the requirement to help them to safety asserts itself as an ethical 'demand'. The moral response thus cannot be prompted by any quasi-physical necessity. In fact the ethical demand, which is perceived to be unconditional by a morally aware being, paradoxically presupposes physical freedom not to act. Freedom itself, Steiner shows, must not be conceived as random undetermined acts, but only as a 'higher law' that shows itself in the presence of more developed consciousness. Such a consciousness is beyond that known in animals, being essentially reflexive: like Goethean 'seeing' of Nature, it manifests itself both in connection with what is seen, but also as drawing attention to the moral catalyst—the human presence. Here indeed, as so often, Steiner was taking further some hints of Goethe.* Moral truth is an objective dimension manifested in the presence of humanity. All Goethean 'seeing', in fact, whether seemingly just of 'Nature' or of a human situation, places us quite objectively in a position of moral awareness. In all of which we have a basis for understanding, or rather concretely perceiving, the spirit in Nature.

'Direction', 'increased complexity', 'higher'—these are all value-saturated concepts (which is why many scientists haver at introducing them). If the world does indeed exhibit that broad 'trunk' of evolution to higher forms (complexity) which de Duve accepts we see, we may best explain it as uncovering higher-level coherence, essentially like that of moral perception. And if its coherence is closely related to our level of consciousness and moral awareness, we may plausibly argue that it shows human emergence making evolutionary sense in just the way Rudolf Steiner suggested. Human complex existence, emerging from that etheric point, would be identical with the drive to bring about the increasing elaborateness and autonomy in the sequence of living things. That drive would have no quasi-physical necessity or determinism, however (that would be the old abandoned

---

* In what one might term his academic days, Steiner had a number of contacts with Max Scheler whom he regarded as a significant ally in his own kind of thinking: the latter developed an approach to moral values which was similarly *objective*, a matter of specific perception—like Steiner decisively turning away from the dominant subjective and Idealist ideas of ethics prevalent in the nineteenth century.

theology, God said and it was so), and it would not be a physical but a value-pressure, or active moral force—the arch upwards of that broad trunk at the core of living diversity. It would be the spiritual law which eventually becomes explicit and acknowledged in us as moral beings.

If this seems to verge perilously on metaphysical hypotheses, it must be stressed that there *is* nothing but the actual evolution which has taken place and is still taking place; by learning to see in a Goethean way, we expand our consciousness to see the moral 'slant' of the events which brought us here. Rudolf Steiner speaks of the 'gesture'—the human gesture of things.

In Steiner's work, therefore, the separation of the animals (and other beings) to occupy their respective niches, flourish, and die, while the human archetype in all its richness remains still-to-be-actualised, and absorbing specific developments on its higher level, is described in terms both of fact and of 'fall' and 'sacrifice'. We can follow him still further in imagination from our glance at the early stages of Earth's development. We reached the stage when a kind of 'fascination' or longing of cosmic consciousness arose around the primaeval fish and similar organisms. Subsequently the changing continents and climates offered further evolutionary opportunities; and alongside their animal existence we must think also of the human archetype thereby coming closer to realisation, not yet attaching itself to them but freeing itself, divining ever greater inward autonomy, even as they assumed greater fixity. These potential animalities are there in human nature, but the one-sided realisations of animal life also liberated the human potential which cannot as yet come into outer expression. Any of the animal tendencies would be a premature narrowing of the autonomous being that is essential to the human. We can follow Goethe's lead once more in according a gratitude to all the 'lower' manifestations of evolution— even a reverence though not quite a religious one. The moral significance of their sacrifice is clear as soon as we remember to put ourselves into the picture.*

---

* Goethe's approach is taken up by Rudolf Steiner in his lectures on the *Gospel of John* (New York 1962) pp. 24ff. This proper reverence for Nature from the viewpoint of human evolution is of course a fundamental attitude in Steiner, and one which we need to put in place of the exploitative understandings of science which have wreaked so much havoc on our planet.

Animals and Atlantis

As we move on to the great animal elaborations of the Mesozoic era, we must suppose these were also a profound enrichment of the human archetype, bringing closer its possible organic evolution as the life-processes of the animal organs reached ever greater complexity. Steiner characterises them and the drawing of the emergent consciousness of humanity toward them, during the phase he calls 'Lemurian' because it took place predominantly on the great southern continent (nowadays termed Gondwana). He has described in particular the significant development associated with the breathing-process. The life-sustaining content of the environment is absorbed directly by the plant or simple organisms. In evolving the ability to breathe, a new life-process arises to meet the further needs of the organism in a novel, rhythmic way. He touched especially on the concomitant crisis of development in consciousness, initiated already with the first advent of this breathing-relationship. In human existence, it lays the foundation of the later ideational and feeling life—a higher-level interchange between inner and outer. (Steiner rightly rejected the artificial camera-model of consciousness and pointed to its organic prototype in the interchange of the atmospheric environment, now outside, now within the organism.) Consciousness intensified with the sense of life and death in their alternation through breathing.[*] This dualistic relationship is different from the plant-archetype which mirrors the cosmic situation of Sun and Earth; it is related more to the duality of Earth and Moon. 'There were ever higher animal forms, which man took into himself,' experiencing them as members of his own being, because they 'contained the germ of the later human.'[†] It is the picture once more of Paradise.

The archetype of humanity is beginning to assume the morphology which will enable it to take on organic existence—separating out into a more vegetative and a more animal (sentient) double structure. The cosmic (or in Steiner's terms astral) 'body, which also still contained the life-body and the physical body within it in a state of dissolution ... separated into a finer astral portion, the sentient soul, and a coarser etheric portion that could be affected by earthly conditions. At this point

---

[*] On this crisis in evolutionary development, Steiner, *Egyptian Myths and Mysteries* pp.58-9.

[†] Steiner, *Egyptian Myths* pp.74-5.

the etheric or life-body, which already existed in prototype, made its appearance ... the coarser portions of the etheric body that were sensitive to sound and light were separating out."* In all of this we continue to see the inner reality corresponding to the embodiment of the animal forms, a radical extension of the fluid metamorphosis which underlies each specific modification of the plant.

So we come once again to the stage which Steiner named from its notable palaeogeographic feature: the continent filling the Atlantic Basin—corresponding in time to the late Cretaceous and the immediately following geological phases (Paleogene). Across that vast continent the Mesozoic flora and fauna including the great sauropods (dinosaurs) rayed out and proliferated. The etheric human was closer than ever to animal existence, especially with the separation of the reptile genera. In Atlantis, the human was closely connected only with those forms which are closest to us in our existence even today, whose intimate biological relationship to us was indicated by Steiner and so brilliantly elucidated in biological terms by Wolfgang Schad.

Steiner is clear, however, that none of the 'human forms' which he describes for this period would have left any fossil record, unlike the creatures which physically populated the Earth such as the dinosaurs or later the giant mammals and birds. Of these latter indeed we have some spectacular fossil evidence. But the human, still etheric-formative forms were still connected more with the cosmic forces of evolution—those astronomical interactions which are the fundamental 'situations' that are captured, so to speak, and internalised in biological evolution. As yet they were not tied-in to any specific ecological niches on Earth.

In Steiner's account, then, the primal relations of light and darkness, of energy and material substance, even as they were already defined in the early stages of the solar system, set the initial parameters for life. The diversity and changing relations among the planetary bodies, for Steiner offer a model for later animal and human developments. Back in the Cretaceous era, the biological 'script' for human existence is still suspended in the chain of those 'nested hierarchies' of structures which reach up to stellar level, but more especially to planetary diversity. We must see if we can briefly put the cosmic setting and the emergence of human diversity out of the animal stages into a coherent living sketch.

---

* Steiner, *Outline of Esoteric Science* p.212.

The realisation that the planetary cosmos has gone through distinctive phases of organisation, establishing models for development in evolution, is not actually unique to Steiner:

> The surfaces of planets and moons cannot be predicted from a few general rules. To understand planetary surfaces, we must learn the particular history of each body as an individual object—the story of its collisions and catastrophes, more than its steady accumulations. The planets and moons are not a repetitive suite, formed under a few simple laws of nature. They are individual bodies with complex histories. Planets are like organisms, not water molecules, they have irreducible personalities built by history.

So wrote Stephen J. Gould in a famous essay. And his words are not just a charming analogy: they are part of the increasingly recognised application of evolutionary thinking to cosmology—mirrored by the increasing recognition of the role of cosmic ideas in biology. It need no longer strain credulity when Rudolf Steiner connects the evolutionary emergence of animals and human beings with the stages of cosmic formation of the solar system. The idea that life's structures resonate within a hierarchy of structures in the larger architecture of the cosmos has been discovered elsewhere, independently. Steiner's clairvoyant account goes considerably beyond the results of ordinary scientific thinking; but it goes beyond ordinary science in the direction that science is itself travelling in some of its most exciting and illuminating insights. Gould's observation that the planetary system is already a level where we witness the crystallising of individuality, exemplifying complexities which are more familiarly played out on the level of animal and human development, comes to meet Rudolf Steiner's investigations and helps us take further direction from them.

In the time of Atlantis, the etheric 'information' for human emergence is still a cosmic patterning that will later form the model for biological realisation. Many of the presuppositions for such realisation to come closer must already have unfolded as cosmic rhythms. For the human archetype is not just generic. Individuality is a crucial element in it, without which we cannot speak of human existence at all; but individuality originates in the cosmos.

Indeed individuality has a long history in cosmic terms, on many levels, but it is central to human development in a quite special sense. Human beings just cannot be conceived at all in terms of general features alone. We have sketched the manner in which, in animal and plant life, species are defined by the specific ecological niches that are occupied—while they last. The species inhabiting them necessarily become extinct every time they change too fundamentally. The animal organism's ability to adapt to change is limited, even in higher animals though they may have an impressive range of instincts to deal with manifold dangers and difficulties. As evolution progresses, more of the environment can be internalised and made part of the life-cycle of a complex animal. Even so, an advanced mammal like a hedgehog has the same range of responses whether it lives in a cloistered wood or near a dangerous motorway, and further adaptation is at best restricted. Animals as sophisticated as chimpanzees (much like the more proverbial dog) can learn new behaviours up to a certain age but cannot go on adapting. Of course humans have their limits too. But in the case of *Homo sapiens* we have societies which live in Arctic tundra, others in rainforests, some in deserts, some in marshlands, or in cities; their social organisations are also strikingly various: unlike the packs in which predators hunt or the protective herds of bovines or elephants, they vary to suit individuals. Still more impressively, in the case of human societies individual visitors can move between them (as permitted) and quite rapidly enter into their appropriate strategies of life. Thus it is the individuals who emerge in societies that define the basis of human life, and not merely different societies who produce separate fixed types. Societies are not mere 'systems of difference' as often alleged, but systems to be inhabited by creative individuals: they foster adaptable members, of which alone they can be constituted. (Where this is not so, we detect decadent or corrupt societies.) People mistakenly suppose that societies 'stay the same': but they do not, as history shows. Enduring cultural norms and values are carried by individuals, who sustain a society's identity not on an outward but on a spiritual level.

To find one's creative place in a society requires a personal history of self-discovery as well as an acceptance of shaping by society's demands; a society by its nature offers 'openings' for particular talents which can be developed. The knocks and blows as well as loving guidance that make us angular and unique are not deformation of a set type, but an enrichment that can be used to raise interactions onto a much higher plane. And our history shows: as a biological unit, a social grouping of individuals is evolution at its most successful. It is this social fluidity—more than abstract intelligence or the like— which is the best candidate for what makes us distinctively human.

The ability to reshape ourselves through our history and integrate socially is obviously just the opposite of fitting into a niche. Contrary to Tolstoy and to many theorists whom we cannot now discuss, human society is not like a bee-hive or an ant-hill. Its members do not dovetail by type, but shape the categories of social life through their needs and talents. The separate individual, in such a setting, seems no longer like a mere product of evolution: rather it seems as though evolution itself were embodied in each active agent.[*]

Each human individual proclaims itself the beginning of a further level of manifestation of the life-process itself. Selfhood or egohood is a new inner principle, and the ever-radical newness characteristic of history its mode of being.[†] The process of course necessarily rests

---

[*] Goethean ideas on organic individualisation again lead in the direction of Steiner's own thought: Steiner, *A Theory of Knowledge based on Goethe's World-Conception* (New York 1968) pp.101-3. Considering human society biologically, J.Z. Young observes: 'The diversity of human beings within a community may well constitute a polymorphism that is maintained by ongoing selection. The pattern of human life is not based upon communities composed of identical individuals. Rather, we depend upon the presence of varied capacities, motivations, satisfactions, and hence opinions. The outstanding example of polymorphism is the difference between men and women. To recognise the importance of variety for homeostasis is fundamental to human welfare. One implication is that the varied types are in principle equally valuable and that societies will be stable only if they treat them so': J.Z. Young, *Introduction to the Study of Man* (New York and Oxford 1971) p.543. Steiner stressed that individuals do not *make up* society, but emerge from it and grow beyond it—sometimes using startling analogies. Society, when organised as the state, is often supposed to contain individuals as its cells. 'But the whole being of man, by virtue of his individuality, is far superior to the state; he enters into spiritual realities which the state cannot encompass.... If therefore you think realistically, you will arrive at the idea of an organism consisting of individual cells, but the cells would everywhere extend beyond the epidermis ... the cells would develop independently of the organism and would be self-contained ... as if "living bristles" which felt themselves to be individuals were everywhere extending beyond its skin!'—Steiner, *Building-Stones for an Understanding of the Mystery of Golgotha* (London 1972) p.200 and generally pp.197-202.

[†] An evolutionary approach finally makes sense of that original 'leap' in Plato's thought to the notion of society as reproducing the structure of human nature (*Republic* IV). In Steiner's thought, human individuals certainly are the key to understanding the nature of society, but his evolutionary perspective certainly does not mean that it consists of selves in unrestrained quasi-Darwinian competition. Individualism is only able to work constructively for the individuals themselves when it is structured by the acknowledgement of a shared framework (rights) and a collaborative pooling of talents (brotherhood): see Steiner, *Towards Social Renewal* (London 1999: previously *The Threefold Social Order* and similar titles, orig. 1919). There will naturally be much more to say about the evolution of society later in this book.

upon the foundation of biology, like a new storey added to a dwelling. And even prior to biology, the tendency to individualisation is manifested in essentials already in cosmology—most especially in the diversification of the planetary system, noted by Gould.

With individuality, therefore, it seems that the potential human meaning of our world, its definitive gesture, at last finds complete fulfilment. Evolution-to-individuality has itself been internalised and incorporated, as were its previous stages and products. The human archetype which has reflected the stages of cosmic and animal evolution is now actualised, and human beings walk the Earth.

The evolutionary history of the Atlantean era is therefore, for Steiner, a mapping of the transition from animal to human existence. It is grounded upon those underlying cosmic and biological levels of individualisation, amongst which Steiner points already to the planetary system. A kind of differentiation from the periphery inwards can be discerned and, in the 'planetary' rhythms of evolution at the Atlantean stage of individualisation given by Steiner, a reflection of their ordering from Saturn (in the first two epochs), through the concentric planets in their sequence. This rhythmic archetype presents a picture, a flowing archetype, for the emergence of human individuality—one could also call it, the archetype of human society. It is as yet all a cosmic patterning that will later show itself in physical human forms. As evolutionists, we should think here less in terms of correspondences one-for-one, and more of the direction of the whole development and the potential for those specific histories which Stephen Gould observed.*

The Atlantean stage in Steiner's terminology bridges the Mesozoic into the Paleogene and Neogene eras. In his first Atlantean epochs we see the process of human evolution still bound up therefore with animal forms. Cosmic conditions belonging to the Saturn and Sun stage, and then the Moon stage (planetary differentiation) predominate. These animal-forms embody in themselves the organic spheres of life—breathing/rhythmic system, nervous system, metabolic

---

The threefold character of human nature was the definitive idea which enabled Steiner's thought to progress beyond the unripe anthropic thinking of his earlier *Fragment*.

* The doubling of Saturn in the sequence may initially look arbitrary, but it actually arises systematically in Steiner's thought about the great cosmic rhythms. Systematics, however, is not our concern here. See for some help with the resolution of these issues Steiner, *Ancient Myths, their Meaning and Connection with Evolution* (London 1971) pp.51-59.

system etc., together with a sort of harmonising of them all that is a sort of animal prototype of humanity. In the myths of Paradise, they are typified by bird, lion-carnivore, ungulate-bull and the pre-individualised man (still a kind of group-soul). Steiner summarised his version of man's Atlantean development thus:

> We have pointed out that he appeared in the four types of his group-soul, in the form of the lion, eagle, bull and man. These four types ... meet us, so to speak, before man descends into the physical, before he is individualised. These four typical forms which man had before he entered into physical corporeality are invisible in the present physical human being. They are in the soul-force, pressed, as it were, like india-rubber. It is in fact the case that when man loses control over himself, when his soul becomes silent either by going to sleep, or otherwise falling into a more or less unconscious condition, then one still sees today how the corresponding animal type comes out ....
>
> We spoke of the seven epochs of Atlantean evolution. These seven epochs comprise a first four and the last three. In the first four man was completely group-soul. Then in the fifth age the first impulse of the 'I' soul originated. Therefore we have four stages of development in Atlantis during which man first progresses as group-soul, and each of the first four Atlantean races corresponds to one of the typical animal forms .... The germ of the spiritual man was there, but he could not develop an individual form.[*]

And elsewhere:

> These are the four types. But in the Atlantean time the human form came to dominate, as the human shape gradually constructed itself out of the eagle, lion and bull natures. These transmuted themselves into the full human form, and this gradually transmuted itself into the shape that was present in the middle of Atlantis. Something else occurred through all these events. Four different elements, four forms, merged harmoniously in man.[†]

The animal-forms mentioned are representative of types—that is, they indicate the preponderance of one or other of the bodily-organic systems.

---

[*] Steiner, *The Apocalypse of St. John* (London 1958) pp.162-3.

[†] Steiner, *Egyptian Myths and Mysteries* p.85.

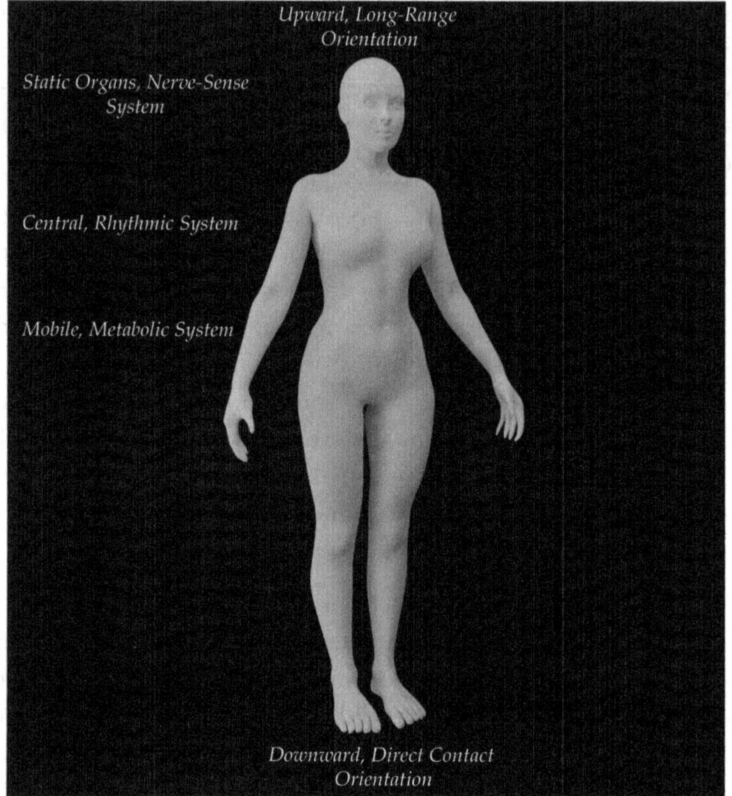

The human form in relation to animal evolution. Wolfgang Schad and following him Martin Lockley have developed Rudolf Steiner's threefold idea into a dynamic understanding of the human form and its relationship to the characteristics of animal evolution. Compare following table:

RODENT	CARNIVORE	UNGULATE
Small, nerve-sense orientation	Central, rhythmic orientation	Large, metabolic orientation
5 toes	4 toes	ungulate
Small body, big tracks	↔	Big body, small tracks
Early evolution	↔	Late evolution

The three orientations which are dovetailed in the human form appear in one-sided dominance in three main groups of mammals. Table after Lockley and Morimoto, How Humanity Came into Being.

Diagrams after Lockley pp.181 & 184 in How Humanity Came into Being, *Edinburgh 2010).*

The three types continue to dominate evolution during the Atlantean phase, corresponding outwardly, in physical life, to the end of the sauria and proliferation of the mammals and megafauna. It was still not possible to keep the divergent tendencies in check before the stage of uniquely dynamic equilibrium achieved in the special human orientation. Moreover, prior to that point, imbalances actually produced within humanity different types. The tendency to individualisation is already at work here—though of course compared to our present-day individuality it is still extremely rudimentary.

The animal tendencies continue as the biological human form is emerging, and several distinct variations emerge—comparable, though to a lesser extent, to the animal adaptation to a particular environment. Steiner calls these (using a term that was widely applied in his time) *races*—though none of them correspond with so-called racial groupings of today.* Because Steiner has been misunderstood in other contexts on the issue, we must avoid reading into them our modern connotations. We shall suggest shortly that we can see an outcome of the tendencies Steiner describes in the difference between ourselves and the Neanderthals (and other hominins).

---

* An exception might seem to be the seventh-stage 'race' in the terminology of Scott-Elliot and Steiner on Atlantis: the Mongolian. Modern anthropology acknowledges a mongoloid genetic grouping, once designated race by nineteenth-century scientific opinion. But the naming of the Atlantean peoples (after the first two otherwise unknowns) reveals that it is based on some affinity to later historical cultures, not races at all. The sequence given is: Rmoahal; Tlavatli; Toltec; First Turanians; Original Semites; Akkadians; Mongolians. Terms like 'primal' or 'original' indicate that these groups are not identical with the historical peoples of that name, known to history. 'Toltec' and 'Akkadian' allude to civilisations of subsequent times; 'Original Semite' indicates a first appearance of features known from the Semitic cultures of history; 'First Turanian' similarly, indicating the truth behind certain Iranian legends. Steiner, *Cosmic Memory* pp.50-58. It is certainly possible to find passages where Steiner does not completely leave behind the racial language of his day. However, Trinkaus and Shipman point out that at that time a looser usage was prevalent: '*race* was then more often used to indicate national rather than strictly biological categories. Attempts to trace the origins of the Welsh as opposed to the English, or the Germans as opposed to the Austrians or French, were common'—Erik Trinkaus and Pat Shipman, *The Neandertals* (London 1993) p.8. The tendency to form different groupings later reappears transformed on the cultural-historical level. The broad characteristics of the three main human populations arise not from biology but out of the history of the diffusion of human beings, with only secondary links to genetic features—see further below pp. 227f.

The different human 'races' in Atlantean times are as yet tendencies contained within the individualisation-process, but remain within the sphere of the human and do not become animal forms. The 'races' were a prior stage to human individualisation in the fully social sense, and had subsequently to be overcome and transformed into social identity. The three tendencies then work socially, as essential aspects of society for every member, not of distinct types but upheld by everyone in varying equilibrations. The three functions must always operate for any society, but not distinct groups. In fact, Rudolf Steiner declared that distinctions of 'race' were valid only back in that Atlantean stage of evolution, and not in subsequent human history.* Later racial labellings he always regarded as regressive manifestations, running counter to the evolutionary emergence of human social *individuality*. Racial differences, in so far as they exist at all, are a fringe phenomenon and the sign of a failure to shape cultural identity out of the legitimate resources of shared human creativity.

There are necessarily divisions and stages in humanity's development through historical change, but on this front it is basically quite striking that Steiner specifically rejected the misleading language of races that was widespread at the time, and insisted from the outset that the epochs in the development of humanity must be understood as 'cultural epochs', and had nothing to do with successive races. Failure to move forward in evolution, and instead to fall back on restrictive notions of who belongs to humanity on inherited or racial grounds of whatever kind, he rejected outright.[†] It is abundantly clear that he did not believe it had ever played a determining, but only a hindering role in the achievements of civilisation, which belong purely to man in his historical, i.e. his culturally creative existence.[‡]

---

* Steiner, *The Gospel of St. John* pp.151-2: 'Human evolution ... upon ancient Atlantis can be divided into seven successive groups. To these seven groups the designation racial development is applicable.... But it cannot be said that the humanity of the earliest post-Atlantean age ... differed sufficiently from ourselves to justify calling it another "race".... False ideas can very easily be created by this word "race" through a failure to see that what produces differences in humanity of the present era is something of a much more inward character.' Thus Steiner points instead to the role of religion, which took a different form and role in each post-Atlantean epoch.

† See for example Steiner, *Outline of Esoteric Science* pp. 256ff; *Theosophy of the Rosicrucian* (London 1981) pp. 136ff.

‡ He anticipates in fact the modern view that the 'racial' characteristics of the large populations of today are the result of cultural and historical decisions—not their

The Pattern of Society

After our lightning streak through the sweep of evolution, as it appears in the perspective of Steiner's thought, we can now turn to his description of Atlantis and see that it treats of the same process—in societal terms.

The same story of humanisation is already told, from a more inward perspective, in the collection of essays by Rudolf Steiner in which he had first taken up Scott-Elliot's researches. In them he focusses wholly on the changes in consciousness which accompany and facilitate the process of becoming human. The seven stages of Atlantis present, as it were, the *beau idéal* of the transition, through the overcoming of the one-sided 'racial' tendencies that are still present, to human-social mode of existence. Recent palaeoanthropology has likewise come to stress that the opportunities of social life created the 'run-away brain'. Humanity's external form was the result of our unique development, not its cause as Darwin had thought.

In Steiner's essays, the first stage consists in the central role of memory, preserving the connection with deeds, places and events. Remembering, and so being able to share experience, is here considered as a first step to developing socialisation and its individualising effects. Yet it is still far from identical with our modern experience. It instigates a sense of identity, though as yet a collective identity grounded in the shared memory-experiences. (We shall try to describe it further in detail later.)

Only at a second stage does the desire arise to be included oneself with the doers of remembered past deeds. One who can number himself with the definitive acts of the community comes thereby to possess their authority over the group in his own right, and these became leaders or charismatic kings as a result. The group itself thereby attains to definition, or active self-assertion, rather than having mere de facto existence. In the third stage, we have the further step of experienced *community*, essentially bound up with the acknowledgement of the remembered ones, considered as still playing a role in the present as well as the past,

---

cause. In the transition from the first two, semi-animal stages, Steiner wrote, 'initially the formation of groups depended wholly on natural forces, on common descent'. The beginnings of *community*, however, were effected when 'powerful personality recruited a number of people for a joint undertaking, and the memory of this action formed the social group' (*Cosmic Memory* p.53). The transmission of the memories through inheritance, since 'the capacities of the father were generally transmitted to the son', thus follows from the defining of the group through social interaction. Cf. J.Z. Young, *Introduction to the Study of Man* pp.588ff.

so giving the community a continuing identity. Initiation into their wisdom becomes the basis of its guidance. However, the potential for rivalry and internal conflict, which has gradually been growing in all this, is played out in the fourth stage (associated with the rebellious Turanians of legend). It was to overcome their competitiveness and mutually destructive rivalries that intelligence develops as submission to judgement, enabling society to channel its members' energies into more constructive forms. Rudolf Steiner makes the critical point that the origins of intelligence belong to developing socialisation—and not to some post-Enlightenment fantasy of man's drive to understand and master the external world.

Individuals also begin to resist some urges in themselves and approve others, on a social model. The evolution of society and individuality are inseparably intertwined. The sixth and seventh stages continue the intellectual and personality trends further, but also show how such processes exhibit a rhythmic rise and fall, and enter a distinct decline after reaching a peak. The notion of human progress simply going on and on is another Enlightenment fantasy. In reality, strong cultural decline and in the space different opportunities arise. From amongst the forms they develop, new modes of incorporating the past are found, with transformational effect. A new phase begins, making history happen—if it were not so, we would be back to mere animal accommodation to a particular time and place, necessitating eventual extinction. Instead, we have the human dimension of history with its tragedies as well as hopes.*

Once civilisation is up and running, the accumulation from all these stages becomes part of the mixture through memory and tradition, and is ever structured anew. But the building-up process is necessarily first of all one stage at a time, step-by-step. And its rhythmic, 'planetary' structures continue to underpin later history. Steiner's threefold understanding of human biology enabled him to follow through an ancient line of thought connecting the structure of society with the materials of human nature—an important feature of his later thought.

This is, however, no Platonic ideal of the state. In all this we are still in the realm of the human project in evolution, as it draws ever nearer to realisation. What we see are assimilations into the human archetype, as one-sided tendencies are brought into balance. Society is something that can only be actualised through individuals as

---

* Steiner, *Cosmic Memory* pp.50-58; a similar account, with references to the planetary types in Steiner, *Outline of Esoteric Science* (New York 1997) pp.240ff.

they evolve. The etheric 'type' of humanity is no abstract map of necessary human features, but a mass of individually different forms with their human family likeness. Steiner affirmed very seriously the philosophical viewpoint which held that each human individual holds the place occupied in the animal world by the species.* His 'clairvoyant' perspective here stays essentially true to evolutionary thinking in terms of actual forerunners, not ideas. It is to be understood therefore that even whilst these features were still being scripted in the etheric type of humanity, human beings were already *souls*, absorbing them inwardly. The formation of society that they experience step-by-step is a coming to expression of their humanity, as spiritually actual, and thence into biological expression in the structure of the evolving environment of the Earth. It all happened when conditions became fitting.

Rudolf Steiner insists that these etheric formative-forces of humanity, 'bodies of information' working into matter as Paul Davies has surmised, left no physical trace until later stages when, as fully human beings, they became biological forms.[†] They did on the other hand already exert some influence on their future environment. They could possess a 'real causal efficacy', acting to organise life 'in the same way that matter can be moved around by physical forces'.[‡] Such forces, though they have never previously been satisfactorily explained, have elsewhere been investigated under the heading of so-called 'psychokinesis' (levitations, etc.) which are sometimes exhibited quite extensively by certain mediums, or in poltergeist phenomena, etc. Steiner considers these as further evidence of atavistic powers which once belonged to the mode of existence of human beings, and most especially so in the Atlantean period. Nowadays they show themselves only as atavisms, when they have not been diverted and modified by our later development (notably into intelligence).

---

* See for example Steiner, *Theosophy* (London 1970) pp. 52ff.

† Steiner, *Wonders of the World, Ordeals of the Soul, Revelations of the Spirit* (London 1963) pp.112-3: 'What then were these Atlantean bodies like? In the satyrs and fauns and in Pan … Greek imaginative intuition has elaborated pictures of normally developed Atlanteans…. Atlantean humanity were so formed that they had no skeletons such as men have today…. For this reason it was incapable of being preserved, and the geology or palaeontology of today will be hard put to it to find any trace of them.'

‡ Davies, *Origin of Life* p.242.

In their archaic form, he connects them explicitly with the environmental relationships of the time, and particularly the life of plants, in which case they could be utilised in a particular way to affect their growth—and in other modes. Even much later, biological human communities were thus actively integrated into Nature. In an Atlantean 'settlement everything was still in alliance with nature ... built of trees with artfully entwined branches. What the work of human hands created at that time grew out of nature. And man felt wholly related to nature.' Human existence was still etheric existence, intimately linked with the plant world: in still more primordial times, Rudolf Steiner adds, human presence was still more closely identical with the actual shaping-process of the environment.[*]

Many mythologies recall an awareness of man's 'giant' ancestors, who are commonly said to have made or shaped the landscape which is home to the tribe. In the Enoch-tradition the Atlanteans are characterised as giants with massive limbs, the children of the fallen angels. Their size was not really a physical extension. Psychokinetic powers are described mythologically as extra arms or other limbs. Though mainly familiar today from Indian iconography (the timeless dance of many-armed Shiva), the idea of primordial semi-divine or giant human figures of enormous energy and (typically) eight arms can be shown to have been known across the world from India to archaic Scandinavia.[†] In parallel, however, the structuring of the archetype of humanity into a socio-cultural process was bound up with a powerful intensification arising out of the dull primal consciousness, now was becoming ever more focussed and awake.

In such ways, then, already by the end of Cretaceous Period with its massive chalky deposits, followed by the great catastrophe or Extinction Event, the fluid archetype of human existence was pressing ever closer toward its realisation on Earth. It was already individualised and ensouled, as the myriad human 'dust' of Steiner's clairvoyant vision.

---

[*] Steiner, *Cosmic Memory* p.47. For the shaping influence of germinal humanity on the environment in still earlier periods, cf. p.74. The landscape thus sculpted was experienced as an equivalent of a temple, making it a reflection of the divine-creative—though not at all like a temple construction of modern times (p.75).

[†] Cf. the brilliant study by G. Dumézil, *Mythe et Épopée I. II. III* (Paris 1995) for the instances of Sisupāla (India) pp.729-30 and Starkadr (Scandinavia) pp.744-46. These heroes had to be lopped of their extra limbs to become acceptably human, but still possess tremendous energy, which can be transmitted upon death to a successor.

Only much later, in the Pleistocene, as we reach the age of human archaeological remains, will we finally arrive at a line of biological human beings (*Homo sapiens*) on Earth, spreading out over the surface of the globe. But there is a strange dislocation to be noted at this very juncture. According to Steiner, these earliest Modern Humans cannot immediately be related to that archetypal crystallisation of the human that we have been describing, and which is the inner reality in any truly evolutionary understanding of man and animal.

Instead, we find these first biological humans emerging on a portion of the old massive supercontinent that has by now broken up into Africa, Antarctica, etc. which Steiner called by the name Lemuria. We may relate them in the first instance to that pre-individualised type of humanity which was nested among the higher animal types. Rudolf Steiner calls them, in his terminology not an Atlantean, but a Lemurian humanity, and from it indeed we are all descended. He essentially concurs in this with modern geneticists, in fact: it is certain that all Modern Humans can be traced back to a single human pair coming out of Africa. But this humanity does not emerge from the trends we have been examining. It is a distinctive feature of Steiner's account that he brings us in our explorations not to one origin of humanity, but in this strange way to two. Or is it a real dichotomy? The solution to this apparent riddle, I shall suggest, when we unravel the links between Steiner's descriptions and of what we know from the latest palaeoanthropology, may actually explain the conundrum of the anatomically-but-not-behaviourally Modern Humans we met in an earlier chapter. And it may elucidate the presence alongside them of another human species.

Chapter 5

LEMURIAN CHILDHOOD

Back to the (New) Beginning

With our broad background awareness of evolution, we must try to assemble a unified narrative of human origins—from fossils, from modern science and from Rudolf Steiner.

By a sort of comprehensive detour-of-ideas we have gained some inkling of the way that cosmic, plant- and animal-evolution was a process that involved the emergence of physical life, but also of dynamic organising structures, endowed with the complex information needed for living things. The process at the same time was suggesting in ever more detail the organic form in which the 'Anthropic' potential of our cosmos could most fully be represented—always given the right evolutionary opportunity. Rather as Goethe trained himself to 'see' the flux of transformations behind the specific plant-forms, so Rudolf Steiner characterised the archetype of human existence coming ever closer to earthly realisation, as seen by his scientific clairvoyance. It is a living form or forms, for it is inseparable from the idea of its actualisation through social life made up of individuals with different features ('polymorphism'), whose evolution can encompass the breadth of spiritual life as they shape the direction of their consciousness.

We sought for such ideas, initially, when we were confronted by a somewhat paradoxical situation in the picture that has emerged of early humans from the fossil record. For we found that contrary to former theoretical expectations, we witness the form of Anatomically Modern Humans on the one hand, without any evidence of characteristic human behaviour. Such behaviour then appears, rather suddenly, as if by a cognitive leap; but alongside all this, we observe a different yet likewise fully human species—*Homo neanderthalensis*—who turns out to be neither a crude forerunner nor a poor cousin of *Homo sapiens*. Our evolution is bound up with an encounter between the two species, after which the Neanderthals rapidly (and mysteriously) became extinct. Now Rudolf Steiner described a critical divide in human evolution, and a subsequent coming together, in ways which quite remarkably recall the modern picture. We are now in a position to explain more fully what he meant, and to see how it helps us understand the evidence from a palaeoanthropology that is increasingly challenging ideas of how we came to be.

The pivotal moment for understanding our evolution comes at the distinctive point when human activities start to look (in most ways) distinctively modern: in the phase of the late Stone Age called Upper Palaeolithic. So impressive and widespread are the evidences for the deployment of abstract thought, fully developed creative language, representational art, organised lifestyle involving planning and diversification of skills, burial practices that express hopes of a life in the Beyond, clothing that does not just cover and insulate but tells onlookers about the person, importance and social standing of the wearer—that we can hardly fail to recognise ourselves in the mirror. The Anatomically Modern people whose fossil remains are associated with these evidences are indeed indistinguishable from ourselves in almost every physical detail. Their cultural advent was sudden and expansive, leading to the widely used expression 'explosion to modernity'. Long-standard Darwinian ideas of apes slowly turning into humans have turned out not to fit the case at all, and have collapsed on all fronts of human development as our knowledge has increased.

We sounded a note of caution, however, about too easily assuming that because Upper Palaeolithic humans could think and speak like us, the content of their minds was exactly similar to ours. Their consciousness—their basic way of envisioning themselves in the world they thought about—may have been significantly different. We noted, too, that the new storyline has uncovered not one but a series of striking paradoxes. Their culture may have arrived in a burst of creativity, or cognitive revolution, but that in turn means that Anatomically Modern people existed before the revolution—modern-looking, but initially without the characteristic activities and attributes they would one day exhibit. How did they come to be suddenly transformed? Darwin's clever idea that increasingly complex people were made by increasingly complex activities, requiring dexterity and thinking-power, has been completely thrown out. What is left is a still unanswered question.

It now appears to scientists that the 'runaway brain' throve, not on outer conquest but on awakening cultural and social life, and its expansion then determined the biology of reproduction, childhood and continuing adaptation which we shall have to explore. Scientists like Richard Leakey nowadays refer not to external changes which gave humans their position of advantage in evolution, but to a novel 'commitment to uprightness': a new attitude, or change of consciousness, rather than just new technical achievements. But how or why did this happen? And equally, why did the cultural explosion of which humans were capable not happen immediately?

To complete the paradox, during the time leading up to the cultural revolution we now know that there were other humans—especially the Neanderthals. To scientists and to the modern mind generally, these remarkable others had long been perceived as a threat to our self-image, and had repeatedly been subject to extraordinary retrospective abuse! We are historically and religiously attuned to think that ours was a unique success story, and ours a unique place in Nature. But while Modern Humans were posing new and difficult questions to researchers, the Neanderthals were emerging as far more than a vestige of some primitive human stage. With every recent discovery made, they have shown over and again that, though different, they were as convincingly human, and perhaps earlier in showing it. With their larger-than-Modern brains, moreover, they may have given us a significant gift. For it is now abundantly clear that they encountered and interbred with the anatomical Moderns and that, just around the time of the Upper Palaeolithic explosion, they helped to shape our own inheritance—just before inexplicably going extinct.

A new approach to the entire question of humanity seemed called for: and it has been the source of some exciting ideas that reach out far beyond the limits of anthropology, or even biology. The notion that humanity was essentially a spin-off from the behaviour of highly developed animal species (the Australopithecine apes) was now flatly contradicted by the evidence. Instead, ideas that had swept through the physical and cosmological sciences were suggesting that 'humanity' was a conundrum which required the vista of the whole universe for its solution. The extraordinary number of fine-tunings and harmonisations in basic laws, events and directions of cosmic development seemed to many scientists to point to a world in which man is an integral phenomenon rather than an accidental product. There may be other (possible or actual) worlds where things work differently, but the world we know has a human meaning—a coherent basis which allows us to evolve, as by a biological imperative, and finally comes to consciousness in the human knowledge which traces it once more to its foundations. Long before this 'Anthropic' turn in science, however, it is remarkable that similar ideas were elaborated by Rudolf Steiner, based on an intensification of modern consciousness that built on Goethe's increasingly acknowledged contribution to scientific method. Most remarkably of all, Steiner's researches brought him to describe a stage when there were two kinds—or even different 'species' (*Menschenart*)—of humanity which interacted so as to create the life-situation of modern man.

Over the last chapters we have seen that for Steiner the series of fine-tunings in the laws of Nature, which enable us to be here, expressed itself in the broad rhythms of cosmic development. Just as a rhythm is present as a whole in every component stroke, yet adds nothing materially to their sum, the developmental rhythm of those 'nested hierarchies' of form, from cosmic ratios to planetary individualities and the phases of the transformations of the Earth's surface, is present in living organisms like a non-material script. Steiner explains that it can be perceived non-materially in modified states of awareness. It forms a 'body of information', or etheric body, which assigns living significance to each element in its make-up (as understood also by astrobiologists like Paul Davies or Lee Smolin). The presence of such information written into the logic of the cosmos means that there is an evolutionary pressure, or biological imperative, which helps to ensure that they come about as and when they can. The progress of cosmic development may be seen as the separating out of the several possibilities that are essential preconditions of the conscious human being, whose existence is the proof of what was there, implicit in the beginning. The script of our own organic being is elaborated and enriched as human existence in biological form becomes a nearer potential reality.

The great rhythmic stages in which the possibility of human evolution is becoming clearer are naturally those of the epochs of life (Proterozoic, Mesozoic, etc.). On the changing continents life was becoming ever more complex, and successive plant- and animal-forms dominated the changing shapes of the continents. Steiner refers to a time when a continent covered the Atlantic Basin (thence called Atlantis) and to an earlier vast continent (or Lemuria) of which Africa, Australia and South America are subsequent divided parts (nowadays often termed Gondwana). In those eras (which are correlates of the geological layers of the Earth), man did not have an existence that was organic and material; yet these phases set the stage for much of what he was subsequently to be, and constitute a spiritual ancestry, ethereal rather than physical in form, writing ever more explicitly the script that would make us what we are. These two primordial epochs project, however, two different images of humanity: I shall contend that the remarkable description given of them by Steiner can solve many of the conundrums which confront us in the fossil evidence concerning the two main human lines. Steiner himself has fascinating things to say about the phase when later, under material conditions, two human types actually walked the Earth.*

An Improbable and Tenuous Humanity

Probably the most serious upset caused by the re-evaluation of fossil evidence from the Modern and Neanderthal inhabitants of the Middle East was the chronological reversal (see above, pp. 35ff). Better dating techniques showed that the conventional assumption had been wrong: the old idea had been that Neanderthals and similar types were primitive forerunners who gave way to the Moderns, or *Homo sapiens*. The Moderns had triumphantly emerged, showering around them all the manifestations of culture and intelligence. But no. It turned out that the Anatomical Moderns (i.e. outwardly resembling ourselves) from that region and, as it was soon shown, equally elsewhere, had been present from vastly earlier times, and existed long before that outburst of culture and mind that brings them fascinatingly close to us.† So the Modern form, anatomically, was not automatically connected with mental development, technology, culture etc., and we are left instead with the puzzling presence of our own predecessors essentially complete but somehow simple, unactivated. We must envisage them lacking those skills and the mental richness which bursts out in the Upper Palaeolithic 'Explosion'.

These early humans equate well with Steiner's Lemurian type of humanity, and Steiner concurs that from them all Modern Humans

---

* Of course Steiner could not refer us to evidences that were not yet available. In the *Outline of Esoteric Science* he describes how forms of some of the 'races' (more what we would term sub-species) from Atlantis became subsequently physical inhabitants alongside the main line of humans and *persisted for considerable periods*. However, they then suffered rapid extinction (p.250). It is hard not to connect this with what Steiner says about the special role of one such group which interbred with us—see further below, pp. 271ff.

† The analysis of a an Anatomically Modern jawbone from the beginning of the Middle Palaeolithic in Israel in 2018 confirmed the rightness of the new picture in more detail (recounted first-hand by members of the team; for the article go to www.nytimes.com/2018/01/25/science/jawbone-fossil-israel.html). The researchers affirmed the evidence as confirming much older Modern Human origins than once supposed, and cautioned that nevertheless 'Early modern humans in many respects were not so modern' (p.2).

have descended. In fact, like the modern geneticists, he is in agreement that everyone alive today is descended from a single pair of them. They do not emerge from that constellation of socially interactive tendencies which he associated with the Atlantean type, however. But they are highly relevant to the question we raised as to the biological basis on which that new layer of human complexity could be built up. The process of individualisation is indeed, as we have seen, a key aspect of evolution generally—including cosmic-planetary evolution. But if human beings were to embody it, there must have been a development out of the ever-increasing complexity of animal existence to enable it to take organic form. This older biological aspect of the human is associated by Steiner with the Lemurian continent,* which means, in the period of human biology, with its surviving African core (the other continents having sheered off and dispersed across the globe).

We are now in a position to understand what is implied in Steiner's picture of the two human kinds which correspond to the encounter that modern science confirms, prior to the creative leap that produced recognisable kin to ourselves in the Upper Palaeolithic. The older kind was the foundation of the biological/anatomical human form; the second one was connected, in a way yet undefined, with the cognitive leap to modernity. The crux of the immediate question posed is: how did the foundation for the realising of the human-social constellation come to be there? What characteristics did it possess on which further levels could develop?

The fossil evidence is that prior to the 'Explosion', these humans came upon the Middle Eastern and European scene where they met the Neanderthals in a streaming 'out of Africa'. They were a further wave in the succession of hominins who had long before originated there. The phrase 'out of Africa' has come to be used for the theory itself. The events we have touched on are usually termed 'Out of Africa II', because it is clear that there had been a prior diffusion of hominins, starting in Africa and then spreading out into Eurasia, much earlier than the period we are discussing, which thus constituted 'Out of Africa I'. Genetic studies have made it apparent that from the diffusion called Out of Africa II, all of the human beings currently inhabiting our globe originated. This striking recognition has swept away the formerly prevalent notion of a gradual humanisation all over the world which went under the name of the 'multi-regional hypothesis'. It followed the Darwinian gradualist

---

* Steiner, *Cosmic Memory* pp. 71ff.

line of thought. Now, instead, we have a radically different history involving the specific African origin of humankind. Chris Stringer of the London Natural History Museum championed the view, and few now disagree with his conviction that the older populations across the world did not gradually get refined into humans—they were replaced by the in-streaming new population. We (their successors) are now, indeed, the sole species representing our genus in existence—which biologically is an unusual pattern, to say the least. With this shift-in-ideas, science immediately came much closer to the thought-patterns we find in Rudolf Steiner, for whom it was clear that there had been successive, overlapping different types of humanity. (It must be remembered, though, that elements from the older ones were always absorbed and furthered in the new.)

Another phrase much associated with the theory is 'African Eve', which denotes the unique mother from whom genetics indicates that we all descended. From her we draw our physical make-up, though of course she had her Adam who affected it too; equally of course, she was not the only one of her kind and nor was he. Yet her offspring are the only ones which went on to multiply and become present-day humanity, even though she and he were part of a general population—one which, her offspring excepted, otherwise died out. The resemblance to Steiner's ideas goes far beyond the general notion of different populations which replaced each other in human evolution. For Steiner emphasises in many aspects the extraordinary difficulty involved in evolving a viable humanity.

When Steiner speaks of that unique pair from whom all present humanity is descended, it is to stress the extreme difficulties of the case. He mentions the difficulties of changing climate, with phases of rapid cooling, and of the difficulty of realising the human type in the bodily-material substance available. Such pressures are the driving force of animalisation, which represents a useful adaptation to particular outer conditions. Animals fit well into their immediate environment, or migrate or hibernate or do other things to be able to stay in it. But the human archetype is that which retains maximum adaptability, so that to changing conditions we find answers out of our greater malleability. We thus face many more challenges. To have so much adaptability, we need extremely sophisticated, refined bodily construction. Until we have artificial shelter, or clothing, we are therefore more than ever vulnerable. In a phase of great outer perturbations, such as Steiner describes for the time of the first origins of humanity, getting everything right is an evolutionary high-risk strategy and likely to involve many failures. Animal solutions are a constant tendency since they involve a more immediate answer. To maintain the inner flexibility of the human is ultimately a better solution, but getting it up-and-running is an evolutionary conundrum: 'so that it may be said, with approximate accuracy, at least, that there was a single couple in existence which retained sufficient strength ... to "hold out", as it were, through the period'. The population to which they belonged was increasingly depleted, and 'became less and less numerous'. Only in the instance he refers to was it possible 'for human substance to become more refined and rendered suitable.... The descendants of this one main pair were able to live in more pliable substance than had been available before.'*

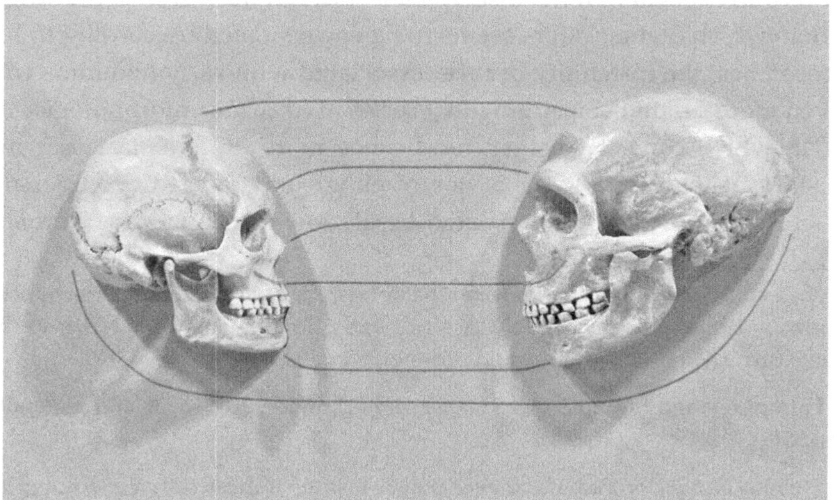

Skulls of different humanities. A comparison of an Anatomically Modern Human (Homo sapiens) *with a Neanderthal exhibits immediately some of the striking differences. The Neanderthal (right) shows the typical larger brain capacity: the male up to 1740 ml though more usually c. 1600 ml, whereas a modern human averages 1200 to 1500 ml. Also noticeable is the beautifully spherical back of the head sometimes likened to an 'occipital bun'. The Neanderthal chin is rather weakly developed, though present in some specimens; and the brow-ridge is noticeable, which led early cartoonists to extravagant 'low brow' parody representations. The Neanderthal nose must have been somewhat large, perhaps to help breathing in Ice Age conditions.*

---

* Steiner, *The Gospel of St. Luke* (London 1964) p.83. It is hardly necessary to point out that, although Steiner is keen to indicate that there is a truth behind the religious traditions of an Adam and Eve (an idea much wider than just the Bible), the reference to the dwindling population from which they derived shows that his approach is in no sense biblically derivative or in any sense fundamentalist. The conditions of matter to which Steiner refers involve more than biological change—but it is the biological aspect alone which concerns us here.

It is hard not to think of the fossil hominins which are sometimes discovered both in *gracile* or more nearly human form, while in other cases the same species is found in a *robust* type. The latter indicates much heavier bony skull and skeletal structure, with massive build.[*] The earlier representatives of the 'Lemurian' type are especially prone to this tendency. *Homo habilis* was an important discovery of the Leakeys: it represents a human type still vastly more primitive than the early Modern representatives, but it had a large brain (essential to that adaptive range of behaviours). Sometimes the skeleton suggests fully human appearance, though with a flat face and larger teeth; but in other specimens the indications are much more ape-like, so that Humphrey and Stringer remark, 'the human status of *H. habilis* is therefore still controversial'.[†] In the discrepancy between these forms, the struggle to which Steiner refers seems to be encapsulated. According to his researches, the instability of form associated with early hominins was even more dramatic, resulting in a series of 'unviable morphologies'.[‡]

In which context we may recall once more the odd fact that the history of the genus *Homo* is one of either unparalleled evolutionary success (ourselves) or of repeated break-downs, extinctions. (Perhaps

---

[*] The evolutionary division between the 'robust' (heavily built and large boned) forms and the more 'gracile' character of early hominins is well shown in the table furnished by Humphrey and Stringer, *Our Human Story* pp. 18-19.

[†] Humphrey and Stringer, *Our Human Story* (London 2019) p.88 and generally pp.85ff.

[‡] Steiner, *Outline of Esoteric Science* p.249. Steiner indicates that a number of these nevertheless persisted for considerable periods. Other 'Lemurian' stages of humanity may be recognised in types such as *Homo erectus* and *Homo heidelbergensis*, the protagonists of 'Out of Africa I', who stand progressively closer to the Modern (*sapiens*) type. Nevertheless, it seems that even *Homo erectus* could slip back into anomalous forms—such is the presumed origin of *Homo floresianus* (popularly nicknamed 'the Hobbit'), a dwarf form which persisted on the Indonesian island of Flores into relatively modern times. The mysterious Denisovans are known only from minimal evidence, but may well indicate a similar story. Steiner stresses that such representatives of the human experiment, on the whole, inevitably died out. More than twenty extinct representatives of the genus *Homo* are now recognised. L. Humphrey and C. Stringer, *Our Human Story* (Natural History Museum London 2019) pp.86-8 note in the case of early *sapiens* the evidence which suggests 'that archaic skull morphologies persisted alongside more modern ones' (p.136), which is in line with Steiner's remarks. The greater success of *Homo erectus*, based on evidence from dispersion as far away as China and Java, also suggests ambiguities—the brains of the 'survivors' are about 50% larger than the older specimens (pp.96-7).

Steiner's curious qualification—'with approximate accuracy'—implies that the unique pair situation happened more than once? Perhaps it is something of an archetypal human situation?)

Finally, we may bear in mind that the Anthropic approach to science is based on the recognition of the extreme improbability that all the conditions, the linked fine-tunings, for full humanity to exist could ever happen. Biological types still require situations in which they can come about. Human existence inherently belongs, therefore, in a situation of extreme unlikelihood or instability, when it is against the odds that so many circumstances should apparently conspire. In a phase of climatic change, and of biological transition to more gracile forms which must reflect significant lifestyle changes, it seems that finally the shift occurred: unable to hold on to any fixed evolutionary niche, the requisite pair instead internalised adaptability itself—the 'human' meaning which stands written at the beginning of all evolution.

Rudolf Steiner's almost heroic description apparently indicates the inner qualities of the primal pair who survived the difficulties of the transition—yet he is plainly talking biologically at the same time: it was in their offspring that they succeeded in developing viable bodies. He is referring, therefore, not to an understanding of their situation yet nonetheless to an inner transformation. Recent researchers have likewise come to speak of a 'commitment' to a mode of upright existence. In one thing Steiner was extraordinarily clear, when so many thinkers thought of man as an extra-intelligent ape, namely, that it was not intelligence which has made humanity distinctive but our special bearing.

Our upright posture enables an utterly different relationship to our environment from the animal's experience. The animal in its evolutionary niche is led by environmental stimuli to behave in appropriate ways which sustain it. The human's loss of connection to their current evolutionary niche at a phase of too-much or too-disruptive change should have meant extinction; instead, humans stepped free and began to relate to their surroundings with a whole range of possible behaviours, endlessly adapted as required. That meant employing consciousness as much or more than cleverness.

Certainly it was not just uprightness physically considered (there had long been partly or largely bipedal apes and hominins), but a *centring of identity* in a permanent upright posture—quite different from the defunct notion of gradually extended animal uprightness. It was a step that decisively re-organised behavioural life, which is perhaps what Richard Leakey means by his 'biological new beginning'. What Steiner is talking about seems to be partly the producing of offspring amongst whom upright posture is fully developed, and so no doubt also the tendency for gracile forms was evolutionarily preferred. In addition it must also have meant the first stages of protecting the young to acquire those multiform patterns of behaviour, which is tied in with many of the biological features that distinctively shape human life. Such extensive learning requires a growth-pattern with a childhood, involving a long immature period, followed by rapid phased growth, maturation, etc.[*] There is a cultural-moral dimension too, since the offspring must have had to learn commitment, acquiring the necessary determination to meet a constant diversity of challenges and also the distinctively human willingness to co-operate with others. Such an evolutionary strategy, separating the life of the organism from its environmental niche and shaping an artificial behavioural life taking account of others, is a kind of dicing with death. Yet in a period of drastic change and strenuous adaptation it also offered the potential of the greatest ever breakthrough and the full realisation of adaptive evolution in each individual. It was in short, humanness as we still know it in its pendulum swing of self-assertion and instability. Steiner points to the young child determinedly lifting itself upright, ever and again falling down until it can stand. Here indeed was something spiritually and biologically new, though it was simultaneously maximum openness to the basic life-situation, and a being-at-home in our own sphere of movement.[†] The paradox of 'new beginning' and return-to-the beginning is fundamental to understanding our place in evolution.

It has only recently come to be clearly understood how much the improbable emergence of humanity is bound up with our evolution during a period of constant climatic fluctuation. In Steiner's terms, this phase reproduces an archetypal 'Lemurian' pattern involving the break-up of a long ('paradisal') phase of stability. The African early phases of *Homo*, still long before the Modern 'explosion' however, recapitulate such a constellation of circumstances. Among

---

[*] Above, pp. 144f.

[†] Cf. Steiner, *Spiritual Guidance of Humanity* (New York 1974) pp.6ff. The forces urging the achievement of uprightness are for Steiner those which make it religiously appropriate to characterise human beings as the 'image of God', 'Son of Man' or Christ-forces.

palaeoanthropologists, Rick Potts of the Smithsonian Institute, has particularly drawn attention to the meteorological link with man's unstable mode of being.[*] One can consider the human quandary in both a positive and a negative light: the ever changing ecology offers a tantalising range of evolutionary opportunities, yet their realisation is only possible after making a radical break with the pursuit of any one in isolation. Equally, by its newness each offers a challenge that may easily become a threat: the fact that a hairless biped of light (gracile) build contrives to prosper through a phase including extreme cold in the glaciations between warmer spells, shows how the human solution is radically opposed to accommodation in a specific evolutionary niche! The decision to stay and adapt must increasingly have become a conscious process.

Spatially considered, human posture is the epitome of instability itself. Rudolf Steiner repeatedly notes that stillness is always an imposition on the inherently dynamic and fluid tendency of the human form. Commenting on his remarks, a recent author connects this with the interesting observations of M. Feldenkrais:

> The human body is badly suited for standing. Statues of human figures have to be strongly connected to a heavy base to prevent them from toppling over at the slightest disturbance. The head, the shoulders and trunk, all the heavy parts, are placed on top, and the base is very small in comparison with the total height. This base becomes larger in all the positions in which the total height of the body is reduced—just the opposite of the requirements for static stability.[†]

We saw earlier that it is just the opposite of the ape's mode of locomotion, in which the animal's much heavier form is carried about requiring maximum contact with its surroundings whether floor or forest boughs.

---

[*] R. Potts, *Humanity's Descent: The Consequences of Ecological Instability* (New York 1997).

[†] Feldenkrais, *The Body and Mature Behaviour* (London 1949) p.68 cited along with Steiner in G. Wilson Knight, *Symbol of Man* (London and New York 1979) p.57. For Steiner's statement see *Eurythmy as Visible Speech* (London 1944) p.14. The unstable construction typical of hominins, with *inward sloping* thigh-bones concentrating all the weight onto the knees, is vividly illustrated in exhibits of the Evolution gallery of the Natural History Museum, London.

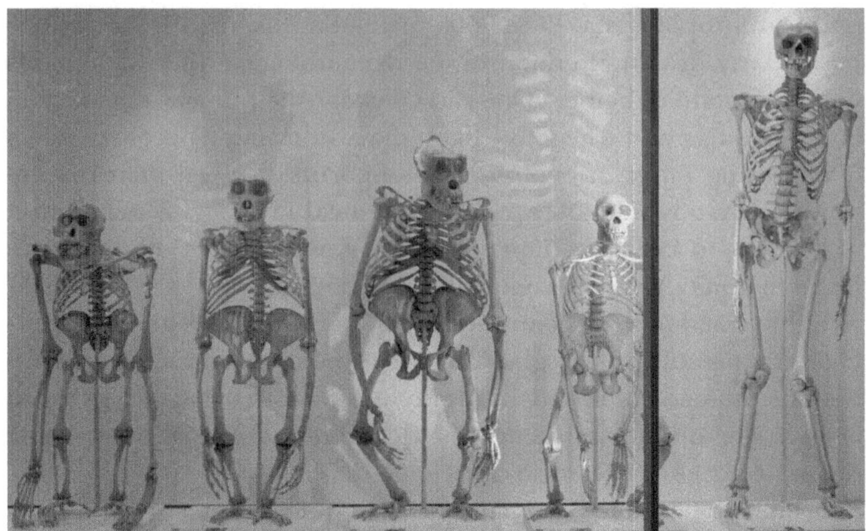

The unique instability of the human standing posture is well illustrated by comparison with the skeletons of the great apes, when these are ranged alongside. Shown here, from left to right, are a Bornean orangutan, a western gorilla, another western gorilla and a chimpanzee, and a human. The apes are firmly supported by spreading their weight, but the tapering leg-bones (femur) of the human mean that standing is precarious and needs constant adjustment. Exhibits from Cambridge museum of Zoology.

In man's case, contact with and dependency on his surroundings is essentially discontinuous and constitutes a moment of transient failure of balance. It is even wrong to say that we stand resting upon the solid ground: in all characteristic positions we are in movement between such moments and avoid them as much as possible; we do necessarily place our foot onto the ground, but only to start immediately to lessen our repose upon it and shift our weight to the other. Human beings move out of themselves, in a way completely unlike the animals, so that their engagement with their world is at every instant elective and directed.

For committedly upright beings like ourselves, even basic activities such as eating are fundamentally transformed: we no longer browse for food, getting it to the mouth direct. We obtain and handle it for consumption using our arms and hands, and normally transfer the whole scene of consumption to a social gathering. Eating is redefined as a communal, meaningful human activity, with shared significance, a meal, not just a natural process. We manufacture, in this and many other still very simple ways, our own environmental niche that is shaped out of our consciously adaptable sphere of activity.

Consciousness thus enters in at every point. Indeed, the human upright stance can only be maintained through the constant intervention of consciousness; it is inherently a *waking* stance, defining our mode of being, and we only abandon it when we need to sleep (and even then we are never inert, like logs—despite the popular simile!). Our basic gesture of maintaining our own sphere of movement is, in fact, the fullest incorporation of consciousness into our whole way of being imaginable. We have mentioned that, from Steiner's perspective, it is utterly wrong to consider consciousness a sort of camera lens or a supposed screen of awareness, related somehow exclusively to the nervous system. It was our adoption of human posture not the size of our brain which drove the heightening of the preliminary-consciousness, implicit in Nature and in organism, to become, gradually, the highly developed faculty we enjoy today—not the making of a sort of intelligent camera. Nor was it ever the solipsistic consciousness which can only register its own point-of-view: consciousness was always social, based on awareness that others were aware too. Ultimately it evolved into individual and reflexive self-consciousness—the fabric of our modern life. What we need still in modern thought is a genuinely evolutionary conception of consciousness. By associating it, and so the human autonomous identity, so directly with upright equilibrium, Steiner suggests a unifying approach that embraces the spiritual and the biological aspects.

On the biological front, the autonomising of our sphere of movement may be considered a further case of the basic phenomenon of evolution: that whereby a portion of the environment is incorporated into increasingly complex forms of life (a warm, watery pond where eggs are laid is replaced by internal gestation, etc.). Instead of thinking of consciousness as our isolated self looking out at the world, we should start with the primal consciousness suffused through the world, and work it in to the self that evolves out of it. In this respect, Richard Dawkins is undoubtedly on the right track when he says, 'Perhaps consciousness arises when the brain's simulation of the world becomes so complete it must include a model of itself.'* He like many other biologists, however, is troubled by the lack of a specific adaptive advantage from consciousness, which would make it relate to a particular niche. (Some scientists in response even break with the fundamentals of evolutionism altogether, and try to 'argue

---

* Discussed in Leakey, *The Origin of Mankind* pp.180-1.

away' consciousness as a mere by-product of the brain, without real significance.) Steiner's idea, whereby consciousness is evolution come full circle, seems better fitted to follow through on Dawkins' insight. It suggests that in conscious activity we touch the fundamental essence of evolution. The biology is explained from that assimilation. The 'runaway brain' is explained, nowadays, as the result of cultural human activity, not its cause. And it suggests the insight that consciousness is fundamental to the origins of the universe. As Roger Penrose is quoted as saying, 'Somehow, our consciousness is the reason the universe is here.'*

Steiner bases one fundamental statement of his approach to human nature on these very ideas, concepts which have come to meet us again from a discussion of modern anthropological discoveries and ideas. The distinctively human stance requires man first to establish his own uprightness, since in contrast to animals, even primates, 'human beings are not given an inborn way of attaining equilibrium in space, but must develop it out of their total being'. He stresses that animal motion is thus fundamentally distinct from ours, since 'animals' orientation in space is determined by their physical organisation, while in human beings it is the soul which establishes the relationship to space and shapes the organisation'. He adds further that language and thinking determine that shaping, e.g. in the link between maturation and the speech-organs, and especially in the forming of the brain as we grow, since 'at the beginning of life the brain is still malleable because we must shape it ourselves to make it an instrument for the thinking appropriate to our essential being'. All this points to the crucial role of a childhood phase when these processes can unfold. They stand in an intimate although enigmatic relationship to consciousness. He was challenging nearly everything in the science of his time, in its attempts to derive human features from animality. Since the discovery of basic 'biological disjunction' between ourselves and even the anthropoid apes, and the uniqueness of genus *Homo*, it is hardly too much to say that these are

---

* See www.mindmatters.ai/2019/08. It may be worth stressing again that Steiner's inclination of thought is here, and in principle, always more deeply *into* the phenomena. It is quite different from the sort of thinking which, when the stage of consciousness is reached, wants to embark upon the unfolding of a 'noosphere' or the like: there, something is added to the reality, and it remains unclear whether it says anything substantively more. Steiner describes the gradual evolution of the organising 'information' body, or etheric, into the human principle of conscious thought, or Intellectual Soul, through the ingredient of reference to self: Steiner, *Theosophy* (London 1970) pp.32-36.

the very ideas which have moved centre-stage in the understanding of human evolution!* Steiner was referring, of course, to people of today, but his remarks still apply because we are here speaking precisely of those first humans who were anatomically identical or near-identical with ourselves.

We must naturally avoid attributing to those first humans our own multi-faceted awareness: that required the accumulation of myriad filterings and discriminations (symbolism, speech, selective highlighting and suppressing) the contents of consciousness. But in their equilibrated activity, they would certainly have experienced the beginnings of psychic self-referral; and they would retain impressions of the world around them with a new intensity. The animal allows itself to be physically controlled by the environment, and so 'earths' the energies that affect it. Human beings, however, retain the impressions from without and confront them in full force of heightened awareness. In their still largely undifferentiated state, perceptions must have had the force of a revelation. We know that young children today experience a fullness of sensation they only later learn to organise. The taking-in of food and consequent sense of sharing in wider existence would have been a cosmic communion. If we accept the supposition that it resembled the fullness of infantile experience, so sensual, blissful, and overwhelmingly powerful that out of it are shaped the very fundamental responses that are the basis of all later psychology, we might indeed speak of the childhood of humanity. In fact, we must reckon it to have been even more potent, and that the child today experiences nothing more than a vestige of archaic ecstasy—for so we must classify early mentality, something Lewis-Williams and others have begun to take seriously. When human beings first confronted the world in self-consciousness, i.e. in the spatial framework of human conscious uprightness, it must have flooded upon their minds and penetrated into them with unimaginable force. Later, shamanic techniques (like those behind the cave-paintings) preserve certain aspects of primal openness, though already under controls.

The great modern student of religions, Mircea Eliade, brilliantly begins his researches with 'hominization' (becoming human) and the step into uprightness. He sees the new awareness-as-space which follows from conscious uprightness as transforming the animal relation to the world. Such a space is 'organized in a structure inaccessible to the prehominians', he comments, and it immediately presents man with

---

* Steiner, *Spiritual Guidance of Man and of Humanity* (New York 1992) pp.5-8.

an experience of infinity. Concrete experience stands within a dimension extending up and down and in all directions to infinity, containing and dwarfing the seer. It would have been dazzling but also potentially overwhelming:

> It is from this original and originating experience—feeling oneself 'thrown' into the middle of an apparently limitless, unknown, and threatening extension—that the different methods of *orientatio* are developed; for it is impossible to survive for any length of time in the vertigo brought on by disorientation. This experience of space oriented around a 'center' explains the importance of ... cosmological symbolism.*

Early humans would necessarily have felt their discovery of the infinite as a god- or giant-like apprehension of the larger reality opening up, and as a teetering on the brink of disintegration. (In all of this we have a graphic intimation of what we clinically describe as the discovery of equilibrium: balance always on the verge of falling. Inwardly, it was the bursting into consciousness of a new world.) An inner answer to the vertiginousness must accordingly be developed—a first assertion of conscious control. Until that evolves, it inevitably leaves us floating in the uncertainty of that primordial, overpowering transcendence.

The myths about Paradise which are one central expression of the human condition comprise both component parts in tense contradiction. For Paradises are always very soon lost. To begin with the myth characterises the sensory bliss of man's primal existence; it forms a 'garden of delights' which is also a cosmological *axis mundi*, linking the earthly to the presence of God. Yet almost as soon as they find themselves there, the first people are shown overwhelmed and threatened with annihilation by the Infinite God whom they had tried to resemble. They have known Paradise, but emerge fragmented and isolated by the discovery of consciousness, imaged as the knowledge of their naked vulnerability. In the legendary elaborations of the story (or are they its suppressed originals?), the first human pair were following in the rash steps of the fallen angel Lucifer, who has intimated to them the god-like implications of the sheer range of possibilities open to them: they will know everything ('the good and the bad' as the Hebrew idiom goes)! But the story warns of the psychic and moral disintegration which follows in both instances. To possess Paradise is beyond humanity's inner strength.

---

* M. Eliade, *A History of Religious Ideas*, vol. I (London 1979) p.3.

The Bible's version is rather severely moralising, but the essential themes are probably the most fundamental and 'primitive' in all spiritual traditions. They are religious and moral intuitions which come close to defining the human condition in their universality and pathos.[*] But is it not an identical drama that is played out in every child's birth and upbringing: the upright human posture in its combination of strength and uncertainty, the plasticity of the brain which can be formed by potent experiences, the absorption of the results into our life-history through childhood, before the 'Fall' of growing up?

So powerful are the formative experiences which work in our primal awakening to life, Steiner explains, that we cannot assimilate under our own control, or even bring back to memory. In the drive to stand upright, in the hammering-out and absorbing of the experiences which plastically mould the brain so it can express our intelligence and enable speech, we are subject to intensities that we can only subsequently ratify and establish as our own. For once we make use of our body externally and fit into a role in the external world, we are no longer able to let all these forces work inwardly upon us as we do in the first three years of life:

> Our constitution as human beings in the physical world is such that, once we are no longer soft and malleable as children, we can no longer stand to have the forces of the spiritual world continue to affect us directly. The forces that underlie our orientation in space and the formation of the larynx and the brain would shatter us if they continued to influence us directly in later life. These forces are so powerful that our organism would waste away beneath their holiness if they continued to work upon us.[†]

The paradox is that consciousness as such depends on the pushing back of these forces, after they have elaborated the foundations of consciousness itself. We must first shape our organism in the child's absorptive way, in order to be able to build our subsequent mental life upon its foundations. We call this pattern childhood. But the pattern was the same in our earliest prehistory: intense perceptions, charged with meaning, belonged at first to ecstatic states which worked formatively on

---

[*] Rudolf Steiner often returns to them, and to the story Adam and Eve our 'first parents'. He more than once remarks that they point to a reality, though he insists it is 'symbolic'—while not always specifying a symbol of exactly what. There is certainly a truth of the genetic origins of humanity.

[†] Steiner, *Spiritual Guidance* p.12.

the brain, and then gradually, over ages of slow development, human beings began to be able to shape their response personally, individually. In short, the irruption of consciousness, like so much in the emergence of humanity, was first a crisis and threatened to overwhelm us before we were able to assimilate it more sedately and 'find our feet'.*

Becoming human, in other words, turns out to have been an extraordinarily risky venture. The in-turning that accompanies the new role of consciousness is profoundly destabilising; we also run the risk of rendering ourselves unfit for external life, since to deal with it we need to find ways of delaying adulthood so as to be able to cope with the extra demands posed by the elaboration of our own inner world. In an already tenuous situation materially and climatically, our strategy for doing so was evidently to prolong the malleable, formative stage of development that in animals usually takes place embryologically, and finishes abruptly when the animal is born. At that stage, the animal is essentially complete, and very quickly able to engage its environment. Humans, however, incorporated the whole evolutionary situation in a much more radical way. By effectively remaining 'embryonic' long after birth, we seized the adaptability to cope with the manifold dimensions of our new life. But existence as a child needed further complex strategies in turn—if we were not to fall prey to external dangers.

Such is the inner story we have hinted at before from a different, more external angle—the story of the unique biological features of genus *Homo*. They entailed extensive alternations in the rhythms of maturation and of physical growth—childhood, with its subsequent phases of rapid growth and consolidation. These must have immediately required new kinds of parental relationship. How did these changes all happen?

It is hard to imagine; but we do know that they did not happen along the once-popular Darwinian lines which envisaged humans getting incrementally more and more upright for longer and longer, getting more brain-power, and more manual skill, more powerful, more dominant. The reality of childhood, as a necessary phase of

---

* The pattern is the same which struck David Lewis-Williams when he observed that the ecstatic experiences of the cave-painters were vivid, shamanic visions— yet they clearly turned into our own inner world of concepts and ideas. It is also the pattern noted by Roger Penrose: knowledge begins in a 'seeing' that can only afterwards be separated out analytically; Owen Barfield noted a similar state of affairs in poetic creativity.

consolidating our complexity, was to make us in many ways more individually vulnerable rather than less—a prolonged extension of the embryonic state was externally dangerous in the extreme. The answer to our individual vulnerability came in the childhood time through accessing the adults' experience: through *learning*, as opposed to taking all the risks oneself. That in turn demanded considerable sophistication of language and communication—consciousness being stretched yet again. The answer to the external vulnerability came in the elaboration of social life, which created a safe space for the growing children. The adults had to make a protective network and be a source of guaranteed provisions. All these modifications, from language to social complexity, were anything but a 'quick fix'. Thus every step of human achievement led in turn to new problems. The Paradise-myths say that we were caught out by the contrary pressures. Steiner implies that by opening ourselves thus consciously to the apprehension of the spiritual at work in us, the infinite, we are liable to be burnt by the intensity of its holiness.

Lapsed from their full Humanity, the divided Adam and Eve take refuge from the flames of holiness when they can no longer sustain the state of Eden. Title page to Blake's Songs of Innocence and Experience.

It is hardly too much to suggest, therefore, that the emergence of Modern Humans and their brilliant leap of adaptation may not have been the unqualified success which is still often rather naively presumed. A success it undoubtedly was—as witness our presence here today. Yet it was such a knife-edge affair that there was surely considerable likelihood that it would follow in the tracks of the previous evolutionary experiments in that direction, which we see in the succession of hominin dispersions, all of which died out. Even the later sudden extinction of the Neanderthals poses the same question. At the very least, we must sadly lay aside the vision of our ancestors being rapidly crowned *sapiens*, taking over the world, triumphally proclaiming their symbolism and technical skill, sweeping aside other species with their cognitive revolution. We are brought face-to-face instead with those more recently revealed early Modern Humans who surprisingly turned out to be older than, not the bright young things who succeeded the Neanderthals in the Middle East. Anatomically Modern, yet not behaviourally precocious as it is so easy for us to wish—how should we envisage them in reality? Widespread traditional intuition considers the first people as half-finished, needing to be completed. Rudolf Steiner indicated that only a very few human beings were able to cope and survive under the conditions of their earthly origins, and it may be that such a situation was played out over a long period, or even repeatedly. It is an inherent human characteristic, in the situation of unaccommodated man (Shakespeare's phrase), to sit on the cusp of life and death; all our great art and spirituality reminds us so, despite our bulwarks of culture and our material paddings. Our survival was tenuous. How then did we build up our defences? How win our victories? Not all at once. The story Steiner tells can help us understand the modern evidence and help take forward modern discoveries.

Early Humans in Search of a Childhood

Describing the 'Lemurian' type of primitive humans, to which those who subsequently spread 'out of Africa' belong, Rudolf Steiner describes some of the earliest stages in human prehistory, and thereby gives a number of clues to the life behind those enigmatic fossil remains.

Before everything else, he stresses that in the earliest phases there was only a small group of full humans—the progeny, as he later recounted, of one ancestral pair. The rest of 'humanity' were still a sort of 'animal men' (hominins, we should say nowadays)—we may think of the still blurred lines of evaluation affecting estimates of species such as *Homo habilis* and the difficulties of tracing the emergence of early *sapiens*.[*] Where we do find the earliest fossil remains which are incontrovertibly *Homo sapiens*, with our distinctive traits, they show that the unique emergence of fully human behaviour and the single line of descent from 'African Eve' was in reality more tenuous. Steiner's emphasis on the extreme difficulty of the conditions which they experienced is nowadays recognised as the evolutionary pressure to maintain maximum adaptability, and must surely also be behind the extraordinary tendency of waves of hominins even before *Homo sapiens*, to spread out across large areas of the globe. Their liberation from the tendency to nest into a particular environment apparently resulted in a restlessness, and perhaps a novel willingness to keep going under different climatic and geographical conditions rather than turn back—perhaps the beginning of that human obsessive with exploration which remains with us still. It has been suggested that following shorelines lured them farther and farther on with the promise of seafood from rocks and pools, taking them even to South East Asia and Australia, but it remains to be explained why they also expanded into areas branching far away from the sea.

Steiner indicates that an important factor, already in the earliest human development, was the emergence of charismatic leaders, and it may have been these which led the way. As Eliade has suggested, the enlarged possibilities of liberated movement and consciousness of infinite space were at once soul-enlarging and intimidating, so that figures who provided a reassuring centre of life and activity would have been especially valued. Steiner takes this insight a step further when he says that the leaders had an intuition of sacral places: and we may suppose by finding centredness there they assured human beings of belonging and gave them boundaries and meaningful patterns of life.[†] We might think of a sort of bonding with the Earth. The quest for such places as spiritual 'centres of the world' would have provided for the first time *homelands*—created by spiritual intuition and 'magical' human action—as opposed to biological niches.

---

[*] Steiner, *Cosmic Memory* p. 76.

[†] The sacred places were certainly places of learning, Steiner says, but 'what later mankind accomplished by reflection, by calculation, at that time still had the character of instinctive activity', though we are not dealing with animal instincts but rather a shaping of the will and indeed psychokinetic faculties. The sacred person 'made' the place sacred by channelling spiritual-natural forces. Nowadays we can feel only an echo of such symbiotic relationship. Individuality was so little developed that it was less personal 'magic' than part of the undifferentiated spiritual-natural formative process: Steiner, *Cosmic Memory* p.75.

The inspired figures provided through their 'cosmic' authority also a centre establishing dominant presence and example; for of course, there was nothing yet in terms of ideals or codes. Steiner is emphatic that there could be no such social principles as yet. Everything was fundamentally based, rather, on the very human trait of imitative behaviour, or training by the will as Steiner says. Certainly authority-figures and spatial centres will have been one of the ways in which human beings began to feel safe and at home, and to feel common allegiance, in the confusingly varied and potentially dislocated world upon which they were embarking. The sheltering power in the cosmic centre, the sacral spot, reassured them all.

The experience of an Earth-mother no doubt goes back to these very first steps in establishing a new relationship to the world as space and authority which is intimated. For it is interesting to note that Steiner presents the leaders of primal human culture, and those who were most utilised by the leaders, as the women. In this he is at one with the growing tendency of modern palaeoanthropologists, for whom 'woman the gatherer' has become at least as important, if not more crucial than the spear-wielding mighty hunters once fantasised to be the direction-setters in human evolution.* Steiner ascribes to the women of earliest mankind the roles of spiritual leadership, or 'priestess', itself a deepening of her role in 'gathering' her family around her. For a modern child, the world is explored as a gradual extension of the mother's love-ensouled body—the same tendencies are surely to be observed in prehistory. A striking description is given by Steiner of a religious scene:

> The priestess, her face turned toward the east, ecstatic, sits upon a seat made of rare natural objects and plants. Slowly in rhythmical sequence, a few strange, constantly repeated sounds stream from her lips. A number of men and women are sitting in circles around her, their faces lost in dreams…. †

Many of the details immediately ring true. The cosmic orientation of the person is a feature of archaic consciousness, indicating that human activity (inner and outer) is transparent to cosmic forces—not yet fully introverted. Nor is the gathering of a kind to stimulate individual responses, but seems rather to be hypnotic in quality. The mention

---

* See the studies collected in Frances Dahlberg (ed.), *Woman the Gatherer* (Yale 1981).

† Steiner, *Cosmic Memory* p.82.

of rhythmical sequences of sounds would reinforce that effect, and is generally an indication of the ready dissociation of primitive consciousness, its tendency to slip into ecstatic states. For the stream of sounds, which Steiner carefully does not call language, is a classic trance phenomenon, still regularly triggered in a variety of religious circles often called misleadingly 'pentecostal'. The whole is, put simply, an ecstatic-spiritual gathering floated by glossolalia or 'speaking in tongues'.* Such ecstasies are the beginning, asserts Steiner, of humanity's intimation of the presence of the gods. In the babbling, repeated sounds, Steiner stresses, there is no question of any specific 'meaning'—so it is not language, which does not yet exist, any more than the social structuring which would enable language to be construed and shared.† Neither was it singing, where sound is organised on definite principles of quite another kind. It was probably very close, however, to baby-language, as even nowadays is babbled by every mother to her still unspeaking child; and it is similarly exciting and infectiously copied, as a child gradually learns through imitation. Nowadays, however,

---

* Note: Steiner is certainly not here expressing any approval for the reviving of charismatic practices of this kind, nor do they have any role in his conception of Christianity. It is well-known in scholarly, as opposed to revivalist, circles that in early Christianity 'speaking in tongues' was not of this kind but involved meaningful words and sense—even if disturbing (cf. Paul in I Corinthians 12:3). In the ancient world oracles, in which utterances were made without human purpose, furnish a closer parallel. The babbling of glossolalia expresses a very primordial, not a Christian conviction of the spirit, even if revived within later religion.

† *Cosmic Memory* p.81: 'For that period there can be no question of meaning in what was spoken. Sound, tone and rhythm were perceived. One did not form a representation of anything accompanying these, but absorbed in the soul the power of what was heard.' Some interesting ideas about glossolalia are put forward in J. Jaynes, *The Origin of Consciousness* (Harmondsworth and New York 1976) pp.357-60. (His version of a different, i.e. what he terms a 'bicameral' mental stage which perceived 'gods', is a still-too-simplistic theory despite some undoubted insights, stressing ecstatic states as part of archaic consciousness.) Steiner's description is certainly related to the mantric use of sounds. Prehistorian Steven Mithen explains that Indian mantras 'are lengthy speech acts and often sound very like spoken language, but they lack any meaning or grammatical structure. They ... require not only correct pronunciation but also the correct rhythm, melody and bodily posture.... They have characteristics of both language and music, but can be defined as neither': Mithen, *The Singing Neanderthals. The Origins of Music, Language, Mind and Body* (London 2006) p.12. Steiner spoke and wrote extensively about the spiritual potency of this level of mantric utterance.

it is soon subordinated to the acquisition of genuine speech. Originally, it was deepened and cultivated in its own right.

Other aspects of woman the gatherer are also intriguingly present. Not only a priestly gatherer of people and leader of 'colonies' to new, oracularly indicated locales, woman was a more literal gatherer of things: rare plants, or other fascinating natural wonders, as an extension of her role as gatherer of food or protective materials. We touch here on a controversial topic—the emergence of gender roles; but prehistory and Steiner have noteworthy contributions to make. Forming protective spaces (literal and metaphorical) was one of the prime needs for earliest humans in need of a childhood. It is in this light that we must understand the remarkable evolution of the human sexes.

Maximising parental care was the evolutionary strategy which was to give humanity the chance of a childhood. Providing it, however, was a development that drew on every aspect of the step into humanity and placed extraordinary demands upon it. Growing up *via* a childhood requires, on the part of the parents, a strong stable relationship which must be maintained at the very least over ten or more years. Animal mating and young-rearing has nothing like it. (Life-long pairings are certainly known, but they do not involve those continual acts of care and the protracted affirmations of the parents' bond.) The child suspends immediate growth toward maturation and in its plasticity first assimilates all that is formative in the first years, and goes on to master multiform ways of dealing with inner and outward challenges. While thus 'growing up', he or she relies on help from a father and mother in conscious and sustained commitment, quite unlike the short-term seasonal behaviours of animals. A large number of distinctly human traits come together: not only physical care and protection, but the provision of activities and roles to copy to arm the offspring for subsequent autonomous life and parenthood in turn, communicating the lessons that can be passed on to stay safe, etc. while steering the young one through successive crises in its development. This hugely enriching the life-experience of the group. It is inconceivable in turn that a family group could be monolithic, furnishing all this and looking out for itself the while. We must presuppose for their success that a larger social group pooled together the resources of others doing likewise, and added in the acquired wisdom of the elder generations, charismatic leaders, etc. from a very early stage.

The intensification of personal relations all around the child, and the provision of contacts with other children, point to the social setting. In this both parents were necessarily drawn in, and as a result their roles and physical development drew close together in a way that is unprecedented in the animal world. In rearing a child, both father and mother were essentially called upon to deepen in an exemplary way the whole human range of behaviour, their specific fatherly or motherly contributions showing only minimal divergences—beyond those directly necessitated by the biological nature of begetting, giving birth, suckling, and so on; beyond these, in sharing all the 'human' tasks of looking after or leading, they are both constantly and closely interacting. The result is another case of biology being shaped by culture. Little of it can be modelled on animal behaviour. After the immediate needs of the young have been met, few animals stay in close contact. The phase of being in heat is usually strictly limited, and only then is the male stimulated. The young even of higher animals are usually soon ready for outward sufficiency through instinctive behaviour, and do not have an extended learning phase.

It is otherwise where the parents represent a fully realised human presence to each other, as well as to their child, and accordingly we see in them the opposite of the highly polarised sexual roles normally found in animals. We are still talking here, in earliest human times, about biological tendencies—not about intellectually conceived education. And the biology manifests itself, in the first instance, in the extraordinarily close physical resemblance of male and female in humans which still holds good today. Jokes and axe-grinding apart, that is the most striking feature of human sexually mature bodies. Both male and female are shaped so as to embody a large percentage of the archetypal human identity. Steiner's clairvoyance sees everyone as basically androgynous. Life in the intensive group containing human pairs did not favour male and female divergence, but rather the ability to share the full range of tasks. Myths of human identity in archaic societies almost always contain a mythology, not only of birth from the sustaining Earth, which we have seen is extremely ancient, but of the first humans' original androgyny. The Paradise myth of the Bible undoubtedly contained both these elements.* Psychologically, we today equally feel our fuller identity as human, and not just male or female. Our fundamental bisexuality is the foundation of both female and male development, psychically and

---

* The creating of the first human being as 'male and female' (Gen. 1:27) is usually blurred in conventional Christian translations. In much of the Jewish tradition it has been rightly interpreted as androgyny. The closely related myth of androgyny and the Fall told in Plato's *Symposium* 189D-195B is explicit confirmation.

physically.* Steiner describes how, clairvoyantly, the bisexual totality of the human is always perceptible surrounding the particular male or female physical individual. Bisexuality is a central feature, both in myth and in evolutionary reality, of the 'Lemurian' human type.

The close resemblance of the female to the male human extends to virtual parity in body-size—only a ten-per-cent size-difference in humans on average; whereas a male gorilla, for example, is a hundred to a hundred-and-fifty per cent larger on average than the female. The prehistoric female australopithecine famous as 'Lucy', whose fossil evidence was discussed earlier, and was once touted as a close antecedent of humans, stood barely three feet high, whilst males of her species would have towered over her at five feet, and were appropriately more massive in build. Frontally, in particular, human males and females are little differentiated. The extreme lack of strong sexual differentiation, biologically, which strikes us in humans even extends to the riddle of the female orgasm—unparalleled in the animal domain and baffling to those who try to justify its biological utility. It is best understood simply as an indication of the extraordinary closeness of male and female constitutions alike to the primal human totality, in all its adaptability and lack of specialisation: the fundamental feature which belongs with all the other aspects of the emergent parenthood/childhood mode of life.†

Providing one's offspring with a childhood permitted primal plasticity to be exploited in long-drawn-out development, then; it also transformed the emotional ties and biological affinity between parents. In the same context we may understand the freeing of human sexual activity from the limitations of the fertility-cycle. Animal sexuality, we have observed, is mostly seasonal and requires the presence of precise stimuli. The evolutionary advantages of being male and female are momentary and

---

* Fundamental human bisexuality is the starting-point of all modern psychologies of sex, and follows from the advances in embryology which in the nineteenth century revealed that male and female emerge fairly late in development, and that the conventional notion of being definitively 'a boy or a girl' has no original physical basis. It should possibly be mentioned here that Steiner was close to the circles of Freud and Breuer, as well as Friedrich Eckstein, in Vienna. He was certainly aware of the latest developments in psychology as well as of Haeckel's embryological investigations.

† See J.Z. Young, *Introduction to the Study of Man* (New York and Oxford 1971) p.483. (The once popular idea that sexual fidelity and freer timing needed to be imposed because the male was often away hunting—a prehistoric version of 'working late at the office'?—is a typical older-style explanation, not now found so convincing.)

basically restricted to the phase of reproduction. Humans, on the other hand, are permanently sexually interested and capable of conceiving at many times of the year. The dangerous gamble of giving birth to vulnerable, partly developed 'children' perhaps required such maximisation.

The evolutionary path to the relative androgyny of humans may yet again be traced directly to the upright posture. Biological female signals for arousal in animals such as primates are based on copulation by the male from behind or with the female prone. But humans, with their commitment to uprightness, relate to each other in the main frontally, so that the scents, swellings and other seasonal signals of animal sex lost their role. Arousal takes place by affectionate behaviour, emotional suggestion and face-to-face relations; disparity of size hinders this kind of encounter, which now instead becomes more equal-sided and personal, independent of seasonal fertility cycles. It already adumbrates the loving-embrace which will enfold the child. But it also enabled sexual fascination to become a long-term force keeping a couple together, and made the matter of complementing each other emotionally and sexually a permanent role, giving a sexual aspect to the whole of their shared life. There was no longer just male and female for the purposes of reproduction, but the emergence of sexualised personalities, i.e. men and women. Relationship to a person replaced simple mating, and relationship with a person was largely based on complementarity overcoming the sense of difference through having the greater part of life-interests in common, though tinged with a fascinatingly tantalising otherness.

Rudolf Steiner has described how the whole human etheric archetype, our biological information forces, can be actualised only in loving complementarity; and that the part which still remains unrealised in each individual forms a powerful drive to seek its representation in another person, presupposing therefore a social situation in which to seek.* Thus a woman has her 'male' aspect just as fully in her etheric body, and likewise the man has his feminine potential. The difference is induced by a polarity of slight holding-back development (in the

---

* See *Cosmic Memory* pp.87-99. For evolutionary aspects see in particular the lecture in Steiner, *The Christ-Impulse and the Evolution of Ego-Consciousness* (New York 1976) pp.93ff. For Steiner this stage of individualisation like the others has a cosmic-structural prototype. He derives the particular account of cosmic development *in this specific detail* from Hegel's interpretation of scientific cosmology—though he emphatically does not employ Hegel's dialectical model: A.V. Miller (ed., trans.), *Hegel's Encyclopaedia of the Philosophical Sciences. Vol. 2 The Philosophy of Nature* (Oxford 1970) pp.92-103.

woman) and slight over-materialisation (in the man). The ultimate androgyny of sexual relations is thus expressed in a continuous quest for ever deepening relationships. It is both an outward search, and also increasingly an inner fulfilment and enrichment. Steiner sees in the deep drive to experience the full range of human stature also a fundamental karmic rhythm—that is to say, the basic pattern of human destinies, alternating male and female lives. The archetype of human existence works in every one of us but, crucially, is not something that can be incarnated in a single quasi-animal form. It is always ongoing transformation. In the myths, sexual determinateness is presented as a failure or a necessary bound set to our fulfilment. Even in sexual rapprochement, the adults cannot find the release into a pure plasticity like that of the child who forms the ideal of their longing.[*]

It need hardly be said that at a fundamental level of our existence, these developments are still the basis of our life and loves. Sexual identities have been a fantastically successful adventure, the original creative trajectory of our path to modernity—though some think that it is approaching its end now. It has permitted an extraordinary diversity of human life to unfold, held out ideals, built cultural and intellectual castles, real and aeriform, for the human spirit. We must beware of supposing, on the other hand, that all the rich social and cultural opportunities for the flowering of relationships existed for the earliest Anatomical Humans, with whom we are physically analogous. Instead, we find a picture which yet again brings out starkly the rudimentary and even more the high-risk nature of these first evolutionary trends.

In line with the logic of human-sexual parity we have just sketched, many palaeoanthropological researchers have concluded that the earliest humans were necessarily monogamous.[†] In animals, the huge disparity in body-size between male and female is a sign of their biologically divergent sexual roles. The massive male gorilla displays his

---

[*] The infant as pre-sexual points to the androgynous human archetype. In many pre-modern and non-Western societies it is normal for a parental couple to bond intensely over the birth of their first child. Nor is their love primarily powered by personal emotion. Only from about the twelfth century AD was the forming of the love-bond pushed to the beginnings of personal attraction, on which the whole subsequent relationship was to be based ('Romantic love'). In highly personalised cultures today, the tendency has been stretched still farther. Prehistorically, the developmental stage of emotional awakening is already 'Atlantean' and thus far ahead of these archetypal identity-forms.

[†] See Richard Leakey, *The Origin of Mankind* pp.23-4.

masculinity to rival males to drive them from his harem of females, and within the group he dominates by his strength and size. The strategy ensures the most virile paternity for the young of all the females he can control, and he will cease to dominate them only when a stronger, younger or more advantaged rival ousts him from his position and takes them over. In our human case, the overriding claim which the vulnerable and growing child exerts arches over both male and female parents, requiring more of a safe space than short-term mating could furnish, means it is their loving bond which keeps the safety net in place and resists incursions from others. However, there is a very human twist in the tale.

The much-needed safe room to develop is necessarily backed-up by having others around in communal life:[*] yet the group has lost its character of a harem dominated by a single male. Instead, there must have been a tolerance of other males in easy proximity—and also females. 'We shall probably never know,' says a comprehensive writer on human nature, 'at what stage of evolution the family emerged from the promiscuous troop.'[†] Did it emerge at all decisively? More likely this phase exhibited the same paradox of strength and weakness, newness and relapse, in human evolution generally. Trance-gatherings of an emotional nature like those Steiner described are associated in the history of religions with unstructured social mixing, like carnival celebrations when the usual restraints are lifted. The liberation of sexual activity from short-term seasonal rhythms, and the emergence of sexualised personalities must in conjunction have created almost as much group sexual instability as monogamous devotion. The group had of its nature to be the field of search for a suitable partner. Yet it must also have constituted a constant source of sexual distraction and temptation, undercutting the evolutionary tendency toward monogamous pairing which alone can furnish the gift of a long and successful childhood.

---

[*] In evolutionary terms, safety became a prime issue already to hominins with life in the open. On this stage Steven Mithen writes: 'Away from the cover of trees, safety can only be found in numbers, which provide more eyes to notice predators.... There is, however, a cost: social tensions leading to conflict can easily arise when large numbers have to live in close proximity to one another'—*The Singing Neanderthals* p.126. He does not specify the causes, but food and sex are no doubt the usual suspects. Uprightness—safety from dangers of the open—stirring of jealous tendencies—yet another version of the human story.

[†] J.Z. Young, *An Introduction to the Study of Man* (New York and Oxford 1971) p.484.

Again we find much to recognise in human behaviour today. In prehistoric times, moreover, we must realise that social and moral-intellectual norms had not yet been invented, such as restrain and guide, however fallibly, the people of today. Perhaps we may just say that the freeing of possibilities which comes about in humans has also made us a species incomparably prone to turn on ourselves in violence or other destructive behaviour. Those intuitive myths which define the human condition in terms of divinely given gifts, catastrophically misused, may be given a definitely evolutionary application. Paradise-and-Fall myths have a central role, invariably, for the human sexes, and afterwards comes the story of Cain and Abel.* In prehistory, Steiner describes how some features of sexual identity were already being formed, and his description certainly does not romanticise the process! It does, however, ring true in many details to what we know of prehistory.

Starting from basic biological similarity, human beings have developed ways of differentiating their appearance. To 'look like a man' we do not just need to be born male, and a 'woman' is much more than femaleness. It is on this level, in fact, that biological ambiguity and incompleteness may actually be overcome. We fashion a way of being and of representing ourselves, a man cultivating hair and beard, for instance, in a manner deliberately contrasting with women, and even contentious alternatives presuppose the same underlying approach. Myriad constructions are possible. All are based, nonetheless, on the basic proximity of human forms and the need to elaborate them in different directions, employing them as a starting-point in order to make a statement of identity. There is no longer just body but a body-image. From this point, where biology leaves us unable to fulfil all our child-potential, the point of the 'Fall' into difference, human beings had henceforth to *make themselves*.

Actual bodily reshaping plays a significant role especially in the very earliest times when the use of symbolism (like ornamentation, clothing, etc.) is not yet available. The practice of circumcision (at puberty) has historically played a hugely important role in human archaic

---

* We have mentioned the Bible and Aristophanes who share similar mythic sources; the Sumerian *Gilgamesh* epic incorporates much of the Paradise/Fall mythology in earlier form: Gilgamesh seeks the Plant of Eternal Life, his friend Enkidu who 'ran with the animals' but has god-like connotations, is seduced by a Temple Prostitute and becomes human—but the 'Cain and Abel' sequel is transmuted into the initial hostility but then heroic friendship of Gilgamesh and Enkidu. How far back do such myths reach?—or are they perpetuated somehow in the human constitution over the ages? We shall later attempt to answer these questions.

history. (Jewish (infant) circumcision is only a rather modern echo of practices which reach back to the first stages of dawning human identity.) It expresses the awareness of the need to alter or complete human existence, for man to achieve wholeness—identity on a higher-than-biological level.* Feminine analogies also exist in archaic societies. Other predominantly male practices include deliberate injuries or distortions, insertion of objects, etc. The alterations commonly assume the aspect of ordeals to be gone through in order to achieve belonging to the group, through making oneself into a fully-formed person. Sexual approximation to the child-potential working in our earliest years, by its very imperfection, thus opens the way to further levels of self-development, which need and which at the same time help to generate a social network. From these practices will evolve Mysteries, which will make more complex roles and human beings to fill them—much of that, however, is still far in the future. Making men and women was their first project.

Steiner characterises the ordeals of 'Lemurian' primal humanity in almost blood-curdling terms. 'The boys,' he wrote,

> had to learn to undergo dangers, to overcome pain, to accomplish daring deeds. Those who could not bear tortures… were not regarded as useful members of mankind. They were left to perish under these exertions … The bearing of heat, even of a searing fire, the piercing of the body with pointed objects, were quite common procedures.

The young men exhibited their prowess, Steiner adds, before the girls who were required to admire their combat-displays.† Rather than being primitive violence, as one might first suppose, their purpose was clearly in some part display itself, i.e. it was a statement of exclusivity, of contrast with the class of non-males. As well as showmanship it was also about demonstrating acceptance into group membership, and thus about the meaningful reshaping of the given individual. Such 'puberty rites' celebrate one version of identity achieved. In practical terms, nevertheless, the boys were thus hardened presumably to be able to protect the group, from other people or from animals.

Girls were developed in quite another fashion. With them, a sort of mental life was cultivated, says Steiner, though without language and even without memory. (When Steiner mentions training of the will as the

---

* For a profound interpretation of the whole complex, see M. Eliade, *Rites and Symbols of Initiation* (New York 1965) pp.21ff ('The Initiatory Ordeals').

† Steiner, *Cosmic Memory* pp.73-4.

group starting-point, we must picture people able to 'pick things up'—as children still learn many things but cannot recall the learning of them.)

> Propensities for dreaming and imagination were developed in the girls, and these were highly valued. Because no memory existed, these propensities could not degenerate. Particular imaginative or dream-conceptions lasted only so long as there was a corresponding external cause; thus they had a real basis in external things …. It was nature's own dreaming and imagination, one might say, which worked in the female soul.*

We note once more the lack of ego-centrism, which in the absence of memory is scarcely able to be present at all. Their feeling of oneness with Nature easily approached the trance-like, and was a sort of dream-vision though an objective one. The most talented in such gifts must have become the charismatic women-leaders mentioned above. Thus the women were trained to enter imaginatively into the environment, which again must have had the practical purposes of making them 'gatherers', able to 'see' hidden valuable foodstuffs, trace sources of water, etc. with a clairvoyant penetration; also, as we have observed, they 'discovered' specially significant objects or things which could be given a quasi-religious meaning, for use in the group séances, or just for their pleasing aesthetic value (so long as we accept Lewis-Williams' caution not to ascribe to early humans our own theories of art-for-art's-sake).† Steiner adds that out of these experiences women were similarly the ones to discover the moral sense of right and wrong. Once again it rings true, since moral intuition is always in archaic times a cosmic intuition, touching on the place of a deed or person in the world-order. Only many ages afterwards would human beings start to reach moral decisions out of themselves, attended with personal feelings.

In these apparently crude first differentiations of gender, the first steps in utilising consciousness to start off an exploration of human potential are enwoven: crude as they seem, they involve going beyond the mere

---

* *Cosmic Memory* p.74. Steiner's assertion that the cultivation of dreamy states was protected from 'degeneration' implies at the same time that they were close to conditions we should call pathological in modern individuals. Do we glimpse a kind of introverted adolescent girl and her obsessions, as seen still today in body-image anxieties, etc.? Essentially they yield psychic control to external forces, though such was the primitive lack of ego-centre that the personality offered little resistance and suffered little damage.

† Lewis-Williams, *The Mind in the Cave* (London 2002) pp. 42ff. (He is of course mainly thinking of rock-art from the much later Upper Palaeolithic, the 'cultural explosion').

given content of the individual, to intuit specific complementarities that have meaning for the group, and the need to achieve one's identity. The centralising role of the ego in constructing experience which we today take for granted, however, hardly existed—it was by confronting the heady emptiness of space, or experiencing the power of cosmic patterns (day/night, seasonal surges) to whirl life along, that such primal explorations were absorbed: traumatically, as ordeals, whether in the women's vivid, trance-like intensities that were Nature's own dreaming, or the violent wildness of 'masculine' physical exercise made into a display, the young people were pushed out into the hostile world, or drawn into dreaming—and survived or not. However, they were exposed to them within a relatively protective familial setting, and thus the children were able to be powerfully reshaped by them as they grew up in their novel biological rhythm through childhood to maturity. Because consciousness was subsumed so strongly in cosmic patterns and forces, the emerging 'gestures' of the sexes would scarcely have been a matter of personal feeling, so important to us moderns, but worked to force their body-awareness and basic attitude of being. They became a pressure to force change upon the body itself, even by violence. The boys hardened by ordeals, pushed uncomfortably out into the world, acquired a bodily hardness and poise that was deliberately different from the girls' being held back in themselves, to dream and let visions lead them. Nowadays these pressures may lead to self-harming or to body-image obsession; these still break out into pathological behaviour, destructive of ego-development, and are overcome in maturity. But in primaeval times they were prime forces of identity-formation.

Through awareness of each sex by the other, each could turn their sense of incompleteness, of being 'thrown' into life, into an intuition of their collective wholeness. The duality highlighted the self-contradictory feelings where biology has left us apparently unfinished. Identity-as-achievement was bound up with sexuality since this highlighted the incompleteness of the individual. As yet early humans embodied it without reflection and even without memory. But they were there—perhaps only just; or a few of them were.

Material Progress?

What then of the popular notion that human qualities emerged in dominance of the environment, primitive technology, use of fire and spears for efficient hunting? What about intellect?

Those things would come; but we have had to find a more fundamental understanding of human emergence, and it is as well to take in for a moment how little it is directly connected with material progress, as worshipped by the Victorian thinkers whose situation necessarily coloured evolutionism.

Steiner stresses that even memory in the real sense did not exist amongst the earliest 'Lemurian' type of humanity, of which it seems that our first African forebears were representative. Steiner habitually used the term memory in a strict philosophical sense for the ability to refer back to a previous experience, locating it in the past, and so being able to compare it with the present. He sometimes pointedly denied that animals have memory *in this sense.* We may protest that our cat greets us affectionately, that our dog even looks out for us—they remember us, surely? But what we mean is essentially that our pets' experience is pervaded and determined by past experiences, of us and of many other things. We do not seriously suppose that our dog has a specific memory of when we were last there in the past, and by comparing us with it realises that we are that same person, as we come down the path at the same time next day. To construct a past, as past, and hold it separate from the present, is memory. A dog or cat fuses present with past experiences and reacts accordingly, hiding under the table or running forward expectantly. Steiner asserts that primal humans still had nothing of memory proper. What then can it mean that they had a crucial childhood time of fluid development and learning?

The growing child was evidently not able, as children do today, to gradually put together a conscious array of knowledge. It was a matter of learning from the adults the appropriate responses to an ever-varying set of circumstances. By going through an extended early-life period of accompanying the adults, he or she was enabled to imitate and absorb a valuable range of expertise to address many of the day-to-day problems of life. Certainly it would have been a knowing-how rather than a knowing-that. They would not have been in a position to recall a past event and conclude that the present one requires the same or a similar response. It would be more like riding a bicycle or sailing a boat: one learns how to take a corner, or to tack against the wind by pulling on the appropriate ropes, without re-evoking the experience of learning how not to fall off or how to keep wind in the sail. Dealing with a snake or frightening away scavengers would become ingrained, but not lead to stored ideas in the mind.

In this way, archaic children would gain a rich store of practical knowledge. Intelligent animals can learn a good deal, too, of course: but they rely on instinctive reactions to a very great extent and much less on learning, and the period in which they can learn is also generally quite limited. Just as we had to say that human posture, and the consciousness it requires, needed a new *commitment*, so the children of the first people were taking part in a life that went over to and relied on learning. In achieving their identity, they found themselves as learners; and found also that to make it the basis of life they needed to learn new things time after time—since learning is precisely freedom from a single setting in which they had to have suitable instinctive reactions.

A necessary basic store of such 'knowledge' had to be gained before adult life—but adults too, unlike animals, went on learning. The parent couple remained united, not just in bringing up their offspring, but in helping each other through new situations. This knowledge too was not abstract knowledge, but in learning about the ever new examples they could give to their child, or in discovering more about the other partner who was almost oneself but fascinatingly different, they had stepped into a potentially infinite series of fluid sharings and appropriations. They had uncovered the boundless scope of such relationships in time, exactly as their human orientation had opened up the infinity of space; in the last analysis, their novel detachment or freedom in both instances is ultimately identical in source. Living out of the diversity in the human archetype, too diverse to be contained within any individual, paradoxically made each self consist of a search directed toward others and a self-transformation to care for and foster others.

In the absence of memory and stored-up ideas, on the other hand, opportunities for their potential transformation remained severely limited; as regards thinking beyond the actual situations of personal and social life, the genie was still firmly in the bottle. The complex and unstable strategy of humanness, moreover, was only successful initially—and perhaps for long, long ages—for small numbers. Only very gradually could it unfold the far-reaching effects we take for granted as human activities and attainments. The twenty extinct hominins are doubtless just a few out of the actual experiments, and perhaps *sapiens* groups often failed, or just managed to keep going. Establishing social organisation in particular must have been a daunting step.

The lack of genuine memory-constructions of the past would mean that maintaining institutions, such as the continuing authority of leadership, over generations could only be of the most rudimentary kind.

We shall see shortly that active companionship with the dead is the inescapable precondition of true society. Without ideas and memories, how could the wise world of those now dead continue to exert itself? Even the family group could not maintain authority once a child had grown up; each generation would simply start again as a new family, not a continuation of the old.

Steiner's most telling comment is that as yet humanity had no stories, and no homes. A story tells of somebody specific and what they did. Without story, a father could not be remembered as having done important things, since these would have rather been absorbed directly into the repertoire of the younger individual who imitated him. Parents lived on in their children in a much more literal way than happens nowadays, where social values and acquiring information take over quite soon in a child's life. It is interesting that, according to Steiner, when memory did begin it was a multi-generational memory—only much later did it become confined to one individual. The absence of separate life-stories will also have meant a sense of belonging probably more intense than we can now recreate. In the most archaic time, there can have been no mental instrumentality for any single individual to conceive of his own bounds, and define the extent of his or her own existence. The mythologies about the first men who lived to astounding ages are an echo of the continuity over many generations that was originally prior even to distinction.* It was first an absolute, even a mystical solidarity: an ultimate expression of that primal 'duty of care' which goes with the project of having a childhood, which defines the human state. Psychic identification among those belonging to the caring-family unit again hints at a somewhat 'ecstatic' consciousness (cf. the ecstasy of mother-baby bonding, or the glamour of sexual attraction), and reminds us in yet another way that humanness was not a possession but a greater-than-self diffusion, of unfathomable reach. In face of biological

---

* Rudolf Steiner linked the Bible's stories of the patriarchs (like Methuselah) to the idea of a long-term shared memory experience. We think much too externally about the role of kinship or common descent in archaic cultures. Note the idea known from earliest Mesopotamian civilisation that the indwelling power (or god) of the father 'passed from the body of the father into the body of the son as generation followed generation.... Since the god who resided in the man's body had earlier been in his father's body and had therefore engendered both him and his sister he is "the son of his god" and his sister is "the daughter of his god"': Thorkild Jacobsen, *The Treasures of Darkness* (New Haven and London 1976) p.159. The usage appears in the Bible (e.g. Ex. 3:15) and is certainly extremely old.

incompleteness, human wholeness must needs be spiritual, aspirational, as care and love. Extended childhood from the start made a bond of love between all involved in creating and sustaining it that must have gone far beyond the range of even the most devoted animal behaviour.

Humanity at this stage lived in the open and in caves.* That means that they did not make dwellings, but sheltered opportunistically in hollows and openings in the Earth. Thus they had territories, in which they felt at one with their 'mothering' environment. Their freedom of movement and apparent curiosity to seek out novel places was complemented by their ability to bond, metaphorically or mystically with the landscapes which received them. The later intensification and adaptation of such feelings into the structuring of life around a 'home' is a more complex act than might first appear, however. We shall try to describe it in a later chapter. One may suggest that the earliest humanity felt at home under the sky and on the Earth, since their freedom of movement placed them in a challenging new relationship with 'open-ended' space. In the stars and in caves, they could come to rest and be carried along by the movements they inspired them to pursue, or escape from the vertiginous. Breaking free from a containing niche which determines an animal's pattern of life must necessarily have resulted in a novel restlessness, which our species' unprecedented spread to all the climes and habitats of the world has still not been able to lay to rest.

With the aid of Rudolf Steiner, it seems we can give a convincing answer to the question confronted by the modern researchers as to the nature of the Anatomically Modern Humans who were as yet devoid of the modern cultural attributes of language, art and basic technology (cf. above, pp. 44ff). It does not throw us back into notions of animal continuity, which the recognition of the 'biological new beginning' with *Homo* and uprightness has now definitively replaced. It acknowledges the unique incorporation of consciousness even into the very gesture of self-balance, i.e. human presence. It acknowledges by the same token the tenuous and knife-edge nature of the human solution to the challenging variety of the world. It acknowledges our way of living in our own sphere of being, and close-sharing with long-term partners and offspring—a way that is precarious and dangerous everywhere but potentially everywhere a triumphant inner adaptation (learning). Yet we did not triumph; we met the world with love and restless aspiration.

---

* Steiner, *Cosmic Memory* p.74.

It may look hardly probable that we should have ever have triumphed. To convert ingenuity at survival into more, the resources of communication by language, of social structure as well as just herding, of tools and established dwellings wherever we settle, will all be needed. Dying out must long have remained an option, and may be termed an established pattern. Infant mortality rates among the large-headed weaklings born to the successors of 'Eve' would have been necessarily high (they remained so even far into the nineteenth century, even in 'advanced' countries). Having many such highly demanding children will have been the death of many of their mothers too. A life based on gathering can support the consumption-needs of physiological high-grade beings through their extreme activity and growing brain-power—but only by a considerable stretch. Hunting may indeed have gradually tipped the balance, even if we must not exaggerate its importance.

It is surely of great significance, therefore, that Steiner stressed the overwhelming difficulties of the human evolutionary project. Many of the first humans, he says, hardly raised themselves above the animal level. Despite the successful step into upright existence and the beginnings of human interaction through care, love and increased consciousness, very few were able to survive in bodies as they had to be on the Earth. Through the tenuous success-story of Lemurian humanity, however, the way was opened for a subsequent evolution, but only through another type of human. We saw in the previous chapter that a pattern of human evolutionary development was already hovering among the cycles of the stars and planets (without which, Earth too could not be). It is what Steiner terms Atlantean humanity, but its story needs another chapter. It expresses more fully the formative-human prototype whose possibility is written into the cosmos with all its amazing convergences and fine-tunings. But it needed very specific circumstances to find a path to actualisation on Earth. The early Modern Humans, however, provided just that. How the development came about is something we can now recognise—but it requires that new chapter to describe.

Chapter 6

NEANDERTHALS AND ATLANTIS

Evolution presents us with an almost unthinkable panorama of complexity—to begin with, life as we are accustomed to think of it, with its types from the microbe to flora and fauna, through all their history, to the world they share with ourselves. Yet beyond that is the intricate order of the environment that is inseparable from life, the finely-tuned balance of the global ecology that scientists increasingly call the living Earth, and the patterns of planetary and solar relations. And beyond that even the explosive history of stars and the formation of the rich diversity of the chemical elements was needed to make up organic life. Living organisms are a realisation of the potential in a uniquely structured world, and their eventual emergence is nowadays acknowledged more and more as the outcome of a 'biological imperative', or 'laws of complexity' already inherent in the universe, and rigorous thinkers scorn the old notion of a haphazard biological accident. These 'nested hierarchies', life-structures dependent on larger structures up to shaping of the galaxies, are not related accidentally to our own existence as observers, in the eyes of many of our acutest scientific thinkers, but already intrinsic to it. The emergent cosmos, with our Sun and Earth, is already an expression of the order which we find in ourselves, enabling an 'Anthropic' or human-centred view of the world that has been one of the most fertile developments in the life-sciences and in recent physical cosmology.

These ideas have been evoked, in earlier chapters, to suggest that Rudolf Steiner's researches into the link between human consciousness and the primal history of the universe may now come into their own in the context of modern thought. In the branchings and differentiations of life during all its evolution, considered in that Anthropic sense, we are uncovering the stages by which our own emergence, as conscious beings, is gradually brought closer. There is nothing metaphysical here. All that is necessary, as Steiner says, is that everything should happen which has happened. Living awareness has not been made to happen in an alien universe, but is an integral part of a living whole. In his 'clairvoyant' account, the human form is written into the world-order from the beginning and its gradual separation appears to enhanced perception as a real process, which he describes. Stages in the prehistoric unfolding of life thus appear already in relation to the possibility of a physically realised humanity, shining out gradually from under the mass of all that is not human in which at first it was buried. At first it was a mere mathematical point or multiplicity of points, as Steiner says. Thus we are able to make sense of his assertion of a connection between the vaster rhythms of evolution, including the emergence of prehistoric animal life on continents which were differently disposed around the globe—on a great southern land mass, or later in the Atlantic Basin—and the human beings who would subsequently live physically on Earth. Their physical possibility was first brought into focus as life separated into contrasting animal types—and here Steiner's ideas have been especially fruitful for biologists like Wolfgang Schad today. But what was clarified in the twists and turns of primordial evolution was no mere 'idea' of the human that might be; it was the spiritual human within the biological imperative of evolution itself, in the laws of its complexity, the pressure of what could be, in an anthropically meaningful universe. And so Rudolf Steiner has described it. He gives to these laws a startling actuality.

One version of the human was in due course to be realised in those early humans who are documented anatomically in Africa—the type which was until recently supposed to be the last step in a sequence of types showing ever greater intellect and technical competence. In other words, they were supposed to lead on directly to ourselves. We saw, however, that the 'Darwinian package' behind such ideas has unravelled in recent decades, conceptually and chronologically. Anatomical modernity turned out to not mean modern behaviour, and the chronology of *Homo sapiens* triumphant was suddenly reversed; rather than the latest stage, it was older than other, alternative humans. It became clear that even the physical human form went back much farther than once was thought—and, to complicate matters, some of those supposedly more 'primitive' hominins, most notably the Neanderthals, turned out to have had their own modes of sophistication which the researchers had previously been unable or unwilling to acknowledge, while the 'Moderns' were still not behaving like us at all. The question of our emergence is starting to assume a totally new—and one might say, more genuinely evolutionary complexion.

The previous chapter suggested that Rudolf Steiner's description of the 'Lemurian' human type which evolved physically in Africa, and whose ultimate origins are connected archetypally with the primordial southern continent of which Africa was once a part, is remarkably apt in the context of modern discoveries. He does not focus initially on intelligence. Rather he enables us to relate the distinctively human form and the cluster of consequences which it entailed in the way of male and female roles, long-term relationships and development, loving care—as well as the more conventional adaptability to many different environments—to humanity's actualisation of the basic principle of earthly evolution itself. By being shaken free from dependence on a unique environment, humanity incarnated the quality of adaptability (transformation, metamorphosis) as such. The relevance in terms of survival and of expansion out into the changing world were fully recognised, but Steiner stressed that those cultural constructions such as language, which would enable the pooling of experience, even of memory which would facilitate the acquiring of skilled techniques and concepts, so shaping social individuality, were still absent. In many ways, one has to envisage early humans perilously torn between conflicting pressures and needs, for which the solutions we know from later history were still far off. It can hardly be accidental that most early versions of *Homo* had a way of dying out. Steiner indeed emphasised that very few steps in the process were successful, and that among the 'human' types many remained essentially animal-like and only a limited number could carry forward the features we recognise as bringing them essentially closer to ourselves. It seems that human mastery of earthly conditions was not the end but only a step with many further developments—necessitating a freedom to project adaptability on quite new levels.

From his clairvoyant perspective, however, those inherent possibilities of human language, memory, culture, etc. already possessed a reality in the 'imperative' of evolution. The emergence of forms necessary to develop them had drawn them ever closer as evolution progressed beyond the Lemurian stage. Realising them in organic human terms actually needed the experiments of 'Lemurian' humanity to have been tried again and again until these could open the way to their gradual adoption. All this was as environmental circumstances altered, echoing large-scale rhythms of earthly development. The acquisition of such features depended, even so, on the perilously unstable 'human situation'. They were for long held back by the difficulties of man's uncertain earthly condition, constantly adapting to changing threats. But just as the emergence of anatomically human forms was made possible by turns in evolutionary history long prior to man's biological appearance, the sketching out of the new enriched human stage belongs in its first potential-for-life to a subsequent large rhythmic division of Earth's history, and new rhythmic disposition of continents following after the Lemurian.

Not just earthly but cosmic-planetary rhythms are involved here, at which we have previously glanced. Steiner's clairvoyant account of these planetary affinities need not nowadays present an insuperable obstacle to scientific understanding: we have already noted that modern science sees in the planetary characteristics a stage of 'individualisation' in the cosmos. These same laws are at work, though nested hierarchically at a different level, in the developments that will bring a next stage in human evolution.[*] Putting it simply, the spiritual prototype of the next phase of humanity was not one but multiform, opening a path to the evolution of a collaborative social way of life. It made possible different characteristic adaptations, each of which realised or brought into a dominant role one set of characteristics or another, in a way that had not previously been possible. The different characteristics of each might then be taken up in some degree into later ones, so that a genuine if rudimentary history arises, enriching each stage with elements still found relevant to the present from the past. Because the emergence of individual characteristics is still bound up with external-biological development, we have seen that it is correct to speak at this juncture of 'races'—so long as we recollect that Steiner saw any regression to this kind of relationship as wholly inappropriate to our own modern period of development. Once, however, it was a necessary and empowering stage.

Steiner ascribes the first indication of these human characteristics, appearing over the evolutionary horizon, as it were, to the time when the Atlantic Basin was occupied by a great land mass. Yet he is clear that there was at that time no human being who could have left a trace in the fossil record. Outwardly, what we observe is the efflorescence of those great reptilian genera which still continue

---

[*] However, Steiner's reference to the connection of the pre-earthly human types with the planets of the solar system does not refer to the planets in their literal modern meaning as material bodies but to qualitative elements in our cosmic system (*Outline of Esoteric Science* (New York 1997) p.221. Any literalistic misunderstandings about Mars-men, Jupiter-men, etc. are a complete misunderstanding. The 'planets' to which Steiner refers are indeed related to our physical solar system through its evolution, in a way which he clarified in his lectures on *Spiritual Beings in the Kingdoms of Nature and the Heavenly Bodies* (Vancouver 1981) esp. pp.95ff.

to furnish the popular idea of prehistory with its amplest sources of fantasy. From Steiner's perspective this non-human proliferation of evolutionary possibilities was at the same time a prerequisite, a necessary precondition for the subsequent working-out of aspects of human evolution, which can be clairvoyantly observed already within the fabric of the world.

These would take their place organically only much later, as the conditions and time of the sauropods (dinosaurs) passed, with its catastrophic conclusion, and the impetus of mammalian evolution took over in its stead. As yet in 'Atlantis' human forms were still, if we will accept the term, 'spiritual'—though at the same time we must think of them, not platonically, but as already undergoing a real history and 'soul-formation'. When external conditions offered the opportunity, these Atlantean human types would be able to shape human development: going beyond the Lemurian humanity which had previously come into being and successfully managed to survive on Earth. The Lemurian type was still essentially to be seen, we suggested, in the first Anatomically Modern Humans whose existence has turned out to be older and more problematic than once seemed the case. The other, Atlantean type can provide us, it may now be added, with a definitive explanation for the evidence concerning that other humanity which appears as the source of other gifts and our advanced consciousness. That is to say, as is becoming increasingly clear, it brings us to the humanity we observe first and foremost in—the Neanderthals.

We shall explore shortly their 'cosmic connection', expressed through music, divination, etc. But Steiner has indicated first of all the fact that this conscious level of human existence was initially and for long ages impossible to realise under earthly conditions, which the 'Lemurian' humanity had adapted to. It remained for those long ages a potential stage for humanity, still written in the cosmos, living in the differentiated qualities which are manifested in planetary relationships. For whereas the Lemurian type succeeded in clinging, however precariously, to earthly existence the Atlantean, which brought the consolidation of human resources through memory, language, etc. was long held back.

In Steiner's rather extensive descriptions of human evolution he is constantly at pains to stress the difficulty human beings experienced incarnating the soul-qualities which belong to man's cosmic affinities. They did not emerge hand-in-hand with anatomical humanness, but quite separately. Here Steiner attained to a point-of-view which has been remarkably confirmed by the recent re-dating of fossil evidence, proving that the 'Moderns' were in fact older than the Neanderthals in time. However, though rooted spiritually in the turn taken by evolution far back in geological time, the Atlantean type he describes certainly must have derived its later physical lineage from the Lemurian-type humanity which was already succeeding, outwardly, in perpetuating itself on the Earth. The human remains we have of course belong to a time-period long after the geological shifts which changed the 'Atlantean' outlook of the continents, with the mass-extinctions which transformed animal life. Rudolf Steiner has pointed, however, to a specific later instance of the interaction between the 'older' type of humanity and the 'younger', more cosmically endowed humans—working on, indeed, into post-Atlantean times; there the new qualitative developments of memory, language and culture play their part. But he makes it clear that the interaction led to a different branch of humanity evolving alongside the original (Anatomically Modern) humanity. A surprisingly exact identification of the events he describes will gradually emerge in our presentation; but our point of departure must be his general insistence that humanity did not, after all, evolve in linear fashion. This was the mistake that in Darwinian form has dogged the whole understanding of prehistory. Rather, a new form of humanity evolved which was shaped to the 'Atlantean', cosmically gifted souls, a type of humanity different in character from those who continued the original earthly line.

When considering this particular evolutionary stage (the question of exact dating must as yet be postponed for detailed later consideration), Steiner remarkably asserted that: 'for a long time the human race continued to consist of these two types' (one might also translate 'two species') 'of human being.'* A different branch of humanity evolved as the

---

* Steiner, *Outline of Esoteric Science* pp.233-4. The term *Menschenart* ('type of human being') must surely be taken as constructed in parallel with '*Tierart*' ('animal species'). A recent translator offered 'human breeds'. In his early writings Rudolf Steiner used terminology which was congenial to the Theosophists and a great many others at the time, speaking of 'root races' (Lemurian, Atlantean etc.) i.e. a diversity greater than that of recognised racial differences but still definitely human. It is unlikely that this way of speaking will appeal nowadays, but it is equally clearly an attempt to distinguish a divergence like that of Neanderthals from Modern Humans, now that we know they were not stages along a single line but a diversity of forms. There is some disagreement currently in scientific terminology, e.g. *Homo neanderthalensis* or *Homo sapiens neanderthalensis* according to theoretical stance.

cosmically orientated souls were increasingly able to take advantage of changing conditions on Earth. Although we must assume an ultimate common lineage, we must stress nonetheless that the new branch was determined by factors quite different from those which shaped the life of the older humans. The latter were essentially southern and African (echoing their archaic Lemurian affinities), and their religious life seems to have been a powerful bonding with the Earth. The Neanderthals in modern perspective, however, fit the picture of Steiner's cosmic orientation, and are in their formation essentially Western and European: no decisively Neanderthal remains have yet been found in continental Africa.* Lemurian humans were inwardly further away from modern assumptions than scientists for a long time wished to believe, in spite of being so intimately related, anatomically and morphologically, to ourselves. Of them Rudolf Steiner gives a remarkably convincing account. He also integrates the second type of humanity into our evolution in a way which makes extraordinary sense of the Neanderthals, whose challenge to our self-image can no longer be ignored.

The 'cosmic orientation' of the Neanderthals can be exemplified from modern research especially in the fields of memory, language and music. These things together will furnish us with a coherent picture of the 'oracular' modality of their consciousness.

Neanderthals: Living with the Dead

Views of the Neanderthals have had to alter profoundly since the old dismissive days, when everything perverse and degenerate was projected upon them. In nothing more significantly can the change be felt than in the admission, now widely held among researchers, that the Neanderthals were the first to bury their dead. The implications of such an act are far-reaching. Claims that they merely 'disposed' of them, or that the evident signs of meaningful symbolic activity around

---

The scientific idea of different human species was of course not directly available to Steiner.

* The equivalent 'race' in the scheme of Scott-Elliot is said to have had its origin in the area now comprising Scotland and Ireland: *Story of Atlantis* p.22. Those regions had, of course, once been part of the Atlantic continent, prior to the opening up of the Atlantic Ocean. In the line of heredity, however, they must have branched off from the Lemurian humanity, as is generally assumed for the Neanderthals.

them were accidental accumulations from seepages, no longer (if the archaeologist will forgive the allusion) hold water. A burial is a deliberate placing and memorialising, the marking of an end and a holding in mind. It may not be evident to the lay reader, however, just how enormous is the inner step involved. We might suppose that beliefs about the dead arose as speculative notions about survival, etc. But the archaic experience of the dead is much more immediate. Burial of their dead is first of all a sign that the Neanderthals had stepped into the dimension of memory and did not just forget one who had died. The person deliberately buried could be found afterwards, so that his or her identity continued to be felt. Much that we know about the Neanderthals indicates that they had recognised the almost infinite consequences for human life.

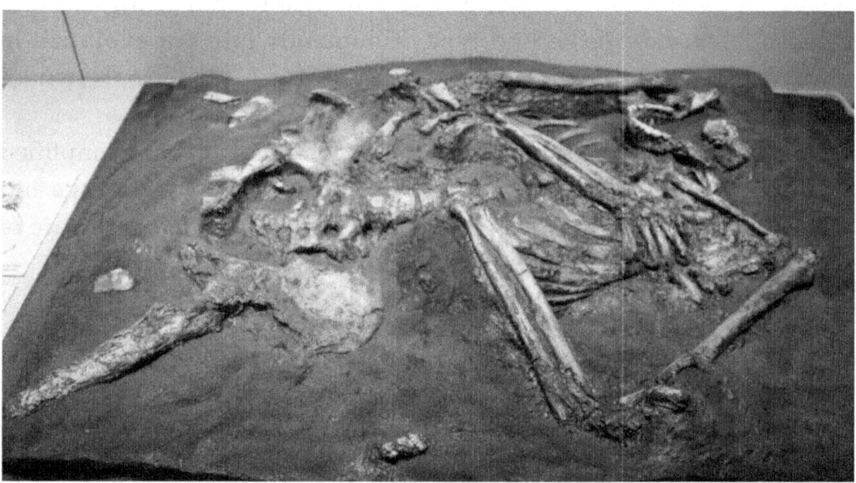

It is widely accepted nowadays that the Neanderthals were the first systematically to bury their dead. Signs of ritual behaviour such as scattering flowers sometimes occur. A grave is a memorial, perpetuating the presence of the dead which is an essential step in human society, since cultural forms are then continued in their name. Religious memory probably took the form (as Rudolf Steiner suggests) of vivid localised presence, in which the person's deeds and thoughts had continuing reality. Many archaic cultures still experience their forebears within the landscape—and memorials make a first map/ record connecting the scenes of the group's history. Neanderthal burial (cast), Israel Museum, Jerusalem.

Burying someone is an act of memory—of not letting that person disappear from the flux of life's events. We noted previously that Rudolf Steiner insists upon the precise definition of memory (distinct from mere absorption of past experiences) as the ability to place a person or event in the past. The decision to bury a dead individual requires just that: it is not merely saying that the affects of that person linger on in the group, but categorically that he or she can be thought of as someone who was part of the human group but is no longer so. And yet being no longer present, he or she can still be contemplated, acknowledged, as the one who is memorialised.

It is worth imagining for a moment, in detail, the process by which something so revolutionary as memory could have come into the range of human experience. It is all too easy just to think that people woke up one day with the ability to remember past people and acts. We so readily take things for granted, but obviously, a little reflection tells us it cannot have been so. There is a leap in the underlying psychology—a radical development from elementary forms. One may say that burying someone enacts objectively the process of leaving someone behind in the past, by giving them a place to which their family or others who knew them—or (later) who can be told about them—can come back. The memorial establishes their relative or hero as belonging to the past, who therefore must be grasped as a spiritual identity which can be acknowledged independently of his literal presence. Human beings who bury and remember their dead inhabit a conscious world which is not dependent on physical contact but constitutes a spiritual reality of other beings, past and present. However, in its character as memory it remains on the same level as physical interactions—making it all the more immediate and direct, so that the dead are as vividly real as the living. Thus it would be wholly wrong to suppose that their continuance would be a speculation or an abstract idea. Yet memory-awareness of the dead is crucial to transcending the limits of the present, and of one person's experience: truly a spiritual awakening for the living. Continuing attachment to the dead is almost certainly the origin of pastness, and opened up a new level of spiritual existence. Naturally it was also a new level of self-awareness.

The well-known Neanderthal 'flower burial' of Shanidar cave in northern Iraq revolutionised the picture of these long-disparaged people. Large quantities of pollen found in the burial made it clear that the individual had been laid to rest with elaborate ceremony, and the fact that analysis of the flowers concerned suggested medicinal usage indicated to many researchers that he may have been a medicine-man. But perhaps flower-culture was simply highly elaborated among his people. At any rate, they placed the deceased individual in a cave, centrally positioned in such a way that the site clearly became a memorial place, not to be used again for any other purpose.* Thus this burial is a memorialising, and the buried person is one who can be found again. Perhaps originally only a few, outstanding individuals were granted such status. Many writers on the subject of prehistory still ransack the evidence for 'ideas' of the afterlife, or 'survival', but that is to import our own questing, uncertain consciousness into an earlier scene: rather, the buried dead *are* the first memories of the tribe, they are themselves the ideas of the afterlife, and their continued reality is as objective as the place where they can still be found and so where the memory lives.

Revisiting the place where someone is buried still carries intense meaning for many people today. At that spot it is possible to remember someone in a quite special way, a way which bears the hallmark of a primary phenomenon in the inner life. (But we must not be tempted by any foolish reductive tendency to suppose that the notion of 'survival' is a mistaken idea whose reality is the dead body preserved in the tomb. For it is to the spiritual reality of the person that the phenomenon bears testimony. We find there the active essence of the person remembered, not a mere left-over. Physically, they have been left behind, as the tomb shows, and it is on the spiritual level that we still relate to them. If we (Moderns) want to fashion an argument, it must surely be based on the primary factor at work here, which is that in human life our most significant experience can be shaped and determined by presences not empirically present. Originally they were unforgotten people who first inhabited a 'past', or in subsequent, more abstract scenarios, 'ideas' which likewise extend beyond the mere physical situation to which they relate. The dead were the first *ideas* of mankind. Even today the more intense cases of remembering still reveal their original form. And finding one's way back to a past as a revisiting of a memorial site was for long the human mode of remembering. Only gradually has the encounter with another world in memory been overlaid by the abstract line of time upon which we hang our experiences out to dry, i.e. when we relate everything primarily to our self. The latter is a form of inner organisation

---

* Leakey, *Origin of Humankind* (London 1995) pp.198-9 comments on the poignancy of this particular burial. For more details see Trinkaus and Shipman, *The Neandertals* (London 1994) pp.336-41. The pollen-analysis was undertaken by Arlette Leroi-Gourhan, and rapidly led to the popularising of the 'flower people' in the 1960s and 1970s. For details of other Neanderthal burials, see Trinkaus and Shipman pp.176-8, 186-9, 254-8, etc.

which could originate only in a much more ego-centric age, in fairly close proximity to our own.

Ancient remembering, even for the great civilisations of antiquity, still had the quasi-spatial character of a journey.* It took people out of themselves, i.e. out of their perceived present, to come face-to-face once more with an objective realm of experience. Later filtering preserved only subjective aspects—shrinking our experience from older modalities as we centre our modern-day focus on the ego alone. Formerly memory truly opened into another country, another yet still continuing reality. If, as it appears, the Neanderthals were the first to stand at that ocean's edge of the mind, one might well say a whole new spiritual dimension was discovered for humanity.

Rudolf Steiner is one who above all realised that to get back to the roots of human experience we have to deconstruct our more modern mental life which stands in its way, in which we conceive our self as the sole theatre of inner processes. To explain how we have reached that point is exactly the issue! The realisation that 'remembering something' was originally a journey back to the place where an event happened enabled Steiner to divine an essential aspect of early human consciousness. It was not subjectivity in our sense, but placed everything in human life on the level of cosmic reality. The head, says Steiner, was not a protected space of 'my own' thoughts, but was a part of the world alongside everything else. It is we who have left behind the original archaic experience of humanity, though it has been long echoed especially in Oriental thought:

> Through this fact we are able to have a conceptual memory, we are able to remember things through thinking. We form pictures in thought of what we have experienced as abstract memories in our head. That could not have been done by a man of ancient times

---

* In primordial historical times, such as those of the Sumerian-Babylonian *Epic of Gilgamesh*, archaic modes of experience still shine through. Thus the events of the Flood and its Noah-figure have been long left behind at the time of the hero, Gilgamesh, King of Uruk. But if he travels far enough away, at the edge of the world, we learn, he can arrive back at those events and consult the venerable Noah-figure Utnapishtim in person. The story constitutes the XI Tablet in N.K. Sandars, *The Epic of Gilgamesh* (Harmondsworth 1960). It is striking that in recent scientific thought the interrelationship of space and time is again taken into account. 'Linear' time is no given reality but an abstraction many times sifted from our primal involvement with the world we experience.

This apparently empty courtyard or theatre is shown blank to make room for the crowd of memory images which can be mentally placed there, so as to be found when needed by a speaker or a sage. The technique is a vestige of a type of memory-experience going back to prehistory—cultivated also in classical times and as the art of memoria *technique at the Renaissance by John Dee and others. Robert Fludd, de* praeternaturali *(1621).*

who did not have thought, but still had his head. He could not form memory pictures. But in those archaic ... regions where people were still conscious of their head, but had as yet no thoughts and hence no memories, we find something else developed to a remarkable degree .... One would see on every side small posts placed in the earth, or here and there a sign.... Such memorials were to be found scattered over all inhabited areas. Whenever anything happened, a man would set up some kind of memorial, and when he came back to the place, he lived through the event over again in the memorial he had made.[*]

Thought has given us the freedom to organise our inner life, and to remember selectively in accordance with our interests and ideas; but

____

[*] Steiner, *World History in the Light of Anthroposophy* (London 1977) p.15. Deconstruction of modern consciousness, pp.7ff.

we must think of archaic memory, in the absence of modern-style ego-representation, as a more unconscious and compulsive power. The drive to revisit the site of some important event or, especially, perhaps, of the memorial of a dominant leading figure who had been buried, would have been an urgent but not very lucid pressure to go back. (Anecdotally, I recall that as a young animal-spotter I was one day startled to come upon a basking snake and felt for days afterwards a hard-to-resist compulsion to return to that same spot, based on the quite irrational conviction that the snake would be there again.) Only gradually has freedom within our inner life been gained. Steiner characterises a subsequent stage of 'rhythmic memory', and only then of conceptual memory which essentially coincided with the developments that brought about archaic and classical Greek culture.* But if we want to understand prehistoric human experience we must avoid that fatal mistake of ransacking prehistory only for features of intelligence, or technical development which remind us of ourselves, and enter into the life of people who did not yet think as we do but whose inner life is the strangely unfamiliar prototype which turned slowly into our own. Memory is one instance.

Another significant, related case is language. Neanderthals it is now agreed certainly had language of some sort. We may be certain of this from their ability to co-operate extensively, which would not be possible without speech. But a recent researcher who has acquired a deep empathy for early humans at various stages suggests that Neanderthal language would have been lacking a feature which all linguistic analysts recognise is the latest modality to emerge. That is the ability to refer to the future.† And what we nowadays ordinarily mean by 'thinking' is more or less exactly that—moving from what has happened to what is going to happen next. Neanderthal language, however, would have been purely an articulation of memory, able to refer solely to the present and the things of the past. We can observe this from their Mousterian culture and artefacts, which were clearly based on repeating a formula from the past as a basic feature.

---

* Steiner, *World History* pp.16-18. The 'rhythmic' stage of memory is attested in all the ancient and modern archaic cultures where mythic events which define their way of life are re-enacted or retold in seasonal or other cyclic forms, such as the Egyptian or Mesopotamian cultures of the Middle East.

† Robert Whallon's conclusion, cited in Chris Stringer and Clive Gamble, *In Search of the Neanderthals* (London 1993) p.217.

A picture is starting to emerge. The compulsion to experience an event again, or return to the place where a person died, must have been the strong force which impelled people first of all to bury people and make a memorial. Such is the first act of memory in Steiner's strong sense—being aware of the past. Remembering what had been done in that way was a powerful basis for repetition, in handiwork or in language. It also brought an ongoing relationship to place. The past reality of a place had as its first content the doings of the person who had lived or at least was buried there. The consciousness of the identity between the past of a place and the buried/remembered person, who remains a defining presence, as an archetypal human experience, leads directly to the idea of the dead living on in the landscape, or under the Earth we tread, and of having shaped the world in which later generations live. Eliade describes how already in very archaic religions like that of the Australian aborigines, sacred knowledge 'is equivalent to learning what *happened* in the primordial Time .... The sacred and secret lore now depends on the mythical Ancestors, no longer on the Gods. It was the mythical Ancestors who lived the primordial drama which established the world in its present form.'[*] Place became the setting for imitative human actions guided by the ancestors. This is not a religious *supposition* when we find it among many archaic peoples, but an immediate reality of consciousness in what Steiner calls the 'pre-religious' time.

It was the Neanderthals who first show signs of this development, after the 'sensing the gods' of the older, Lemurian human type. But it still belongs to the 'given' of early human consciousness in its development, not reflective thought. Since it was as yet untrammelled by the filters and constrictions of advanced ego-awareness, it would have approached more to what we might call psychometry (where events and circumstances can be actually perceived by a psychic in connection with a place or thing). And Steiner affirms that it is the authentic way of conceiving the spiritual presence of the dead and their relation to the living.[†]

---

[*] Mircea Eliade, *Rites and Symbols of Initiation* (New York 1958) p.40.

[†] For the concept of the pre-religious stage of consciousness, cf. Steiner, *The Apocalypse of St. John* (London 1958) pp.16-17: 'We must look back into times of the prehistoric past, the pre-religious age of mankind, in order to envisage the origin of religion. Is there a pre-religious age of humanity? Yes, a time existed when there was no religion; this is acknowledged by spiritual science though in a very different sense from the assertions of materialism .... Times

Developmental researchers in child-psychology independently of Steiner's approach have observed that their young subjects quite spontaneously find their way to a strikingly similar way of thought. Children relate the features of their environment, such as mountains and lakes, to prior human beings who lived there and made it for them, such as their parents. Their psychologically consistent constructions can be proved to be independent of 'misunderstandings' based on what they have been taught about human industry, etc. As a result, researchers have noted, even an educated child is not usually inclined at first to take up formal religious notions of creation, but 'spontaneously

---

in which humanity had a inkling rather than a definite vision ... of the spiritual reality around him—these are the religious ages of mankind. And before these ages were others when man did not need such a sense of aspiring to the supersensible spiritual world, because he knew that world as today he knows the things of the sense-world.' For the dead as a presence in the landscape, Steiner, *The Inner Nature of Man between Death and Rebirth* (London 1959) gives an in-depth account which we can only indicate here. 'Up to his death, man ... has become accustomed to standing on the solid, material earth and to see upon it beings of the mineral, plant and animal kingdoms' like a sphere around him (p.80). But after death that sphere of consciousness is seen *from outside*, and the dead person is now within the world around. There he is a creative presence, whereas after death we can no longer work transformatively on ourselves. 'When man is born physically, he confronts an incomprehensible outer world' but cannot directly see himself, whereas 'after death, after birth into the spiritual world ... he begins to be in the environment, to which he himself has given birth, has himself built up around him. He has given birth to the world, whereas when he is born into the physical, the world has given birth to him' (p.85). Ultimately, for Steiner, this is a karmic process preparing conditions for a future life—one of the ways in which humanity is working to transform the Earth itself. The kind of closed 'sphere' of external perception we nowadays have during life has itself evolved fairly recently, so that a less ego-centred ancient humanity was naturally much closer to the dead. Clairvoyant perception on Steiner's model, as we have seen, is to be understood as part of the evolution of our consciousness, rather than as a special faculty or 'gift'. In this context it is interesting to note that actors or speakers who use imagined 'theatres' from which they can recover items of information or speeches can thereby remember much more than is accessible to linear memory, which would soon be overloaded. Thus *memoria technica*, which is an artificial continuation of the old local memory into modern mentality, shows itself to be an *extension of consciousness*, still pointing to its clairvoyant origin. On the fascinating history of the technique, cf. Frances Yates, *The Art of Memory* (Harmondsworth 1969). For memories, especially formative childhood memories, in the landscape, cf. Steiner, *Mystery Knowledge and Mystery Centres* (London 1997) pp.56-9.

attributes to his parents the perfections and attributes he will later transfer to God ... it is, therefore, man who is thought to be omniscient and all-powerful, and it is he who has created all things'.[*] It has also been established that young children regard learning in a distinctive way, different from adults: they describe it always as remembering what they already knew.

The dominance of memory, rather than the thinking we know today, is taken by Steiner as the key to the consciousness of the Atlantean human type, implanting the faculty later elaborated through the successive stages of local, rhythmic and linear consciousness we have mentioned. And he connects it directly with the emergence of a cult of the ancestors, just as we find implied by the Neanderthal burials that clearly bear the meaning of memorials. There is evidence too that Neanderthals carried their dead around on their migrations, indicating their importance to the continuing identity of the group. 'The memory, the remembrance of the ancestors or those who had acquired status during life, developed', says Steiner. 'From this there emerged among some tribes a kind of religious veneration of the deceased, a sort of ancestor cult.' From it also derived the first hierarchical structuring of society, he says, since anyone who could connect themselves with the remembered deeds of a great one strove in turn to copy them and participated in the kudos of the awe-inspiring deceased. In the earliest periods, a successor's deeds were only felt to be significant in so far they chimed with the exploits of the remembered forerunner. To gain recognition such a one continually 'had to demonstrate by new deeds that he still possessed his old power. He had to recall the old works to memory by means of new ones. What was done was not esteemed for its own sake.'[†] The leader who could evoke the charisma of the past by his acts, however, gained a sort of 'regal status'. We noted that charismatic figures like this were probably those who were first to obtain memorials, like the burial at Shanidar.

---

[*] Jean Piaget, *The Child's Conception of the World* (St. Albans 1973) p.400 and generally pp.396ff pulling together the evidence of the book. Piaget's reconstruction of the formation of the idea is valuable but still has a somewhat artificial feel, based on purely hypothetical 'confusions'; is not Steiner's acceptance of the primary truth of the child's thought also intellectually the more cogent? It is our consciousness which needs explaining, is it not—on the basis of the child's prior form?

[†] Steiner, *Cosmic Memory* p.52.

Those who could not aspire to leadership were nevertheless drawn into the web of memories, and felt their communal identity in the wake of the events. 'Previously the formation of groups depended wholly upon natural forces, upon common descent. Man did not add anything through his own mind to what nature had made of him. Now a powerful personality recruited a number of people for a joint undertaking, and the memory of this joint action formed a social group.'* He did so in the name of a venerated and semi-divine progenitor. Such is the power of memory, and the ancestor cult which played such an enormous part in spiritual history. Such primal memories were not yet the property of individuals—memory consisted firstly of the deeds of the deceased originator, or gradually of the accumulated layers of ancestors who repeated and augmented his acts. Only very much later did 'linear' memories serve to construct our personal life. Memories are originally the groundwork of community. In fact, recognition of the many dimensions of human life which are activated around the apparently simple act of memorialising a past figure can furnish us with a key to Steiner's crucially illuminating conception of an oracular culture—the central idea in his portrayal of Atlantean humanity. Much that we know of the Neanderthals will thereby also stand out in bold relief.

In Rudolf Steiner's earliest researches in prehistory, he speaks conventionally of the spiritual 'mysteries' which played the role in Atlantis of guiding in human lives; but very quickly he adopted a more specific language, and referred to them rather as 'oracles'. Even in more recent antiquity, including Greek times, mysteries and oracles have been related but distinct phenomena. In a mystery, we have to do with cultic actions which transform ('mould') the elements of human nature in accordance with a spiritual ideal: in mystery-rites, the transforming presence of a spiritual power is directly perceived and sets those who have experienced it apart from the ordinary population, the 'profane'. Often, they have worked on themselves intensively in preparation. When they address their fellow-men, it is as teachers from a higher, divine perspective. They recognise others on the same level as forming a 'brotherhood'. In an oracle, on the other hand, a god communicates with human seekers who have come to a sacred locality where special qualities can be found. Often such a place suggests the confluence of elemental powers. It is specific needs or some personal crisis which brings them, not the wish to stay and form a higher group. The divinity

---

* Steiner, *Cosmic Memory* p.53.

solves a problem or answers questions which puzzle mortals, but the knowledge imparted is for essentially practical use, offering problem resolution, cure from disease, divination of battle outcomes, etc. Then it sends the seeker back into the thick of life.* One who has been favoured by an oracle is raised to leadership in an immediate practical way—but without a supporting group, and without learning lessons (for the god will speak again if needed). All that is necessary is to remember where to go for answers.

Oracles produce people with an answer to a situation, and others who will follow as directed—a primitive but decisive kind of leadership. The relatively small social groups documented among the Neanderthals would be the result. Mysteries created a special group of people who carried a higher-divine awareness, which would develop society in light of a higher ideal. Oracles were directed at a single social situation and a human need, although they did need an exertion of human individual effort to respond to them. Indeed a human intermediary is needed to make human sense of the god's utterance, for as Chaucer inimitably put it, 'goddes speken in amphibologyes'. They cannot be readily understood in the normal way. The utterance of a god expands human meaning to reveal a richer, god-discerned meaning that unlocks the situation when once we see what is implied. Thus it is finally for a human interpreter to realise what the god is trying to get through to him, and without the ability to spot the relevance of divine insight the oracular meaning is frittered away in ambiguities.† But if it is conveyed to the right person, with the specific problem that is addressed, then *heureka!*

Oracular knowledge, as Steiner saw, belongs to the earliest type of spiritual experience of humanity, and this concept can bring together much that we have already touched on above. The act of memorialising a leader or an ancestor meant that human beings entered into a world which transcended the flux of present life. The memories of the things that had

---

* Thus an ancient town or community or an individual would send a representative to an oracle, such as Apollo's at Delphi, when it had a problem; the Mysteries, however, brought new initiatives and led developments with divine impulsion, rather than waiting to be asked. A human question could be taken to an oracle; the mystery-god would call those whom chose to be his own and impart their mission.

† Sybils were oracular interpreters in classical Greece, and read the divine meaning in patterns of scattered leaves; in archaic China, sticks were thrown and the patterns used to divine the significance of the 'moment' (*I Ching*). But the wind scattered the 'Sibylline leaves' once the moment had passed. The sticks interpreted one significant moment.

marked out such a figure were the first 'ideas' which humans took up, and the place where the memorial lay was thereby made a gateway into a world of higher significance. The history of the place was vividly, clairvoyantly experienced, and it redefined the present reality to which seeker or returning descendant joined it by his pilgrimage to the now 'sacred' spot. But the experience did not lead to the bestowing of special religious insight (or teachings, a much later idea), so much as to the empowering of a new individual who himself took up the definitive acts of the remembered one, the ancestor or divinity, sending him or her back into the community to resolve its problems and tensions as an empowered, god-deputed figure.

The compelling power which drew people back to these memorial sites, or originally to bury their dead there, is the primal need we have recognised—the struggle to remember. It can still be felt in the wish to go back to the burial places of people who have been important to us, or to where we remember special events taking place. Its mystique is compounded of the quality of the place, and of the person—originally there could scarcely be any difference. The dead person was now at one with the natural forces which shaped the locality, which would not be distinguished from the human activities there. Oracles of the dead as such are known, but in an oracle a cosmic power spoke (an earthly, or a sky-divinity) for which there was perhaps also an ancestor as human intermediary, and a living enactor of its message. At the moment of revelation such a one was fused with the presence of the god and was equally divine. The ancestor was one with the shaping power in the sacred spot. Under quite other conditions, young children still spontaneously live in terms of oracular power; they feel the power to give meaning to their world as radiating from their parents, or other impressive persons, and feel that they 'remember' what they learn from them. For much of human history and in many societies, deceased ancestral figures have loomed even larger as authoritative models. But in all this, it is earthly life which receives spiritual clarification and ordering—oracles very rarely addressed transcendent goals or new beginnings. The living memories of prehistoric humanity gave them a divinely sanctioned pattern of life, long before direction could be given by formal 'ideas'. Already in Neanderthal times, we must reckon that a memorial site like that of the flower-burial must have been an oracular centre around which subsequent life could be organised, based on the continuing presence of the dead (researchers often use the term 'shaman').

Pundits often make play with the fact that in all supposed communications with the dead, nothing of outstanding brilliance or originality tends to be recorded (as though entering into the afterlife should give you a degree in particle physics or philosophy). But according to Steiner, the dead play exactly the role in human life which early humanity intimated. If we make ourselves sensitive to the presence of the departed, he asserts, we will hear them prompting everyday impulses and ideas. It is in the web of ordinary life that they quietly infiltrate the suggestions which actually determine our lives.[*] Oracles were a way of opening up such awareness. Latter-day individualism has taught us to connect thoughts only with our own self. Many surviving archaic cultures, however, retain a sense that their dead remain amongst them on a day-to-day basis, and prompt the decisions of the community or the family.[†] In many ancient cultures it was common to sleep near the resting-place of a departed relative if one wanted to obtain an oracular dream for guidance. In the earliest prehistory when the dead were given burial, and lived on in the consciousness of the Neanderthals, we must also I think conclude that the immediate presence of the dead enabled the fundamental steps not only of memory-formation but even of society itself.

One who could appropriate the deeds of a past figure, one who had become oracular gained a like authority, and so was able to establish a pattern for future generations. Let us imagine such a person among his followers, and ask what significance that might have at the point in evolution dividing animal and familial groups from *society*. Already in an animal setting, of course, it repeatedly happens that patterns of behaviour which give an advantage to a group can emerge. Members of the group imitate them and thrive accordingly. Evolutionists have frequently been tempted to extrapolate thence a notion that here we can find prototypes of 'culture': they would be transitional between animal and human life. But the model soon collapses: for even among intelligent primates the variant activity can hardly survive from one generation to the next, and there is a return to general instinctive behaviour. That is for the obvious reason that there are

---

[*] Steiner, *The Dead are With Us* (London 2006) pp. 15ff.

[†] Arthur Grimble in his well-known reminiscences of life among the Pacific islanders gives some particularly vivid accounts of the homely relations between the living and the dead: e.g. *A Pattern of Islands* (London 1952) pp.154ff. Handling the skull of an ancestor, for instance, involved a gift-offering and an intimate loving conversation (p.155); the helping presence of the ancestors in everyday activities, pp.156ff ('there was not a cranny of a man's or woman's life into which it did not penetrate', p.160); etc. Grimble later became an academic authority on Gilbertese culture. See also pp.189ff; 242ff; 262ff.

no means of transmission of acquired practices, so as to hand them on to future members of the group. Equally, however, the rise of human 'society' cannot have waited upon the development of abstract thought of the complexity needed to specify interactions and choices such as are needed to perpetuate even the most rudimentary culture. But the remembered dead are the crucial transition—part of what was meant by saying that the dead were the first 'ideas' of humanity.

The oracular dead who were celebrated for their life-experience and achievement did not hand their wisdom down in precepts or even in heroic narratives like the later epics (requiring elaborate techniques of transmission). In prehistoric times, being in touch with the dead and their wisdom meant that as living memories the dead continued to be present and themselves perpetuated the insights and actions they had attained, mingling in the following generations and enriching them with guidance and instruction. The experienced presence of the dead is the transition from animal group to human society. Only in that way could the organising of social relations accumulate a knowledge of the values and decisions which had established its cultural uniqueness, which need no longer be lost in a fall-back upon common instinct. Those who shared in the wisdom of the dead transmitted the practical content and the authority to maintain social structures and involve or enforce participation by others. Human achievements were incorporated into the success of the social unit through the presence of the dead.

We are all constrained by the thought of our time. Many today will urge that prehistoric humanity (and historic humanity too!) merely had the illusion that the dead moved among them and made the continuity of society a living reality. They imagined or vividly remembered them. Yet we must concede that without that illusion there would not have been that consolidation of human experience we call social life. Moreover, those faculties of imagining and abstractly remembering come later, not as a first stage: it is we Moderns who have overlaid past presences with our ego so that they become mere shadows, who have put imagination at the service of merely personal fantasy. In actuality, we could not have these facilities either without the primal stage of recognising the reality of the past and its bearing on our present in the full-blooded form which ancient and archaic cultures knew and relied on.

It was long the convention that Neanderthal culture should be interpreted minimalistically. But as perspectives begin to clear, their ability to organise complex seasonal patterns of living, and to show signs of a more structured social organisation dissipates old romantic notions of prehistory such as the 'democratic' hunting fraternity. The most recent research on the Neanderthals points rather to 'sophisticated economic and social organization that included detailed planning of hunting activities at specific times of the year', convincing Clive Finlayson 'that they were our cognitive equals and that they were not'—as the older view maintained—'replaced by cognitively superior humans'.* They were just as much pioneers of the human. There was clearly leadership, familiarity with patterns derived from the past, and a recognition of hierarchical social roles—all of which Steiner convincingly derived from the Atlantean mentality. All these things are in the gift of the dead. Gradually, ancestral voices turned into thinking, consolidating and socialising past experience. Decisive signs once as divinatory puzzles turned into connections of understanding. As the ritualised burials of the Neanderthals show in essential symbolism, theirs was a life lived out in the light of the past—in the great theatre of memory called human society, where these things all occurred.

Memory and Music

We are having to accustom ourselves to the idea that human prehistory was not the story of a single trajectory. The Neanderthal line was elaborating cultural life (memory, economic organisation, symbolism) already in the Middle Palaeolithic, whilst the predecessors of Modern Anatomical Humans as yet showed no signs of that Upper Palaeolithic awakening. Rudolf Steiner's story of two human types, ultimately coming together into one, has become dramatically relevant. So too have his indications concerning thought and speech. There too a revised and novel narrative is starting to take shape. Language was once treated as a closed domain which evolved from simple grunts to more complex speech, all with a severely practical aim; in reality, it now fits into a richer, more surprising account—revolving more around music than Darwinian achievements.

The standard approach until recently contrasted music and speech as black with white: music was of no cognitive significance, and in many researchers' minds of no real significance at all, just an entertaining

---

* Clive Finlayson, *The Smart Neanderthal* (2019) pp.195-6. Compare this recent view with the more grudging recognition from Clive Gamble and Chris Stringer, *In Search of the Neanderthals* pp.216-8.

waste of time; whereas speech was meaningfulness and collaboration, the cradle of thought, and developing it was a central requirement of human culture. This dualistic view continued to be held tenaciously even when the Darwinian knot tying thought-advances closely to technological (hunting) skill began to fray apart. Recently, however, Steven Mithen has brilliantly challenged the scientific status quo with a far-ranging study of the role of music in prehistoric culture, and in particular the significance of 'the singing Neanderthals'.[*] Music, he argues, is far too important and integral to every known human society to be merely a diversion, and has far too much in common or overlapping with speech to be without significance for the rise of language. Dancing and singing are primary human activities, without which society is almost unimaginable, and in particular with a range from holy to incidental, from heartfelt to gymnastic, which unifies much of cultural life. Music's central role demands recognition, and its prehistory cannot be unimportant.

His work resonates in many respects with Steiner's conclusions about the Atlantean focus in human development: Steiner emphatically associates this line with music and music with the emergence of language (i.e. the use of uttered sounds to convey meaning). He indicates specifically that language went through a phase when it was essentially singing before it became the purely spoken medium we now know.[†] This in essence is also Mithen's conclusion; but Steiner is able to show how it evolves on the basis of the Atlantean consciousness and its defining feature of *memory*. For music and memory are connected, and indeed come extraordinarily close to being ultimately identical. Thus memory is a link both to music and to thought: after all, popular speech talks of remembering a tune as *knowing* the music.

In the Atlantean mentality, we have seen, memories do not stimulate the working-out of further connections, pointing to future outcomes. With their inner resources they could not yet extrapolate from an experience a 'thought' in that sense. The re-enlivening of the significant past in memory was, however, a vivid, clairvoyant evocation and could be taken up and imitated—effectively re-lived. This could be by revisiting the oracular site where a person or a memory-experience had been memorialised. However, though early humans could not yet

---

[*] Steven Mithen, *The Singing Neanderthals* (London 2006) esp. pp.221ff.

[†] See for instance Steiner's lectures, *Art in the Light of Mystery-Wisdom* (London 1970) pp. 79ff.

'think' about a memory, a qualitative response could be formulated, and expressed: musically. Such a musical response may be considered a step in the direction of thought-memory, as we shall see.

A *melos*, whether a tune or other grouping of sounds, is basically a combination of sounds heard or made that together produce a distinct impression; and making or especially remembering (repeating) them is a musical experience. Indeed it is repetition which establishes the sequence as *meant*. We are most familiar with music in linear combinations (i.e. melody) but other idioms are possible, and indeed melody is rarely heard without other musical effects being added (percussion, chords etc.). Although the sounds may initially be taken from, or in some way suggested by Nature, they are music only when taken over and enjoyed in themselves. Therefore *melos* has no existence, essentially, except as memory—taking for granted the lack until historic times of artificial methods of notating or recording sounds. When taken up as music, it becomes something that can be re-run through being sung, played or just recollected. In a social setting, the singing or playing of the tune will normally bring a surge of recognition that often spreads like a ripple-effect through others who hear. (When a popular singer comes to a well-known item, this effect is particularly striking.) Even if it copies something from Nature, a melody can never be found again except by recollection/recognition, but when it is enunciated it asserts itself powerfully and, unless sophisticated conventions suppress the urge, brings with it a powerful urge to join in. Still more, it brings an urge to move to the music in the audience, and for attention to become absorbed in it. A definite emotional resonance is also liable to spread through the listening group. Music is distinctive for the way it engages us on so many levels at once.

Our modern psychologies of memory have probably given far too much attention to pictorial imagining—an emphasis bequeathed to them by the philosophers' eighteenth-century fascination with the prototype camera, and continued in misguided comparisons with video-recording today. Yet a very little reflection will assure us that remembering things is not at all like watching a film of past happenings. Recalling what we did last summer is not like sifting through a file of old photographs, re-living each scene. It is more like when we are asked to sing or whistle a melody, and, as the phrase is, 'knowing how it goes'. That does not mean that we sing out the tune, or even that it is actually played out fully in our heads, any more than we need summon up any actual mental picture of that disastrous holiday to 'remember what happened'.

Being able to get back a tune is a rather pure instance of memory in action, and it may then readily be accompanied by images if we allow them. Music often vividly conjures up outward associations of landscape, or the ballroom, or other settings or private suggestions of the hearer's own. To those who know it, just to take one instance, Sibelius' music easily prompts ideas of bleak northern forests and mountains and frosty air. Interestingly, the music effectively conjures images of the landscape, but not the reverse: visiting mountains does not, so I find, as compellingly summon up the well-known music.

Being able to recall a tune and knowing how to continue it is an act which reveals the essence of remembering, and reveals the way that particular images and specifics can be carried along with it.* The aspect of recognition is an overture to conceptualisation ('Ah, that's what he's singing'), but the musical experience remains firmly orientated toward the past and does not invite anyone to continue the tune in a different way—that would just spoil the melody. Contrastingly, with a conceptual memory or a language-based account of an event, reviving it frequently prompts further comments or continuations. Note too that if one hears a spoken description, it is most likely that it will leave an impression of the event, which may then be discussed in other words, whereas with a musical impression we reproduce the medium, and repeat the melody note-for-note.

Music has in common with language, however, that it is based on inner connections, and so has to be grasped as a complete system or not at all. To hear even one melodic interval (e.g. a rising fourth) as music is to experience that interval as implicitly part of a set of tonal/modal relationships which make a system, however simple (or complex). Similarly language requires the grasping of every unit and inflection as part of a whole, which can become ever more elaborate. Both language and music have to be learned, for the most part unconsciously, during those earliest creative years. Music is encountered first, perhaps, in mother's crooning over her baby, which

---

* The association of music with memory is anything but a novel idea, of course— but perhaps the oracular connection will also need to become clear. The chief oracular god of the classical Greek world, Apollo, was at the same time 'leader of the Muses' (*mousagetes*), goddesses who gave their name to music. The goddesses were mythologically the daughters of Memory (*Mnemosyne*). Whereas we moderns tend to contrast fixed *memory* and free *inspiration*, the ancients thought otherwise. Homer derives his poetry from the Muse or Muses, but not in our sense of liberated language: rather the Muses 'enlarge the recollection of the poet' by bringing before him the great events and numberless heroes of the legendary past; cf. Bruno Snell, *The Discovery of the Mind* (New York 1982) pp.136ff (p.137) and see further below.

already also hints at language; but neither speech nor music can be acquired by external imitation. Linguists have now long recognised that from very early on a child creates its own statements, and is never merely copying adult speech. Even before that, children can creatively take up the elements of music.[*] Children are once again drawing upon those same resources mentioned by Rudolf Steiner (above, pp. 173ff), which earliest humanity already used to shape their faculties of expression and understanding.

If music is so intimately related to memory, we should expect a mode of human life which was fundamentally memory-based to have a deep relation to music. And as far as the Neanderthals are concerned, we would almost certainly be making a correct supposition. Indeed if we correlate what we know of them with Steiner's account we may say that they evinced what we may term a musical consciousness.

Ironically, Mithen himself is unconvinced by the evidence of the bone flute found some twenty-five years ago at a Neanderthal site in Slovenia. Yet many others find it hard not to recognise in the pierced holes and signs of further modification at the end of the fragmentary object a viable musical instrument (replicas of which have been successfully played by musicologists).[†]

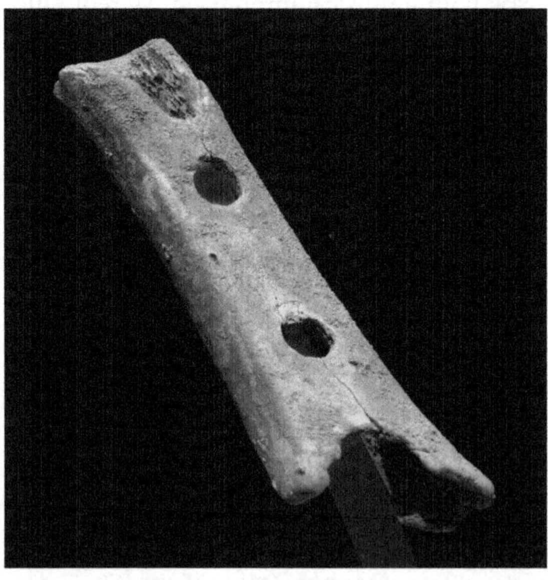

Some researchers believe that Neanderthal consciousness was not focussed on images, like the Modern Humans of the Upper Palaeolithic, but was deeply musical. The evidence from this apparent bone flute, though of undoubted Neanderthal provenance, remains controversial however. Most telling in its favour, perhaps, is the fact that musicologists have been able to reconstruct a playable instrument based on the surviving fragment. Bone flute, Divje Babe, Slovenia

---

[*] Mithen, *Singing Neanderthals* pp.69-70 gives the arguments for the priority of musical structures in infant development, and concludes that the cognitive 'networks for language are built upon or replicate those of music', not the other way around.

[†] Mithen, op. cit., pp.243-4.

And yet, after decades of work to understand the prehistoric mind he has come to the conclusion that music provides the long-awaited answer to the riddle of the Neanderthals. Their culture remained inscrutable, and therefore liable to be put down in favour of the later *Homo sapiens* triumph, because they left so few material signs (figures, symbols, etc.) to show their mindset. Why did their sophisticated method of chipping flint tools never lead them to explore sculpture of other kinds? Now he has proposed that we have missed their significance, because they did not express themselves through objects, or images, but through sounds—through music. And sounds, of course, leave no archaeological trace. Their achievements in social organisation, cultural stability and long-term survival can all be put in a new context and become brilliantly revealing, once we grasp that their lifestyle and spirituality was propelled by their musical activity. And is it so improbable, in that case, to suppose that amongst the non-representational, enigmatic objects they might produce would be a musical instrument?

The lack of visual artefacts means only that Neanderthal interaction and co-ordination were governed not by pictorial symbols (clothes, jewels, power-signifiers) but by the power of musical expression. Mithen sketches for us a wonderful picture of a Neanderthal life which used to the full all the resources of those larger-than-modern brains:

> The Neanderthals used their brains for a sophisticated communication system that was Holistic, manipulative, multi-modal, musical and mimetic in character.... They were 'singing Neanderthals'—although their songs lacked any words—and were also intensely emotional beings: happy Neanderthals, sad Neanderthals, angry Neanderthals, disgusted Neanderthals, envious Neanderthals, guilty Neanderthals, grief-stricken Neanderthals, and Neanderthals in love. Such emotions were present because their lifestyle required intelligent decision-making and extensive social cooperation.[*]

Sympathetic emotions are also evidenced by the long-term care of the injured or incapacitated which is proven by several Neanderthal remains. From their remains, some of these individuals must have needed intensive therapy to survive and be rehabilitated, and music played a role here too:

> I think it most likely that 'music therapy' was used to reduce stress, improve healing and facilitate the return of motor coordination....

---

[*] Mithen, op. cit., p.221.

> I can readily imagine Neanderthals singing gently to calm the distressed and those in pain, and tapping out rhythms to help the movements of those who were injured or disabled.[*]

Working within a shared range of experience, i.e. strictly within what was remembered, he adds:

> There would have been limited, if any, need for the type of novel utterances that could only have been produced by compositional language. To put it plainly, they didn't have much to say that had not either been said many times before or could not be said by changing the intonation, rhythm, pitch, melody and accompanying gestures.[†]

Such a reconstruction fits well with Steiner's indications of an Atlantean type of past-orientated, memory-based, musical mentality. Indeed, whereas it has long been exposed as a fallacy that higher intelligence correlates with a larger brain, there is recent evidence that extensive memory does so.[‡] The mystery of the Neanderthal brain may finally have found a solution.

The stress on emotional development also rings true to Steiner's perspective. He noted that Atlantean humanity were the first that 'developed feelings, which their Lemurian ancestors did not yet know'; he once more drew the link with memory, which in the emotional sphere becomes attachment.[§] We may suspect, moreover, that love was also in origin a love for the remembered dead, whose continuing presence in the social order gradually wove people together in a loving web of attachments encompassing also the living. Love and death remain archetypally linked in myth and romance through to the mythical romances of modern psychological theory. Through death, love becomes a pure intensity which transcends external connections.

Mithen's musical reconstruction of Neanderthal existence is for the most part extremely convincing. Yet his wordless songs pose a

---

[*] Mithen, op. cit., p.236.

[†] Mithen, op. cit., p.228.

[‡] Cf. J. Pearson in the online academic journal *The Conversation*—https://theconversation.com/brain-size-matters-when-it-comes-to-remembering-30728. Memories are stored and preserved; yet despite using some distinctly quantitative metaphors he emphasises that the memories are not recordings and are retained only by 'voluntary effort' and active 'processing'.

[§] Steiner, *Cosmic Memory* p.50.

riddle. From mother-and-baby song onward, singing and babbling involves a sort of word-formation, even if it be goo-goo or da-da-da. It seems Mithen presses the point too much that there is 'no evidence of Neanderthal language'. Most other researchers are convinced that they had speech. It is true that anatomical reconstruction seemed to show that the Neanderthal larynx was placed differently to ours and would have struggled to produce articulate sounds, but this view has since been modified by better evidence.* Rudolf Steiner suggests a less artificial bifurcation, and he attributes to the ancient Atlantean human type both music and speech—at first closely intertwined, only later separate:

> During the evolution of humanity, speech has gradually emerged from what was originally an element of song. The further back we go into prehistoric ages, the more speech resembles recitative and ultimately song. In very ancient times, when humanity revealed itself through sound, tone was not differentiated into song and speech; these were identical. ... The primordial speech of human beings is at the same time song.†

We may take up Mithen's beautiful picture of Neanderthal life, therefore, with a less rigid division of ideas—a more evolutionary approach. Mithen himself had earlier cited evidence from modern child-research that music (song) is an independent and prior development in infants, not derived from speech. Nevertheless, that research concluded, the two need not be unrelated. Rather, the evidence suggested that the developmental 'networks for language are built upon or replicate those of music' (above, pp. 207ff). We may envisage that the prehistoric song of the Neanderthals was not yet speech, but that it contained components (such as sound differentiations like vowels and consonants) which later became independent of the song-element, and were able to develop in a new direction as spoken language. Music, in a complementary way, was able to further its non-vocal potential which we know from its later history (that bone flute comes to mind once more). Exactly how far the differentiation

---

* Stringer and Gamble, *In Search of the Neanderthals* pp.88-91; Mithen, pp.226-8. Mithen's reconstruction would give us humming Neanderthals, but I find it hard to square this with the supposed expressive range which he claims: lusty- bitter-grief-stricken humming? Hence my dislike for his generic term Hmmmm.

† Steiner, *Art in the Light of Mystery Wisdom* (London 1970) p.80; more on Atlantean music, pp.118ff.

proceeded in the Neanderthal phase of human evolution remains, like so much else, still far from finally determined.

Mithen was over defensive, perhaps, because some anthropologists have employed a notion of language-development which assumed that words were initially disjunct utterances, with narrow fixed meanings, later strung together. But language-development is not really comprehensible in these terms. Thinkers like Owen Barfield (with Steiner's thought in the background) have shown that linguistic meaning is originally holistic just as much as Mithen's musical 'meaning'. Only afterwards do unitary meanings get divided; the dictionary is the grave, not the cradle of language.[*] Steiner himself pointed to an important historical evidence of thought-structures rooted in the archaic Atlantean mentality, to which he repeatedly drew the analogy, in the oldest layers of the culture of ancient China.[†] There he found concepts which belong to the holistic phase of language, and to its musical and oracular significance. Of course, Chinese culture also evidences many thousands of years of history, both in terms of its people and of their language and thought. It is essentially to what we mean by culture, however, to maintain some long-term values beneath the surface of change—and China has a commitment to preserving the treasured elements of its past which is exemplary.

One absolutely central notion, of great antiquity, is the concept of the *tao* (or *dao*). Modern interpreters of Chinese thought such as Richard Wilhelm have employed equivalents such as *Sinn* ('meaning'). He quotes Lao-tzu (Lao-tse), the sixth-century BC founder of Taoist philosophy, to the effect that the enlightened man sees through to the 'immutable law' present behind all change: 'This law is the *tao* of Lao-tse.'[‡] But Steiner emphasised that in its original significance, the Chinese *tao* which guides judgements is by no means something equivalent to the 'reason' of today. Its connotation in Chinese thought still captures an echo of a primal, holistic mode of thought which is an active divine presence in the cosmos:

> The Atlantean did not raise himself to his God through concepts and
> ideas. He discerned something holy in nature as the keynote of the

---

[*] Owen Barfield, *Speaker's Meaning* (London 2011); and many other works.

[†] Steiner, *The Mission of Folk-Souls* (New York 2023).

[‡] R. Wilhelm, *I Ching. The Book of Changes* trans. C.F. Baynes (London and Henley 1951) p.lv.

divine; it was as though he breathed in and breathed out his God. If he wished to express what he heard in this way, he would embody it in a sound similar to the Chinese TAO. For the Atlantean this was the sound which permeated the whole of nature. When he touched a leaf or saw a flash of lightning, he was aware that part of the Godhead was revealed before him; it was as if he were touching the robe of the divine … the body of the Godhead.[*]

The all-pervading sound that is not yet articulated as word is the medium of prehistoric consciousness. We certainly know that Lao-tzu was already drawing on older sources for his ideas, and in particular that he was familiar with the ancient oracular source of knowledge, or *I Ching*. 'Lao-tse knew this book,' says Wilhelm, 'and some of his profoundest aphorisms were inspired by it. Indeed, his whole thought is permeated by its teachings.'[†]

Chinese tradition considers this oracle to be the original fount of human culture. It was said to have been devised by a mythical culture-hero. 'Fu Hsi is a legendary figure,' we learn, 'representing the era of hunting and fishing and the invention of cooking. The fact that he is designated as the inventor of the linear signs of the Book of Changes means that they have been held to be of such antiquity that they antedate historical memory. Moreover, the eight trigrams have names that do not occur in any other connection in the Chinese language'[‡] and are already mentioned in connection with dynasties supposed to have a date c.2000 BC. Its basic principles have a profound affinity with the origins of the concept of *tao*, which Wilhelm further paraphrases as:

> the course of things, the principle of the one in the many. That it may become manifest, a decision, a postulate, is necessary … the 'great primal beginning' of all that exists, *t'ai chi*—the ridgepole, the line. With this line, which itself represents oneness, duality comes into the world, for the line at the same time posits an above and a below, a right and a left, front and back—in a word, the world of opposites. These opposites became known under the names yin and yang and created a great stir ….[§]

---

[*] Steiner, *Founding a Science of the Spirit* (London 1999) p.93.

[†] Wilhelm, *I Ching* p.liv.

[‡] Wilhelm, *I Ching* p.lviii.

[§] Wilhelm, *I Ching* p.lv.

Now, to make a 'decision' about the truth and arrive at the principle of things, we moderns reckon to haul up the necessary conception out of ourselves by our own activity. The ancient oracular cultures, however, did not alienate thought from the level of all other things, and the postulated 'line' is reminiscent of the lines which, in various combinations, the *I Ching* employs to represent the first principle and the space which receives it, the above and the below, heaven and Earth, and all those opposites which made such a stir. The 'decision' is the throwing of the yarrow-sticks, which the one who knows how to see into the structure of the moment can interpret within the wholeness of the *tao* itself.*

It is as though all of Chinese thought is generated by that line in its multifarious combinations. Though unfathomably enriched by the millennia-long history of the Middle Kingdom, something going back to an original oracular culture that is still more archaic has come down in the primary concept, or sound, the *tao*. Steiner connects it with prehistoric consciousness and the Atlantean human type.

The elements of the oracle almost certainly had musical significance too, though this has become obscure over the ages. It is noteworthy that we also find surviving among the Chinese people a form of language where pitch still plays a part, and an attitude of veneration of the ancestors, and many other features which we now know cohere together from prehistoric times. It is certainly possible, therefore, that Chinese civilisation preserved aspects of a still older mentality. But of course we also know that children can evince features of archaic culture simply through their development. Chinese people share in the inheritance of the Anatomically Modern 'Eve', who is the genetic source of everyone's biological make-up among the peoples on Earth today. It is also the case, however, that several populations have received marked gifts from the Neanderthals, especially outside Africa. But is there any evidence that Steiner could be right when he suggests that Atlantean biological features are preserved among the Chinese people in a special way—that they are in some hereditary sense related to that still more archaic cultural source biologically?

In 2010 it became clear that Neanderthals and Anatomically Modern Humans had interbred in the Middle East, which had given to most of the modern population of the world a 1% to 4% admixture

---

* 'All individuals are not equally fitted to consult the oracle. It requires a clear and tranquil mind, receptive to the cosmic influences hidden in the humble divining stalks'—Wilhelm, op. cit., p.liv; cf. p.lvii.

of Neanderthal descent (Africans less so). The expansion of Modern Humans across the world carried that mixture right across Asia. But then in 2013 a complicating fact emerged. The East Asian populations which centrally include ethnic Chinese, studied by a follow-up team of researchers, did not merely confirm that they had inherited some Neanderthal features: they turned out to have *more* such features than the Europeans who were supposedly closer to the interbreeding-event. In other words, they were not just transporting across the world their human-Neanderthal biological make-up, but had somewhere in their archaic history a further, more significant such development in common with the Neanderthals. 'At least two separate episodes of admixture between Neanderthals and Modern Humans must have occurred,' the researchers concluded, 'and at least one of these episodes must have occurred after the separation of the ancestors of modern Europeans and East Asians.'[*]

How was it that one arm of advancing humanity encountered a Neanderthal-type population, resulting in a strain that has left its mark only in the Far East? Some have postulated instead a prolonged interaction between humans and Neanderthals from which different populations 'dropped out' at different stages; but there is the difficulty of the rapid disappearance of the Neanderthals in the aftermath of their decisive contact. At any rate, Steiner's indication of the Atlantean affinities of the East Asian peoples certainly takes on a potential new significance. And the picture strikingly confirms his fundamental account of the threefold distribution of humanity coming down to modern times. We are all Modern Humans, descended as he already pointed out, from a single original pair. But the history of our development is still discernible in the broad division into the African peoples from whom we descended ('Lemurian type'), the Asians with the largest inheritance from the Neanderthals ('Atlantean' type), and the European group (Indo-European type).[†] We need not conceive of these in nineteenth-century terms as continuing 'races', a concept which

---

[*] J.D. Wall et al., 'Higher Levels of Neanderthal Ancestry in East Asians than Europeans', published online 14.2.2013, at www.10.1534/genetics.112.148213, p.26.

[†] Steiner, *Cosmic Memory* p. 48; cf. p.58—which refers to a 'Mongol' Atlantean race. Though Atlantean in type, however, Scott-Elliot, *Story of Atlantis and The Lost Lemuria* p.23 says that it had its origin on the plains of Tartary where it 'multiplied exceedingly, and even at the present day a majority of the earth's inhabitants technically belong to it' though mingled with other types (cf. p.37).

Steiner rejected and instead emphasised human cultural evolution. But they do rightly indicate the presence in our prehistory of other genetic groups than the single line of *Homo sapiens*. If our interpretation of the Neanderthals as belonging to the Atlantean type in human evolution is correct, Steiner's indications can still chime creatively with the latest research into the increasingly complex and fascinating question of human ancestry.

Chapter 7

ORACLES AND FLIGHT-PATHS:
THE NEANDERTHAL WORLD

Oracular Networks

Rudolf Steiner's perspective of the 'evolution of consciousness' enables us to see in the extraordinary musical development of the Neanderthals the emergence of human characteristics, in embryonic form, that have since become our inalienable inheritance. We feel our kinship and continuity with these extraordinary people, strange and baffling though their archaic consciousness may nevertheless still appear to us.

They lived out of music and memory, though their recollection always needed to be triggered by an external memorial. The awakened memories also powerfully addressed their present needs in an oracular way, answering their questions in a way not essentially different from the devices of the *I Ching* or the Sibylline leaves. At the site of a burial, they could experience once again the presence and archetypal deeds of a quasi-parental, oracular predecessor, and renew them before his monitoring gaze as a model for the next generation—or even coming generations, accumulating layers of memory. The qualities of such deeds could also be sung, by Neanderthal voices adept in capturing the emotional pathos, grandeur or glory of a situation in sounds, and intensify the memory as a force shaping community action. For the community must have assembled at the oracular scene so as to have been swayed by the inspired memorialist, thus constituted as their leader. In primal stages of society, imitation of the actions would then continue but the memory would fade; gradually, however, ways must have been found to extend awareness of the basis of actions in a way that could integrate society more fully. One of the most potent was language: the speech-sounds that were originally included in song-utterances would eventually be developed into an 'objective' way of referring to the heroic proto-type or his acts. The activating of memories at the oracular site could be repeated through 'speaking' about them, though the linguistic complexities we take for granted today evolved very slowly. Expert opinion inclines to the view that Neanderthal language never developed a future tense. It remained, like the consciousness of Steiner's description of Atlantean humanity, orientated to the past.

Personal memory like ours was as yet still far, far down the line. If the empathy of modern researchers is correct, the extension of memory was not initially in that direction. It did not collect memories around a single 'self' but consisted more in a quasi-spatial assembling, an arranging in localities of the mind (like a memory-theatre) of the sources of imitative knowledge. The effect was to bind the environing world and the community inhabiting it together in a first cosmic order. Separate blocks of memory and knowledge were not pooled so as to give the sense of personal continuity like ours today, but rather were marked out on the sacred landscape. The Neanderthal consciousness remained 'domain-specific', says Mithen, accessing different kinds of knowledge and expertise not only in different places but storing them in separate 'chapels' of the mind. The interrelations between them were not grasped by a central consciousness—rather, their prototype was a networked landscape staked out by separate oracular sources. Different knowledge was available in a different place, linked together not on the human level but as part of the cosmic wholeness of life. Only afterwards was the whole landscape internalised so as to indicate a central 'self'.

Memorial sites, especially burials, create for the first time a linked series of pathways which the dead and the living traverse. Grave-markers are also way-markers joining remembered sites. The landscape becomes a known locality, in a way different from the older relationship to the movements of migrating prey or to the changing sources of fruits or water. We know that the Neanderthals established precisely such distinctive finite territories. The technique of memory-theatre made everything part of an organised system, enabling stronger economic organisation too. Individual memorial sites stood in relationship to overarching oracular locales, at which the different segments of the community must have periodically united for the kind of enactments and song-celebrations which we have just tried to indicate. One or more standing stones marked out a 'sacred way' bringing people together and opening up communication, dominating the landscape. Ultimately, each oracle-site stood in relation to a 'heavenly' prototype, since the phenomenon of the cognitive landscape opened the understanding of space extending up into the sky, as also, of course, down into underground realm—of memories and the dead.*

An 'intriguing feature of the Neanderthal archaeological record', remarks Steven Mithen, is the presence of carefully cleared 'performing areas' in a number of caves, areas too small for any obvious practical gathering, sleeping etc. but strongly suggesting that one person acted, sang or danced while the others sat around the body of the cave.† Whether these sites were also associated with burials, or grew up in association with burials nearby we do not know—we are of course at the mercy of chance when it comes to data from so long ago. The line between shelters for the living and the dead in prehistoric times is anyway a very blurred one: Rudolf Steiner pointed out that the first 'dwellings' were actually for the dead, and that the living took shelter with them and gradually extended their practice of building homes. In many archaic cultures, as we have noted, the dead still find a place either quite literally, or in an ancestor-shrine, in everybody's home. Indeed, he argues, architecture came into being as an extension of the 'laws' of the physical body—the body as it is apart from the spiritual life previously informing it. The ritual placing/burial of the body and its artistic contemplation is thus quite literally the starting-point of human building activity, prompting an externalisation of creative forces based on intuitive affinity with its form.‡ The memorial stone was simply a substitute for the interred body which furnished the original centre of social life, and the spirituality of dolmens and similar structures makes permanent in archaic consciousness that continuing presence.

In that context, Mithen also discusses the evidence for what may have been a more than usually significant site. In fact it must have been a major Neanderthal oracle-centre: the cave at Bruniquel in southern France, near Montauban.§ The entire region is rich in prehistoric sites,

---

* Cf. the brilliant discussion in M.P. Nilsson, *History of Greek Religion* (Oxford 1925) pp.109ff. See also M. Eliade, *Patterns in Comparative Religion* (London 1971) pp.216ff.

† Mithen, *Singing Neanderthals* (London 2005) p.242 mentions such features in Neanderthal sites at the Grottes Vaufrey in the Dordogne, the Grotte du Lazaret near Nice, and Les Canalettes in the Languedoc.

‡ Steiner, *Art in the Light of Mystery Wisdom* (London 1970) pp.19-20. Temporary encampments there certainly also were; but the permanent centres were originally only for the dead and these alone would survive in the archaeological record.

§ Mithen, op. cit., pp.242-3.

and must have been of deep significance to palaeolithic people, or possibly even earlier.

The cave itself extends hundreds of metres into the rock, and then, in utter darkness, has been shaped into a chamber with extensive signs of building activity. There are cleared areas and extraordinarily constructed walls incorporating hundreds of broken-off pieces from stalagmites in the cave. Such elaborate wall-construction and complex re-forming of the natural chamber goes well beyond anything that had been expected of Neanderthals, and the chamber is far underground by any prehistoric standards. As Mithen remarks, it must have been constructed using artificial light from torches, and anything which took place there must likewise have been done by fire- and torch-light. The cleared area indicates to him that it was a dancing-ground for the purposes of communication among the Neanderthals, and he believes that they exploited the magnifying effect of singing in the resonant enclosed space. But the reasons for situating their jamboree so far underground are not thereby sufficiently accounted for.

A further striking aspect of the cave is the lack of wall-painting, such as adorns so many prehistoric sites in France and elsewhere. This, he remarks, is typical of the non-visual Neanderthals—although the evidence about that has become a little more complicated. I propose that cave with its dancing-ground and lack of visual presences is actually typical of Neanderthal culture, specifically, in being an *oracular centre*—rather than a Mystery-site such as we find in the palaeolithic phase, where (as at Lascaux) animal and cosmic symbolism guided the procedures and inner processes which unfolded there.

It is not that the Neanderthals knew nothing of this kind of spirituality; but it is not the purpose of the cave at Bruniquel. There it is oracular knowledge which was sought, through a medium who 'performed' it, and conveyed it in the form which shaped their basic mode of life. At Bruniquel, if we may develop Mithen's idea, we are not embarking on an ecstatic path through the cosmos but communing with the dead deep under the Earth, and dancing their wisdom in oracular song. And that despite the sky-attributions of the marker-stones to which we shall soon advert. The implied presence of an audience indicates this too, in contrast to secret ecstatic rites which in vision take place before the gods, or at most a favoured few. Here, we must imagine the solemnity of a visit to find solidarity with their

This chamber deep underground at Bruniquel is one of several that can be shown to be Neanderthal sites, and features elaborate circular structures made using stalactites as building-blocks. Such building activity was once thought far beyond the capabilities of Neanderthal humanity. Striking, however, is the absence of rock painting—even though it has been commented that the surfaces in the cave are highly suitable. Steven Mithen has interpreted Neanderthal mentality as fundamentally musical, and believes the acoustic properties of the site were exploited for gatherings. Building underground would have needed such enormous determination and effort that these must have had great importance—presumably of a religious nature, probably of an oracular type. Later oracle-gods like Apollo at Delphi were closely associated with music and with cave-utterances. Neanderthal structures in the cave of Bruniquel.

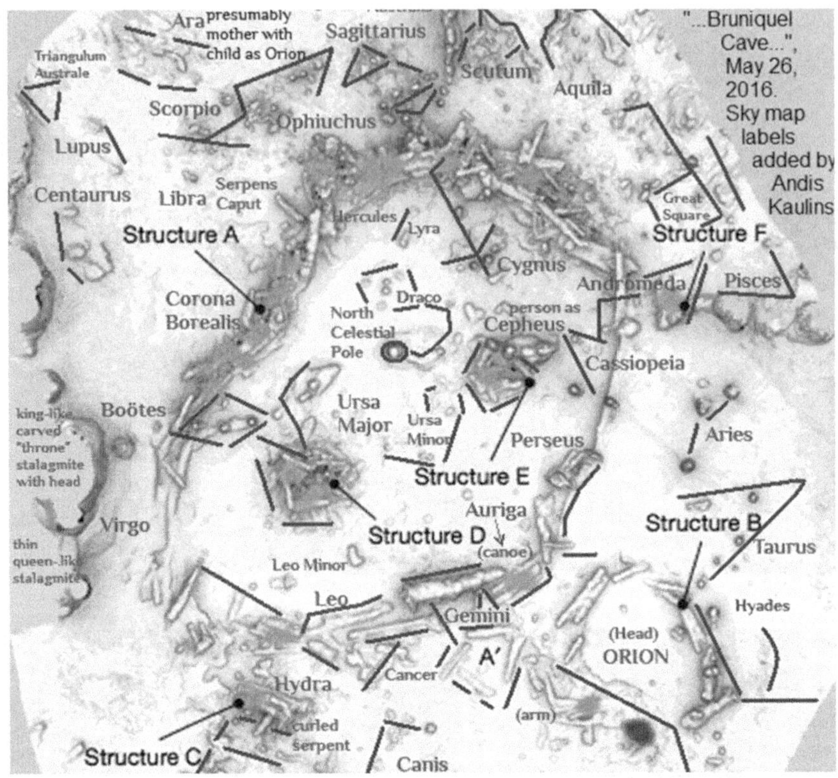

Neanderthal underground space at Bruniquel, interpreted as an experience of the night sky centred on the Great Bear. Despite the total lack of images, the main attributions by Andis Kullin (2016) may well be basically correct if the detail is sometimes carried into speculative exaggeration.

ancestors and knowledge of their *mores* which serves to draw the people together into a community.

The extraordinary effort needed to construct the underground chamber suggests that it is an exceptionally important site, and further work at the cave from 2014 opened up a further level of understanding. Detailed excavation revealed a complex ring-monument, again going far beyond what had previously been expected of Neanderthal competence.* Now detailed symbolism and composition were unde-

---

* Jaubert, Jacques; Verheyden, Sophie; Genty, Dominique; Soulier, Michel; Cheng, Hai; Blamart, Dominique; Burlet, Christian; Camus, Hubert; Delaby, Serge; Deldicque, Damien; Edwards, R. Lawrence; Ferrier, Catherine; Lacrampe-Cuyaubère, François; Lévêque, François; Maksud, Frédéric; Mora, Pascal; Muth, Xavier; Régnier, Édouard; Rouzaud, Jean-Noël; Santos, Frédéric (2 June 2016) [online 25 May 2016]. 'Early Neanderthal Construc-

niable. Enthusiasts have been quick to find symbolic interpretations everywhere, going far beyond the initial assessment by the experts. We must treat their speculations with a certain amount of caution. But the impressive ring form of the monument does suggest an allusion to the sky, and it is possible to see references to the striking star-patterns of certain prominent constellations in the shapes found in the stone—especially the Great Bear near the centre (reflecting its place near the celestial pole).

So there may after all be cosmic allusions at Bruniquel: not in the sense of a shamanic journey, but in establishing a network of oracular sites which were considered to reflect a heavenly totality. The ancestors, as we mentioned in the last chapter, replaced the direct presence of gods in human experience. Now they were the ones who, having known the gods directly, could pass on their wisdom in a human way. Steiner referred to seven major oracles which were the basis of the Atlantean mentality, and connects their qualities with the planetary individualities. Yet he also indicates that actual planetary cosmology was a later development of human consciousness, and thus is not the origin of the conception.[*] The seven sources or qualities of wisdom are a more primary manifestation in human life. The seven stars of the Bear which never set, but seem to watch unsleepingly over the world below, were certainly an earlier image for the primal spiritual powers.

The presence of Neanderthal humans at the Bruniquel cave is archaeologically assured by a burnt bear-bone—one of a scattering of indications of a bear-cult among the Neanderthals. Moreover, the association of the seven stars of *Ursa major* (which we often call the Plough) with a bear or bears is a fact of cultural prehistory manifested right across the globe. It is documented from ancient Greece, whence derives our version, to India where the Seven Sages, sources of all wisdom, are the 'seven Bears', to native North America where the constellation may represent God himself who sometimes takes the form of a bear. Very widespread too is the connection of bears with the original ancestors of humanity in myths and legends.[†] To

---

tions deep in Bruniquel Cave in Southwestern France'. *Nature.* 534 (7605): 111–114. doi:10.1038/nature18291. ISSN 0028-0836. PMID 27251286

[*] See Steiner's lectures, *The Gospel of St. Matthew* pp. 38ff and Christ and the Spiritual World (London 2006) lecture 4.

[†] There are some penetrating comments on this mythology and its basis in Karl König, *Elephants, Bears, Horses, Cats and Dogs* (Edinburgh 1992) pp.35-56.

be found in so many places, so far apart, the conception of the seven bears as ancestors or wisdom-figures and as divine beings must have been very archaic. If the ring-monument at Bruniquel does allude to the Great Bear, it may be one of the oldest pieces of evidence for the conception of the seven starry sources of knowledge, mediated by the ancestors who were themselves approached in some manner through a cult of the bear.*

The Indian mythology of the 'seven bears' or *Rishis*, who are also human ancestors, holds a secret that can be understood more fully if we follow through on Rudolf Steiner's ideas. The latter often indicated the continuity of the most archaic India with prehistoric, 'Atlantean' ideas—in the sense of a visionary ability to recreate the ancient mentality, and live in the pictures of a primordial time, to which the *Rishis* belong par excellence. The seven Sages represent spiritual qualities that were afterwards associated with the planets, but which were still experienced in an older configuration. When we speak of the prehistoric humanity of the Neanderthals as 'Atlantean', we mean specifically that they are determined biologically by the archetype that was formed in the evolutionary phase when a great continent filled the Atlantic Basin; that archetype was not then physically realised, but held back among the stars, still ethereal and implicit in the cosmic order. The seven qualities emerged from the rhythm of its primal elaboration (see above). When humanity was lifted in oracular consultations into those creative source-realms, the constellation of

---

* Most of the evidence for the conception naturally comes from later mythological traditions which may, however, point back to this source. The bear-cult of the Neanderthals is often treated nowadays with great scepticism, after the excesses of Bächler and others. It is partly just a prejudice against the primitive Neanderthals having 'religion', however, and the argument generally used, namely that bear materials are found in caves but not where there are elsewhere signs of human activities, seemingly forgets that sacred events have their special places. Perhaps the sacrality of the caves was influenced at an early stage by the presence of bears? So much is changing in our understanding, that we need to keep an open mind to the evidence here. A Neanderthal site in southern Poland (Dziadowa Skata Cave) in 1953 yielded a bear-bone with markings that have recently been re-examined: the seventeen parallel lines are now considered to have been deliberately made by careful carving action. 'It is one of the quite rare Neanderthal objects of symbolic nature,' says Tomasz Płonka, professor of archaeology at the University of Wrocław, and proves the craftsman's cognitive faculty. Bones from cooking were not employed—to choose a bear-bone might indicate special significance?

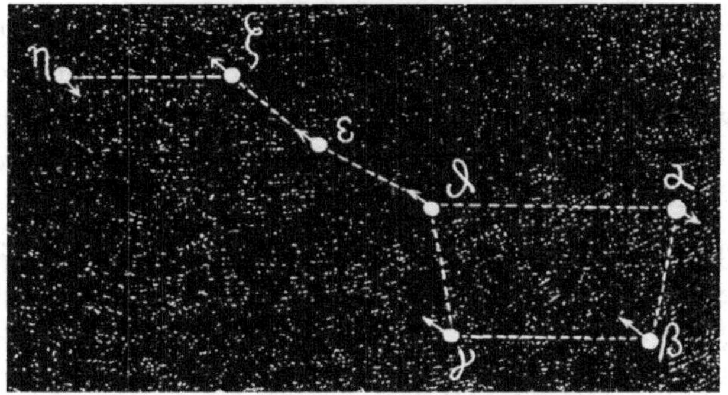

The Seven Sages (Saptarishi) are the oracular sources of pre-Vedic Indian religion. For reasons never explained they are identified mythologically with the main stars of the Great Bear (the Plough or Big Dipper). Rudolf Steiner associates them with the Manu- traditions and Atlantis, and it may be plausible to suggest that the constellation where they can be found was chosen because it is a picture of the ancient continent. In mythology the seven Pleiades are the wives of the Rishis, seeming to confirm that it is the shape of the star-group which is important. Oracles are usually at marginal sites, where mountain meets sea, etc. They would thus pick out the outline of the land mass and its 'tail'. Important sacred sites are still found there today.

stars visible in the Great Bear was somehow experienced as a map of the cosmic structure of that time.

The exact shape of the constellation is of first importance to the significance of the *Rishis*, each one of whom was identified with a bright star from the body or the arching tail of the Bear. We know that the shape was important, because in mythology each of the *Rishis* was provided with a wife, and each of the wives also had a starry identity as one of the seven Pleiades, a tight grouping of stars in the constellation we call Taurus. A comparison of the star-cluster with the stars of the Great Bear immediately reveals the striking similarity of their disposition. It was because of their identical pattern that the Pleiades too evoked the qualities of the primal age. Each of the goddesses of the Pleiades exists also as a *chakram* or centre of spiritual energy within the human form, that can be activated by yoga training. But the *Rishis* themselves have a basically macrocosmic signification—a feature which recalls the old oracular consciousness, for which different qualities and ideas were located in specific places and centres, in this instance to be found, evidently, spread out in space in a pattern corresponding to that shape of the Great Bear and the Pleiades. What does that pattern originally represent?

If we gaze closely at the map of Atlantis which we have recovered from modern scientific and from visionary sources, there is a striking geographical feature to be observed. The great body of the continent stretches westward from the point of contiguity with present-day Ireland across the vast area now ocean, occupying a huge slightly irregular rectangle. Further west still is the long chain of what will later be the Rockies, then a separate slightly curving land mass some hundreds of miles across the sea from the Atlantic mainland. Do we have in the Indian mythic picture of the *Rishis* the heavenly projection, so to speak, of the map of Atlantis? The great somewhat square body, and the arching tail would represent our earliest version of the map that was envisioned again in more modern times by John Dee, or whoever lies behind the map whose externals Steiner confirmed. If so, then the *Rishis* would represent oracular centres of creative power at the margins of the ancient continent; such breakthrough places, where land and sea or mountain and sky meet are quite typical for oracular sites. A sequence of oracular sites along the peaks of Laramidia, now the Rockies, would make plausible locations. The region of overlap with modern-day Ireland has been a site of extreme holiness for countless ages of history, and would have been the easternmost oracle of the Atlantean network. The southernmost oracle would have been in the area of present-day Florida: a region that may well be associated with an asteroid catastrophe such as is probably connected with the K-T extinction, or end of Atlantis if we interpret it in the light of the link between Phaethon's crashing of the heavenly-chariot and Plato's myth.

When Indian seers went to visit the *Rishis*, therefore, we must conceive them to be returning to a long-lost world of Atlantis that had physically ceased to exist. They were making an inner pilgrimage to places long distances apart, where the creative energies could be found: or rather they were journeying in the sky, where they knew that the human reality overshadowing what lay below had essentially resided. In the topography of the Great Bear they found a map of the heavenly region where the archetype of humanity was formed, with its seven qualities. The star-goddesses of the Pleiades planted the seven qualities, to one of which each was 'married', within earthly humanity as the *chakras*.

The probable reference to the Great Bear at Bruniquel, and the scattered indications of a bear-cult among the Neanderthals, suggest that they already, as human beings of Atlantean type, experienced their oracle-centres as conforming to a spatial distribution based on the same pattern. They were inhabiting prehistoric Europe, but it generally happens that mythic patterns are transposed to apply to new locations where correspondences in spiritual significance are found. It is to be suspected that the Bruniquel oracular site may turn out to be part of a network of such sacred places recreating the Atlantean pattern over Europe. We are painfully limited in our knowledge of the full significance of so much in the Neanderthal world. Using Steiner's insights, however, we will try to show later in this chapter that the mythological themes of the 'end of Atlantis' were present in their consciousness, and are a key to the further stages of human dispersion and its association with the Upper Palaeolithic pictorial consciousness—in its contrast to the Neanderthal musicality a fundamental polarity in early human evolution.

The seven oracles were undoubtedly 'harmonised' in another way, quite literally, by the musical Neanderthals: as the seven fundamental notes of a mode (something like our scale).* That would be a natural way

---

* The ordering-principles of 12 and 7 in music arise from the natural working of vibration in the cosmos. A note resonates and produces other notes in harmonic relations to it, most powerfully the octave (such as C—C'), and the musical fifth

of conceiving them for Neanderthals. Within their musical conscious-ness each of the oracular tones would remain distinct, corresponding to those unitary 'chapels', each one related to a different aspect of mental life. Presumably, different regions would be dominated by a particular oracle, but rather than feeding into a complex consciousness, people's thoughts and ideas would be laid out like a map. And each would have its unique tone. Mental life moved about, so to speak, as within a memory-theatre, and a scale of being. People were helped too by some further oracular intermediaries—the birds.

The Dream of Flight

It was once supposed that the Neanderthals were too slow and stupid to be able to catch birds. Only *Homo sapiens* could have done so! We now know that Neanderthal culture had a strong connection to birds and feathers.* Markings have been found which show that feathers were removed that cannot have served a practical purpose in preparing food, which leaves the implication that they were used for decoration—pre-sumably self-decoration. How exactly they arranged the feathers, and whether they appreciated their colour or their lightness or their fine tapering, we can have no way of knowing. The fact of their involvement with bird life, however, indicates an extensive adaptation of behaviour, which Finlayson has recently compared with that of North American tribes who know how to lie in wait and to trap even large powerful birds such as eagles.

Given the difficulties (once thought too great for any species except ourselves) involved in any contact with creatures who in a moment can command control of the air, any culture which related to birds can only have done so by a serious, not an incidental effort: it must have been in effect a bird-culture. It is hard to believe, too, that a musical-cul-ture like that of the Neanderthals would not have been fascinated by

---

(such as C—G). In heard terms, seven octaves can be brought to coincide with twelve fifths. The notes thus generated constitute a complete inner world of rela-tions (intervals). On the spiritual basis of the inner musical relationships, cf. Steiner, *Art in the Light of Mystery Wisdom* pp.25-28. The musicologist Deryck Cooke made a profound attempt to derive the expressive range of all music from the basic inter-vals: see his *The Language of Music* (Oxford 1959). See also Heiner Ruland, *Expand-ing Tonal Awareness. A Musical Exploration of the Evolution of Consciousness Guided by the Monochord* (London 1992) esp. pp.64-68 and generally pp.56ff.

* Finlayson, *The Smart Neanderthal* (Oxford 2019) pp.11ff.

bird-calls and bird-song. Surely they imitated it, and perhaps the Slovenian bone flute (assuming it really is a musical instrument) developed as an extension of such mimetic efforts, encouraged by the use of piping as a lure?

The researcher Clive Finlayson has shown that the Neanderthals had an extensive relationship to birds, though evidence that they decorated themselves with feathers remains contentious. Looking out for birds implies a structured approach to sky-zones and directions, and interpretations of times and migrations: their interest would likely have been oracular and divinatory. The advanced civilisation of the Romans based all important decisions on bird-auguries, and preserved an elaborate structural science to interpret them. The founder of Rome, Romulus, receives the auguries: Italian pen and wash drawing, Metropolitan Museum of Art, New York.

Interest in birds would necessarily have involved an intensive turning of awareness to the sky. A remembered landscape with markers would already open out a fully conceptual space, with further extensions up and down—rather than a space of *ad hoc* encounter generated by the lateral mode of locomotion, such as was known before. Envisaged space, with remembered points and a cosmic orientation up and down would offer a far more satisfying solution to the 'vertigo' of the earliest humans. We have suggested that the 'Lemurian' phase found a solution by taking refuge, so to speak, at the bosom of an Earth-mother, bonding with their environment as their primal religious experience. A more potent way of dispelling vertigo, however, would be the kind of cosmos which the Neanderthals must have inhabited, with earthly depths below (caves, the dead, memory) and an organised space above. We may say organised, because a relationship to birds means something quite different from tracking game or seeking until one finds a place of abundance. Birds in their flights and migrations have to be recognised according to season and calculated direction. In days prior to agriculture, before waiting for the next crop gave an idea of the rolling seasons, it was surely looking out for avian migrations which would have given a sense of long-term time, and a meaningfulness to the different quarters of the sky. Birds reappeared at a 'remembered' time and from a specific direction. The world was thereby further structured by an analogy with space and music rather than conceptual thought. The birds were messengers of a cosmic order, though a little nearer than the stars. In addition to their natural fascination, they were an expression before conceptual thinking of dynamic oracular space and time. With the ability to interpret their comings and directions a dimension of control over the world came into human life that is universally associated with sky-divinities.

Birds, in short, must from the start have been auguries—analogised, perhaps, from the ancestor-cults. The significance of avian appearances and behaviour has for most of history been treated in terms of omens and signs. A memory of past actions, or the fixing of an idea, could be determined by such a system just as effectively as by memorial 'posts' or thrown sticks. Just as a burial was an opening into the world of memory, so a bird was an aperture into the world of superhuman order, the determining of destinies, for those who could find within themselves the truth answering the oracular sign.

The sky-world of the birds only impinged on human life in exceptional ways, as when a bird was ensnared, or when an augury was taken. The wearing of feathers by the Neanderthals, as Finlayson suggests, however, indicates that there was a further dimension to the connection. It also poses a problem. Wearing feathers as a sign of status, for instance, is more suggestive of the image-consciousness of our

Upper Palaeolithic near relatives. It goes against what we know of the Neanderthals—which is what led Mithen to seek rather in music for the ordering factor in their culture. It may well be that, rather than thinking of the Neanderthals taking something from the birds to enhance their own lives, we should consider the gesture rather as assimilating them to the world of the birds.

Did the Neanderthals dream of flying like the birds? Turning to Rudolf Steiner, we may note that among the numerous perceptive remarks about 'Atlantean' humanity which have been helpful to understanding the Neanderthals, he observes that their waking and sleeping was not characterised in the same way as ours. Sleeping was a much more vividly conscious state, and not a blank time of unconsciousness as now.* Compared to this kaleidoscopic dreaming consciousness, day-consciousness was much less clearly felt than our modern-day intensity of self. Physical perception was still partly undifferentiated. Even nowadays, sleep-deprivation may easily result in dream-images popping up within conscious awareness. In ancient times, the primary sense (the sense of presence, one might say) which has since been sharpened into different modes of objective perception still enveloped them all, felt like a sort of dreamy aura around what was seen, for example, rather than being concentrated within a spatial form.† Anything perceived carried a penumbra of meaning around it, like a symbol—a portent or sign. It may well be that we need to think of the Neanderthals as much more involved with the dream-world than we are. That even earlier humanity, of the 'Lemurian' type, still lived in states akin to ecstasy. The intense dreaming stage is in that sense a step toward more modern consciousness. Steiner's reference to their less specific modes of sense-perception, with less sharply focussed vision, may one day help us in understanding the larger and differently-shaped eyes of the Neanderthals.‡ We must also be ready to comprehend the way impressions and patterns from day-perception would have been actively elaborated, transfigured (one might say) in the night

---

* Steiner, *Outline of Esoteric Science* (New York 1997) pp.240-241.

† Steiner, *The Apocalypse of St. John* (London 1958) pp.112ff. For Steiner's extensive work on the senses and their differentiation, which he considered the necessary first stage of anthroposophical-scientific understanding, cf. Steiner, *The Wisdom of Man, of the Soul and of the Spirit* (New York 1971) pp.11ff.

‡ See the details in Finlayson, op. cit., pp.92-4.

in dream into spiritual perceptions. Such elaborations would then readily work back into day-life consciousness.

Intensified awareness of bird flights and movements was surely not only a way of obtaining food or decoration, or even divination but a mode of assimilating to the bird-dimension, or taking it into one's dreams. In days when catching things was by pursuit, or lying-in-wait, and by bare hands, 'getting inside' the behaviour of the prey was the starting-point of the hunter and the arduous process of getting to know the characteristic movements and ruses of the animal pursued often became a mystical identification. In their dream-life, it is likely that flying like the birds was a step into the mastery of the heaven-space which in their day-consciousness arched over the ordered landscape. Even today, such flight in dreams is a widely familiar experience, and unless one is a convinced Freudian its origin and derivation in the building-up of psychological layers remains enigmatic.* Our dream-life has shrunk into a shadowy affair compared to that of archaic humanity; but does it retain some features of a once more potent experience?

We might think of magical or inward 'flying' as alluding to the shamanic ecstasy: but ecstatic liberation of the kind shamans most characteristically describe tells of material limitations falling away, or becoming transparent or 'levelled' down so that they can move anywhere unhindered in spirit.† The shaman has to learn through spiritual exercises to be able to 'stand' firm in this new domain without external-perceptible support. Dream-flying does not usually carry such connotations, and dream-experience does not normally involve simultaneous awareness of one's perceptual state and ability to contrast it systematically with the ordinary world. The dream-experience just 'seems to be real'. That might be because it continues a mode of perception that for the Neanderthals was a real-world experience. Their consciousness would put this along with what we call waking experiences all on the same level: they would not have singled out day-consciousness as the sole reality. Oracles are often said to be grasped in a dream (the original meaning of *incubation*).

There may be something primary here of which the shamanic transcendence is a further stage. Shamans themselves preserve an idea that the inner flight, or ability to step free as the world dissolves around

---

* Freud, *Interpretation of Dreams* (New York 1965) esp. pp.428-30.

† See further below.

them, was once a more literal flight. The ascent through cosmic spheres out-of-the-body was once, they report, an event in full reality involving man's whole being. They consider that their type of spiritual detachment is in some sense a form of what was once a more total and holistic mode of being.*

Rudolf Steiner may provide a deeper perspective here, though it is one which is compatible with the approach of modern science. The basic consciousness of Nature underlies all the more specialised versions of consciousness in living things and humans. Even in quite advanced human experience, before quite recent times, consciousness was still 'objective', or actually spread out in nature. Children even nowadays go through early phases when they spontaneously fuse themselves with things that engage their attention—feeling themselves to be a wave, or a car swerving, or ball bouncing. In man's early evolution, with the Atlantean stage, the creative source which brings about our development was certainly more enclosed within us, even if not in the intensely conscious, self-determining manner of today. The cosmic forces which are our specifically mental life were for the Atlanteans becoming less purely cosmic-objective and more a dream consciousness: but they were a true dream, shaping experiences around picture-reflections of the decisive events of our evolution. That meant at the same time an expansion out of the merely earthly environment.†

When we come to actual biologically realised humans of the 'Atlantean' type, the meaning of their existence is still not something that could be grasped as an idea-of-self which gives them their identity, but is still 'outside' them as cosmic music, ancestral memory, and dream-vision. Their emerging day-consciousness is only a part of their life, which would have lain alongside the dream-world. Finally, today

---

* M. Eliade, *Myths, Dreams and Mysteries* (London 1968) notes the 'great difference between the situation of primordial man and that which the shaman recovers during an ecstasy: the shaman bridges only temporarily the gulf dividing Heaven from Earth; he goes up to Heaven in the spirit but no longer, like the primordial man, *in concreto*'—p.64;. and see also pp.99ff.

† The ability of archaic humans to dream into the real environment may be illustrated by the famous account, given by Arthur Grimble from his experiences in the Gilbert Islands, of the hunting of porpoises through collective dream-techniques: Grimble, *A Pattern of Islands* (London 1954) pp.172ff. The outcome was fully real. The episode is suggestive for understanding the magical hunting techniques of early humanity.

our sharp consciousness has suppressed our ability to actually merge with things in dream-life, which has faded into a secondary and often ignored aspect of our psychology. The return to the undivided beginning, and to the cosmic heights, now remains only as an enigmatic dream of flight.

Our evolutionary past is for Steiner not just like a film recording of past conditions, but a continuing active spiritual dimension within us. And its disparity with our actual stage of development (crudely, our body) is what we mean by consciousness. By its means we embody evolution itself. We have made of ourselves a being with a nature partly achieved, partly still consisting of all those conditions we may meet and assimilate. Materialism tries to take the fully conscious, onlooker stage as our starting-point, but history suggests the exact opposite: it shows that early humans lived in a rich engagement with the environment that was less conscious, more dream-like. Far on into historical times, indeed until quite recent centuries, assimilating experience in most cultures was based upon flooding it with a pattern of dream or myth which represented the primordial past. Myth-patterns, as is well-known to psychologists, still recur in many ordinary people's dreams today. But in the past they were the major force in interpreting and landscaping the mentality of whole cultures. It is likely that the Neanderthals lived even more strongly in the dream-world than the ancient cultures we know from the ancient East. Their ancestral memories, indeed, made dream-consciousness and creative for practical life. They were in all probability the pioneers of this fundamentally human pattern of life.

In Steiner's terms, they were living in the mythical picture of the Atlantean stage of evolution. It was by utilising these, Steiner indicates, that a kind of flight was still associated with the archetype of Atlantean humanity. It was a sort of dreaming that really happened. Many primitive traditions, of course, insist that man is able to levitate through a yoga-style spiritual discipline. Shamans 'fly' seated upon their special drum, to which they chant; it is supposed to be made from the wood of the world-tree, i.e. the axis joining Earth and sky.* How far such things were more than a part of the dream-consciousness

---

* Eliade notes that these ideas of the wooden drum and flight are distinct in origin from the actual ecstasies of the shamans: *Myths, Dreams and Mysteries* p.63; cf. pp.102ff. The theme of this state in which weight is abolished, he comments, is of universal diffusion; it is prominent, too, in the accounts of the earliest Chinese emperors.

by the time of the Neanderthals it is hard to say. As a matter of fact, Steiner is rather inclined to play down the stories of the Atlantean airships, as compared to the graphic fantasy of Indian mythology.[*] But when they absorbed their day-experiences into the magnificent dream-world of sensations and harmonies at night, it was indeed a real experience of flying. The cosmic tree whose branches are the sky, and the magic bird in its branches, were probably already familiar themes. Their bird-culture whether or not they bedecked themselves with feathers suggests that they saw human life in terms of the perspective of flight.

The dream-conquest of the skies laid the foundations for the unfolding of an inner sky-consciousness that we now experience in a more abstract manner: the ability to locate detailed perceptions within a large view and oversee the component parts of life, modelled on past experience. We have gradually learned to become our own oracles.

Territory and Conflict: Mythic Resolutions

The Neanderthals already combined earthly-oracular with sky-knowledge. Sky-gods are nearly always masculine and are nearly always associated with organisation and planning. They are often oracular, too, though they may have to share the oracle site with the Earth-powers to whom it aboriginally belonged.[†] The polarity of sky-masculinity and subterranean Earth-mystery is likely to have coincided with the Neanderthal awareness of organised landscape, and bird-dominated skyscape. What kind of leadership emerged from a consciousness straddling these two poles?

Firstly, the factors aimed at stronger pulling-together of social group seem to have been very effective: we know that Neanderthals occupied small territories, and must have had a much sharper sense of regional identity, based partly on an economic sophistication to make use of all

---

[*] Steiner, *Cosmic Memory* p.45: they 'floated a short distance above the ground' and had to manoeuvre around protruding outcrops..

[†] In classical times Apollo, associated with light and sunshine, inherited his oracle at Delphi from the Earth-mother and it retained many subterranean features. It stood at the mythic centre or navel of the Earth, but the sky-father Zeus established its validity by loosing two of his eagles from the opposite ends of the world; they met over Delphi and dropped the *omphalos* or navel-stone there. Zeus himself had his celebrated oracle-sites, e.g. at Dodona with its rustling oaks.

the resources of a limited area, rather than constant wandering. It might also suggest defensiveness and hostility to outsiders and rivalries such as accompany the sense of a greater power base.* Mithen has stressed the evidence for emotional development among the Neanderthals, and their aroused feelings could obviously fuel passionate conflicts, territorial or personal. Their relationship to oracular centres with different qualitative values would suggest that groups with different attitudes and orientations would form, and indeed Steiner has indicated that 'racial' identities based on excluding those of other affiliations emerged as a strong Atlantean feature. Violence as well as care may well have been a familiar part of Neanderthal life, as the high levels of injuries that are an aspect of their fossil record may indicate.† An aristocratic and virile society may lie at their root. Against this, however, we noted the evidence of peaceful coexistence and interbreeding with Modern Humans in the Middle East.

Racial warfare is the essence of Atlantean history as we find it in Plato, and still more so in Scott-Elliot or his sources.‡ In the latter's history, however, one feels the intrusion of nineteenth-century colonial assumptions, when a contention to be the 'superior race' was quite openly advocated and admired. The putting down of less favoured races seemed perhaps the meaning of history itself. Such ideas, however, are out of place in modern understanding. Rudolf Steiner reaches out in a quite different direction, and a surprising one—the annals of ancient Rome. The stories about Rome under its original kings, he asserts, are really revelations from prehistory.§ They are myths of a pattern widely

---

* Finlayson, *Smart Neanderthal* pp.115ff and on strategic planning, p.112. Finlayson also believes the Neanderthals were notable fighters, e.g. who fought to defend Europe by opposing the spread of early African humans, but not everyone accepts his view. Some also think that the Neanderthals finally went extinct after an all-out struggle for survival against those same (superior) humans, but that idea has been effectively demolished by the discovery of evidence showing extensive peaceful co-existence in the Near East and intermarriage. Overall we should not idealise the peaceful aspects of Neanderthals, though avoiding the old parody-view of them as 'savages'.

† Clive Gamble and Chris Stringer, *In Search of the Neanderthals* (London 1993) pp.93ff.

‡ W. Scott-Elliot, *The Story of Atlantis and The Lost Lemuria* (London 1968) pp.21ff, esp.p.29.

§ Steiner, *Apocalypse of St. John* pp.66-7 and generally pp.60ff: though written down later in the celebrated books called Sibylline, the oracular stories referring

documented, dealing with social-racial tensions—and their resolution. They are thus essentially oracular. The oracle-based culture of Greece and Rome is for Steiner the real successor of the ancient Neanderthal world.

It is an extraordinary claim; but advances in the modern understanding of mythology have essentially proved him right. The stories of ancient Rome are a version of oracular tales which are found in similar form among all the so-called Indo-European peoples, and must therefore in their essentials go back to the times of the earliest dispersion of those peoples, or even beyond. They reveal certain recurring patterns of society and cosmic vision. Many of the oldest civilisations among them long continued the practice of basing all important decisions upon oracles, so that the mechanism for the perpetuating of oracular structures and ideas extended their usage for centuries or millennia. Rome itself consulted the Sibylline oracles at every significant juncture, before they were destroyed in a fire on the Capitoline hill in 83 BC. The great comparative mythologist Georges Dumézil devoted his life to uncovering the way such archaic basic ideas at the root of society have been embodied imaginatively in myth, ritual and epic.* We find them, for example, at the foundation of social organisation and religious life in India and Iran, as well as in Rome.

It would be a diversion here to enter into Dumézil's analysis of their scope in detail. We may say, however, that in purely mythological (not philosophical) terms they set forth a kind of dialectic of society, arising as a sort of digest from its earliest stages of development. The first 'king', Romulus, is not a literal historical figure but embodies in all its ambiguity the basic tension of communal life. He is the quasi-divine founder, and his role is to make things happen, to bind and to loosen, to bring people together both good and bad, nobles and criminals, slaves and

---

to the Roman kings are claimed to antedate the epochs of historical events and foretell them. They thus represent the spiritual-cultural forces which shape the successive civilisations of later times. Their oracular character means in effect that they go back to consciousness of the Atlantean type. According to Steiner, the Christian Apocalypse is not literal future prophecy but a work with a like intent of revealing the pattern of society and the world, and so a sort of successor to their Sibylline role. Hence many similar sequences can be found there.

* G. Dumézil, *Mythe et epopée* (Paris 1995, collecting studies from 1968-1986), especially pp.289-312 and pt. III ('histoires romaines') pp.1283ff. The main stories are preserved as 'history' in the early parts of Livy's *Rome from its Foundation (Ab Urbe condita)*, published under the Emperor Augustus to define Rome's place in the world.

immigrants. Society does not originate in some high-minded contract but in a dynamic of interaction. Romulus is the mix of human motives that make us social animals, and is immortalised as the god of the ever regenerating Roman youth, ever resurgent in its energy, society always starting over again. In some sense he is the eternal mixed-up nature of humanity, since he and his brother resemble the old myth of Man and Twin, first to live and die respectively (Remus is killed in a dispute by his brother Romulus). The 'next' king is Numa, who embodies contrastingly the aspect of society which aspires to regularity, codification, fixed modes of behaviour, conformity and respect. His reign is a mythical time of peace and plenty, and he himself is married to the divine powers ruling in Nature, in the guise of the nymph Egeria who whispered to him her elemental (oracular) wisdom. The third king is Tullus Hostilius, in whom the warlike urge breaks out, now more sharply defined, in an age of constant fighting. This time, however, martial energy is channelled and organised to benefit society, conquering enemies to take away the threat of attack and subduing the factors which undermine civilised life. In this mythology, society does not proceed simplistically from primitive roughness to civilised stasis: rather, the tensions inherent in the founding figures have to reach the point at which they fight it out, and then discover how their opposing qualities can creatively co-exist. The wisdom which emerges out of the dynamic process is embodied in the fourth king, Ancus Marcius: on the foundation of dynamic social order he organises the industry and prosperity of Rome.

Dumézil has brilliantly shown, in fact, that the contending and harmonising forces involved in the myths are those of the threefold social order of the Indo-Europeans, from Rome to India: its three classes were based on an ordering sovereignty (Romulus, Numa), the warrior or individualistic ethos (Tullus Hostilius) and the economic sphere of fecundity or prosperity (Ancus Marcius). (The other Roman kings making up the seven are a further working out of the derived relationships.) To be viable, any society needs to have these three functions operating and interacting, and to have the balance between them correlated to suit particular times or stages of society. Rudolf Steiner has a modern version of them, though it must be stressed that his 'functions' are not parcelled out among different castes, but all undertaken to some extent by every individual citizen. In archaic society, separate tribes played each role. The old oracular myths, at any rate, encapsulate a profound understanding of the forces working in a healthy society. They can be shown to underlie the great Indian national epos, old Norse sagas and many other mythologies, notably the ancient Iranian legends about the great dynasties who successively ruled the world: experts in Iranian studies have been able to uncover the sequence of ideas (presented as time-periods): legitimate sovereignty is first established, then overthrown by a violent usurper (represented as 'Turan', or anti-Iran), then triumphantly restored—a parable of society essentially similar to the stories from Rome. An extremely archaic common source must therefore be presupposed. And one may add that a basically similar construct is to be found in China.* Steiner's idea that it stems from an oracular, prehistoric model which laid down how society had to be, thus seems not implausible. His own sequence describing society's stages in the Atlantean epochs could easily be seen in a comparable light (above, pp. 159ff).

Of course societies must originally emerge in sequence through historical stages. Myths, however, are a sort of digest of many and varied efforts at forming society, streamlined together for subsequent use. Their aim is to reinforce the achievement of a viable society, not to take it apart historically. Prehistoric humans must already have sketched out the physiology of the earliest societies. There is of course still a vast time-gap to be bridged from even the oldest high cultures back to the Neanderthals and comparably early humans: but the supposed upheavals of races are increasingly being replaced nowadays by a 'continuity paradigm', positing that languages and ideas are more deeply ingrained and often go right back to prehistory. Certainly the oracular pattern is a sort of mythic charter of extremely archaic origin, indicating in the form of a digest precisely the sort of history which Steiner spells out in his description of the successive Atlantean epochs. In his account the earlier, pre-individualised 'races' have names which were already used and had been published by Scott-Elliot: Rmoahals, Tlavatli. All the rest, however, have names which link them to well-known cultures and peoples of much later times—not necessarily naïve identity, but with a significance nonetheless, as we saw in the case of the seventh or Mongol Atlanteans, who lie behind the strong genetic link of the Far East

---

* Dumézil, *Mythe et epopée* pp.67ff for the heroes of Indian epic (and generally pt. I); Iranian and Norse, pp.263ff; China, pp.289ff. The Iranian sequence of rulers was analysed by Stig Wikander, who concluded that 'the Iranian legend is an heritage from Indo-European antiquity'—his view is summarised in J. Duchesne-Guillemin, *Religion of Ancient Iran* (Bombay 1973) pp.225-6. Yima, the first king, described first as 'shining' then fallen to Earth, is linked to the First Man-and-Twin mythology we noted when discussing Romulus/Remus; Indian myth calls him Yama, first-to-die and so king of the dead. The name is related to Latin *geminus*, 'twin'.

populations with the Neanderthals. Others may reveal their secrets too. The Tlavatli were the first who could 'look back upon certain deeds' and 'demand recognition of them from his fellow men',[*] establishing continuity with the dead as the example for social life. 'Toltecs', the next group, turned this foundation into the first 'forming of a state', based on the principle that 'what an ancestor had done was esteemed by his descendants'[†], i.e. a kind of inner consent rather than the compulsive power of the earlier stage. One may well recognise the stages of Romulus (binding and loosening, leadership, vigour) and of Numa (settled modes of behaviour, peaceful acceptance). 'Kings and leaders' came into existence; 'enormous power was in their hands, and they were greatly venerated'.[‡] But in the fourth phase power went to the heads of emergent dominant, so that 'man wanted to count for something through his power. The greater the power became, the more he wanted to exploit it for himself' and violent misuse, especially of the power over living plant-forces, was the result.[§] In the Roman legend comes Tullus Hostilius; however, the story reminds us also of the Iranian world-dynastic mythology and it is no surprise that the designation given to the fourth Atlantean 'race' as transmitted by Rudolf Steiner is First Turanians,[¶] i.e. the name taken from the usurping, violent people who come to ascendancy after the original archetypal rule.

The story of the origins of society in Steiner is as realistic and as archetypal as the old oracular myths, out of which gradually emerged the threefold understanding of the social order. The social elaboration of human potential was another case of an evolutionary process that threatened to become self-entangled, leading to its own destruction. With early socialisation, as Steiner says in a vivid metaphor, 'it is as if the feet were stubbornly to carry a man forward, while his torso wanted to go backward'. 'Such a destructive effect could only be halted through the development of a higher faculty in man. This was the faculty of thought.... If previously people had given themselves over to every desire, now they first asked whether reflection could

---

[*] Steiner, *Cosmic Memory* p.52.

[†] Steiner, op. cit., p.53.

[‡] Steiner, op. cit., p.54.

[§] Steiner, op. cit., pp.54-5.

[¶] So styled in W. Scott-Elliot, *The Story of Atlantis and The Lost Lemuria* (London 1925) p.18. The terminology has sometimes become slightly distorted from retranslation of Steiner's German renderings back into English.

approve of this desire.'* Thus rationality did not emerge by any sort of inevitable rise of *Homo sapiens*, but was the evolutionary solution to the life-and-death issues that beset society, as it struggled to contend with the powerful contradictory forces within it. (One of Rudolf Steiner's key social insights was that human problems can hardly ever be properly resolved within a dualistically conceived order—rulers and ruled, managers and workers, or whatever.) In his view rationality itself has a primarily social meaning, and by rationality he evidently means the practical and human understanding of how to make things work.

The symbolic Roman 'king' for this resolution-stage is Ancus Marcius, who organises everything that enables Rome to engage in trade, to build, and cultivate and consolidate. He is the representative of the third function in the Indo-European model, just as Steiner's version of prehistory requires a third term to be added to the conflicting elements of order and violence that emerge in archaic society. In the Iranian dynastic myth, the name of the hero who overcomes the usurper, the Dragon-power, and restores the true kingship, is Faridun. His nature is appropriate to his role, since as a hero he is explicitly to be contrasted with the violent nature of the warrior caste, whose exemplar in myth is the formidable Kereshasp. The warrior-nature has its place in society, or in a sense only outside it: the military might of society must only be deployed against its external enemies. (Societies which use their army against their own people are rightly called corrupt.) Faridun is the exemplary mild and friendly hero, more of a community policeman than a death-dealing killer. But his is the power to beat the Dragon. In accordance with the model, however, the Dragon he defeats can never be wholly neutralised but only ever and again be pushed back—marginalised, 'dragged to the edge of the world'. Society will always generate problems, and have to solve them, to push back the usurping forces. And so we see that Faridun in the Iranian version too is essentially the hero of the third function, fecundity, industry and human problem-solving generally.

Iranian legend draws the gardening-parallel with weeding and protecting the crops. The reference to agriculture is obviously anachronistic when we are dealing with the original prehistory. Myths are always enriching themselves as they are retold. The idea that the arrival of reason on the scene was the coming of mediation between otherwise con-

---

* Steiner, op.cit., p.55.

tending factions, however, is absolutely pertinent to understanding the forming of three functional spheres, which continually enabled human life to flourish in all those innovative ways. We can still see the link to our society today. We still, ultimately depend on those primal 'heroes' of humanity who, for instance, demanded that kings and warriors 'listen to reason'. The gods of the third function establish the pattern of Mystery-deities, in which life and death are integrated in a higher life. They are typically represented dying and rising to new life, uniting people in brotherhood, bringing co-operation.[*]

Community was then at a stage when it could (and needed to) move on beyond the purely 'oracular' model, when everybody lived after the same remembered pattern of life. The solution to emerging tensions was to admit a variety of socially legitimate needs. Such a step probably needed more elaborate developments in language above all, such as many recent researchers have surmised at least for the later Neanderthals. The origin of reason was certainly in part a shift from the old musical language to a form that was more inwardly articulate. Even before the threefold ideal was mythologised by the archaic Indo-Europeans, Steiner indicates that it had already established itself in a more basic 'Atlantean' form, which takes it back into deep prehistory. It may be epitomised as: oracles—lawgivers—Mysteries. For society to maintain a living, evolving reality, there had to be interplay between these three: oracles, where diverse questions are brought to a diversity of absolute authorities, empowering individuals; the conformity of 'the state' represented by the lawgiver like Numa, which treats everyone on equal terms and treats of the group as a whole; and Mystery-association, where change and accommodation enables working together, requiring the life-and-death of adaptation to others for practical co-operation, shaping diversity. Such a threefold prototype may well have roots very far back indeed.[†]

The reader may object that in all this I am simply projecting conceptions from later history and elaborate mythology onto the flimsy

---

[*] The life-and-death drama of Mystery-rites may not sound very rational, but Rudolf Steiner was a pioneer in arguing that it was exactly such Mystery-conceptions which eventually led to Greek philosophy itself—cf. his *Christianity as Mystical Fact* (New York 1997) pp. 22ff.

[†] I have proposed this framework for understanding Steiner's ideas along these lines in my Introduction to his *Mystery-Knowledge and Mystery-Centres* (London 1973) pp.5-14. We shall explore more of the material in due course in connection with 'post-Atlantean' prehistory.

THE THREEFOLD STRUCTURE OF SOCIETY

Social spheres of activity and their mythic dramatisation following Dumézil and Rudolf Steiner

Dumézil:

First Social Function:	Second Function:	Third Function:
'Sovereignty'	'Warrior'	'Fecundity'
Gods/heroes of order,	Confrontation,	Gods-men interact,
agreements, rights	self-sacrifice, clemency	regeneration/ healing

Two types:

a) magic compulsion

b) received law

Rudolf Steiner:

'Rights' sphere	'Individual'	'Brotherhood'
Equality, State	Initiative, taking-on-self	Interaction, economics
LIMB-SYSTEM	HEAD-SYSTEM	BREATHING-SYSTEM

Archaic activity:

Hunting/sharing	Oracles	Mysteries
Acceptance of role	onfronting the unknown (threat)	Transformation, death/life

Supposed Kings of Rome = prehistoric social-origins myth, showing emergence of conflicting tendencies and threefold resolution

a) Romulus, binding/ loosing	Tullus Hostilius	Ancus Martius
	violence/	Wealth, favour
b) Numa, law-giver	confrontation	of the people

evidences of Neanderthal life. Perhaps the steps of developing co-operation, reasoned language, etc. took place, but did not their realisation in economic and spiritual terms belong in reality to much later ages? To this one can only reply that the minimalist interpretation of everything Neanderthal has proved increasingly inadequate. Many researchers consider that we should now give them the status they deserve in the record of human development. Who had thought they could create underground structures like the one at Bruniquel? And we should remember that what we call 'primitive' societies very often enjoy a vivid and far-reaching life of myth and symbol. The imaginative building up of these aspects had consequences just as great for human evolution as the once over-valued steps in 'technology' that were supposed to define the rise of prehistoric savages toward man's existence. Monuments like the Bruniquel cave indicate that we hardly know the full scope of Neanderthal building, for no obvious utilitarian purpose. In general, moreover, if Steiner's 'Atlantean' typology is taken seriously, we may expect an ever-increasing recognition of Neanderthal sophistication in many spheres of their cultural life. The truth is that Neanderthal culture clearly evinces that ability to overcome social conflicts and synthesise solutions that all genuine civilisations require.

In other words, it clearly advanced, despite its emphasis on memory. Reluctance to recognise its achievements still makes some researchers reject the evidence of the Chatelperronian industries, in which Neanderthals seem to rival the expertise of the Upper Palaeolithic.[*] We also know now that Neanderthals could produce visual art, closely associated with the Mystery-sphere and its techniques, though they did not give it the prominence of the post-Explosion phase, highlighting their oracular and musical side instead. The synthetic power of reason could ultimately also be expressed in abstract symbols. Thus we have evidence for the dynamic interplay of Steiner's three social components, and a resulting civilisation which could perpetuate and renew itself over long periods of time. It is not difficult to see already in such a prehistoric cultural model the social self, embodying the power of evolution and change. It is working with the three structures of higher-animal development,

---

[*] E. Trinkaus and P. Shipman, *Neandertals* (London 1993) pp.376-9 describe the strong evidence for the Neanderthal context of the Chatelperronian, and the way it polarised opinion about the general nature of the Neanderthals. Even today, experts such as Slimak are unwilling to admit the connection because it does not fit their conception of Neanderthal abilities: Slimak, *The Naked Neanderthal* (London 2023) pp.146-8.

present in human beings as the three complexes, head, breathing and metabolism (picturable as eagle, lion, bull). The free ego is enabled to balance and combine them, in an extension of man's being into communality. However various in expression, it is the same interaction of elements in human nature that was working already among the Neanderthals, and still in our contemporary social order. Our unstable posture in its many permutations thus finds its fullest range of expression in social enactments, reflected even in the most ancient times in the complementary role-play of the myths.

Intermezzo: Myths of Atlantis

Myths bring us back again to Atlantis. But Plato's literary myth of Atlantis has so queered the pitch that we need to start afresh with an understanding of what other myths might be relevant. Not that it helps to assemble every story about a vanished land or a sunken continent. Some may indeed be apposite, but we need to come to them in the framework of a perspective in which to interpret their implications. How did ancient peoples really experience these myths? Rudolf Steiner's history of consciousness provides exactly what we need. In it the myths and legends take on an important new role—comparable, as he says, to the geology which interprets the fossil record. At this crucial juncture, we need to discover the additional dimensions that are present in the original, more authentic myths of Atlantis and the Flood, prior to Plato. Rudolf Steiner enables us to connect them with mankind's evolution in a still more thoroughgoing way. It is hardly too much to say that these myths are the foundation of our whole modern consciousness. But they also reveal how our modern consciousness is grounded in the events and struggles of prehistory.

In his view human beings bring to manifestation the evolutionary nature of the world—which is essentially another way of putting the Anthropic proposition that we become conscious of the very world which is 'finely-tuned' to produce us. To Steiner's higher perception, the evolutionary formation of that world is also the honing of our emerging human constitution. We speak the *logos* which was the creative power that made us; we dream the connection to our surroundings which we have subsequently suppressed in gaining ego-awareness. So that in short our consciousness is the *other side*, the spiritual resevoir, of our physical evolution. Left behind physically, the totality from which we emerged still hovers around us, layered in images and significances. 'Humanity' so conceived is the potential for change which cannot be contained in us all at once, but floats dynamically around us as consciousness (or what we are currently able to experience of it), the cosmic side of whose history is related in graphic descriptions by Steiner in his fundamental works. With the help of Steiner's method we can find again within us those levels of consciousness that still interpenetrated the world around. Ultimately they submerge us in the dim panpsychism of the primordial substance of the world. The Atlantean dream-consciousness to which we referred was still less egocentric than our own. It was still able to give our remote ancestors a vision of their place in the cosmic order, not yet in ideas but by means of tones and images, harmonies and deeds to be imitated. Neanderthals seem to belong to such an inner world, with their musical consciousness and fascination with the sky that was led by the birds. Our own dream-world still has archetypal qualities, but has long been eclipsed by our conscious mind.

The atavistic dream-consciousness did not fade all at once but, according to Steiner's understanding, still finds an echo on another level at a later stage of human development. Once again we see a manifestation of those great evolutionary rhythms he discerned. We like to think of the Greeks discovering rationality, the modern mind and science. But there is another side to them, still almost as important. With them the dream-images have become less vivid and compulsive, but they still remain as ideal presences transfiguring their thoughts, the vivid world of the gods still hovering over all their achievements. In fact, that strange fascination with which prehistory-researchers Clive Gamble and Chris Stringer note the coincidence of the Neanderthal and the Graeco-Roman worlds (above, p. 32), becomes for Rudolf Steiner a clear insight: the world of 'Atlantis' (not in Plato's sense but our real prehistory) was lived over again imaginatively in the religious consciousness of the Greeks. Steiner thereby arrives at a fascinating analysis of a feature of the Greek mind which is normally taken for granted, or at any rate passed over without comment—namely, the transcendent quality of their gods, who for all their vivid human traits tread a path of existence which is distinct from ours, and is never impinged upon by death, suffering, failure, weakness or indecision. Yet they present themselves as 'forms more real than living men', not mere echoes of human thought; the universe is theirs to possess and not humans', and even Plato thought we were no more than their slaves. An intense reality, which now seems incomprehensible to us in our self-centredness, still hovered over them.

The oracle of Apollo told mortals to 'know themselves'—meaning primarily to know that they could not be gods. Men aspire to be great and glorious, powerful or impassioned. In this they fitfully resemble the gods, but Greek art and Greek tragic poetry expose the gulf that opens before all efforts to possess a state like theirs. Gods and men go separate ways. The Delphic theology moulded Greek life and thought over centuries—still showing the power of oracular authority over men's minds. Nowadays we know the Greek viewpoint best from art and literature, though we must not therefore suppose that their gods were just beautiful inventions. Greek art was pervaded, however, by the Delphic attitude, as in the famous 'division of styles' which decreed that heroic deeds should be represented in a 'high' style, avoiding reference to commonplace or weak things, which required a low or comic treatment. Writers and artists were not just creating individual artworks but intimating a whole world of tragic breadth and intensity: a world of consciousness that was not contained within human bounds or temporal limitations. Their heroes such as Achilles or Jason seem just as timeless as their gods. Their pastoral realm of Arcadia has a beautiful simplicity that can never cloy, and will never fade. Like the gods and the heroes it belongs to a past that never goes away, so that we can see it always before our inner eye, yet only to know that we can never make it a present reality.

Rudolf Steiner offers a brilliant analysis of the Greek attitude in connection with his concept of rhythmic phases of development. The mythic figures of the gods, such as Zeus who wields the thunderbolt, or Artemis the magic huntress, have their reality in the time of Atlantis when there was a less separative consciousness. People like the Neanderthals, as we have argued, still echoed that sense of participation in Nature in a dream-like way: they would have lived less in their own separateness, and more in the great deeds of an ancestral hero who merged with or was a link to the divine forces in Nature. They could still feel the reality of those deeds and make themselves the protagonists of the attitude and achievements they remembered. Memory was not just in their own heads, but a presence that had power to compel and to fill the minds of a whole community. The Greek consciousness with its myths, explains Steiner, was partly still a continuation of that older dream-like 'clairvoyance', of sharing in the divine-creative forces. But the reality of it had faded. Humanity had in the meantime cultivated the sense of the ego, and Greek culture was at the forefront of that venture in evolution: as a result, the sense of unity with Nature was atrophied. The old dream still hovered over humanity, but no longer with the old immediacy.

It had once been a true dream, or clairvoyance, in Atlantis, and the Neanderthals can be seen in that light, especially those charismatic leaders who could be lifted above themselves in oracular inspiration. It was their dream-consciousness which continued to survive, though disconnected, as humanity continued to evolve—especially where oracles continued to be authoritative. The idealised 'gods' were an older humanity's self-experience, preserved in later times by the oracular world-view.

We must conceive, then, that 'during the time of Atlantis these gods were human beings, with souls which had a macrocosmic significance'.* We have seen how they were not modern ego-people in their consciousness all through prehistory, but we can say of them in Steiner's words, 'if clairvoyant observation were directed to this Atlantean man who is Zeus'—for instance—'it would acknowledge: As I observe this soul, it becomes larger and larger, it broadens out and is in actuality the macrocosmic counterpart of human soul-forces .... The same thing applied to those other Atlanteans who were really Greek gods.'† Such a condition of consciousness belonged in reality to a time in the distant past. Yet human beings of classical Greek times still experienced that consciousness over and above their emergent ego-consciousness, even though they could no longer ground it in a reality which, in fact, had long ceased to exist. The gods belonged to a consciousness that could not be connected with present earthly conditions, nor with history, time and change. Yet these gods were not an abstraction based on the present-day world, but a more primary truth which demanded recognition—indeed worship. It was also significant that it could be fixed in the form of a fascinatingly perfect *image*.

Steiner took from another mythology an appropriate term to characterise a consciousness which is in this way out-of-step with changing conditions of life: he called it 'luciferic'. He regarded it as a tendency of consciousness per se to hold on to past states in this manner, even whilst acknowledging change. Consciousness is not simply a matter of opening itself in response to the world, it also has a tendency to hold on to that content of awareness and prolong it beyond the conditions which prompted it. (Steiner thought of both tendencies

---

* Steiner, *Wonders of the World. Ordeals of the Soul. Revelations of the Spirit* (London 1963) p.88. This course (or 'cycle') of lectures is fundamental for Steiner's understanding of the development of Greek consciousness, itself crucial to humanity's subsequent historical development.

† Steiner, op. cit., p.86.

as basic evolutionary processes, which always involve both adaptation and stabilisation (homoeostasis). Even when realised mentally they are still not subjective but cosmic and biological in nature.) Consciousness by its very nature involves a damming up or check on the pure flow of experience. It is our holding-on to a state that generates awareness. But when cut off from reality in that fashion, a content of consciousness by no means remains a purely objective picture of the reality which prompted it. It acquires many features that are humanly fascinating, as well as dangerous and illusory.

Though not in origin delusional, it represents an engagement with the world that has lost its connection with the changing context of knowledge. That may be because of outer changes, but very often it is through a further internal development of consciousness as it assumes a more ego-centric character. Thus, for example, we remember childhood feelings of enjoyment or innocence. Such a set of representations presents itself therefore as intimating a paradisal state, for which it sets up an intense longing or nostalgia, since it represents a fuller reality which has had in some measure to be sacrificed or left behind as we grow and develop. We feel that other beings evidently may still possess this state, but humans are excluded from it by their very nature as changing and ageing—their inability to remain ever young. Our affinity and our estrangement from the state which we divine in connection with it generates a sensation of unsurpassed poignancy. The history of humanity cannot seriously be studied without a clear understanding of tendencies such as Steiner indicates under the term 'luciferic', which is not mere fantasy but inherent in the evolution of complex consciousness. It is intrinsic to all the great cultures and civilisations, in their aspirations and their sense of human limitations.

'All that arises in the course of history,' says Steiner, 'in the guise of wonderful programmes, marvellously beautiful ideas, by which it is always believed that somehow or other a return can be made to the Golden Age—all this has its origin in the Luciferic tendencies which flow into man. Everything by which he tries to loosen his connection with reality, to soar above his actual circumstances—all this points to the Luciferic.'* It is one of the most important aspects of Steiner's life's work that he restored and clarified the understanding of the luciferic urge, and showed how it is an integral part of human evolution. The term has its origin in the myth of Lucifer. In its vulgar form it tells how

---

* Steiner, *Three Streams in Human Evolution* (London 1965) pp.16-17.

Lucifer was once a glorious archangel, but is now so no more; it tells of his aspiration to full Godhead but how the sense of his own glory and power, leading the angels in heaven astray, proved empty and landed him in Hell. He and his angels fell, yet not before out of spite or envy he had infected the newly God-created humans with his vain desire and sinful self-glorification. By Steiner's time it had become a Sunday-school fable. In that well-known form it long post-dates the Bible, and is essentially mediaeval or later. Steiner often lamented that even great poets like Milton had failed to delve clearly enough into its significance. However, advances in understanding the ancient Near East have shown that the allusions we do find in the Bible were already part of a far-reaching mythological complex. We now know that from very early times it was connected with ideas about knowledge, rulership, and spiritual aspiration, and was based on a sophisticated time-perspective. Even prior to the biblical materials, we find mythologies which do indeed show a profound comprehension of the significance which Steiner imputes to the luciferic tendency itself, humanly and cosmically.[*] The myths revolve around a celestial phenomenon which can be witnessed on many a (clear) morning before the Sun rises.

The planet Venus is closer to the Sun in comparison with the Earth and therefore, when it appears in our skies, precedes the Sun's rising and may be seen shining out with the first hint of dawn in the East.[†] Nowadays we know that it is a planet relatively close to our Earth as well as to the Sun, and also that it is wrapped in reflective cloud, two factors which make it reflect the Sun's light to us with unparalleled intensity: after the Sun and Moon themselves, it is the brightest object in the sky. When it shines in the half-light of the breaking day, its dazzling appearance speaks to the observer both of its own glorious brilliance and, as it were, prophetically of the full daylight that is starting to dawn. Because of its brilliant appearance and timing the planet was called therefore the 'Light-bringer' (Greek: *Phōs-phoros*; Latin: *Luci-fer*), and names with a similar connotation. A picture can readily form in the mind. Since the Sun has not yet appeared on the scene, it seems initially that the day dances

---

[*] A comprehensive and brilliant literary appreciation of the mythic sources in the light of modern historical research is furnished in Neil Forsyth, *The Old Enemy. Satan and the Combat Myth* (Princeton 1987).

[†] Of course it shines in the West at evening, but quickly disappears from the night-sky because of its proximity to the Sun, disappearing soon after it sets. Whether or not the identity of Morning and Evening Star was realised, the phenomenon of its first-light appearance was mythologised quite independently.

attendance on the bright star which never shows itself to us except in the context of the opening dawn. It is a deeply stirring sight. Even as its light swells, many of the night-time stars are still visible too, especially the brighter amongst them; but even they now seem humbled, even subservient, before the fascinating brilliance of the Morning Star. Then, however, a strange and disturbing drama plays itself out. Just as the radiant Light-bringer reaches the height of its opulence, the glow of the sunshine irradiating the atmosphere pushes up from the horizon and floods the sky, until at last the Sun's disk appears. At once the Light-bringer starts to look second-best, and its light to be increasingly lost in the breaking day. Its hosts of ministering stars quickly fade from the sky, disappearing as if they had fallen out of heaven altogether. And in the all-engulfing brightness of full daylight the Light-bringer himself cringes and shrinks, soon vanishing altogether, not by any external agency but by the dismal failure of his own glory in the greater glory of the day. Such at least is the suggestion which evoked a potent myth.

Outside the biblical sphere, we know that many ancient cultures long preserved features of the 'cosmic-consciousness' such as we have suggested can alone explain the monuments and culture of the Neanderthals. Such 'oracular' consciousness lifted human beings at least temporarily to the level of the divine authority—especially leaders and kings. The prophets of Israel, who set the tone of the Bible, on the other hand, consistently disparaged and mocked such pretensions, considering their own God to be exalted far above the level that could be reached by any mortal. With their conviction of Israel's supremacy, great prophets like Isaiah (Ch. 14) and Ezekiel (Ch. 28) both taunt the kings of the nations for their claim to have ascended into heaven, to have been in the 'garden of God' (the night sky) and moved among the stars. Well!, they say derisively, 'Daystar, Son of Dawn, how you have fallen from heaven!'* From their point of view, the knowledge of the true God has come like the 'dayspring from on high' which shows up the inner weakness and failure of the briefly glorious Morning Star, as claimed by a gentile ruler. Thus in the Bible previous attempts to claim divine authority are denounced as arrogance and emptiness, stolen glory which should

---

* F.H. Borsch. *The Son of Man in Myth and History* (London 1968) pp. 96ff. for a presentation and analysis of the biblical material. For the crucial evidence of Ezekiel 28 see pp.104-6 and p.104n6 for further studies. For Isaiah 14 see further Forsyth, *Old Enemy* pp.134ff. Specifically for the Canaanite (Ugaritic) background to 'the planet Venus, attempting every morning to ascend the heavens but always thrust down again by the rising of the sun,' see Forsyth pp.130-2.

be attributed only to the one true God whom they proclaim. It is a curious feature of the myths and their denouncers, on the other hand, that neither of them really quarrels with the other's analysis. The Bible's prophets do not deny that the rulers did ascend into heaven—they only taunt them that they were unable to remain there, i.e. to hold on permanently to their godlike knowledge and power, which it is not for humans to possess. The myths of the old Near East, conversely, admitted frankly that mankind (personified for instance in Adapa, a sort of Babylonian Adam) quite understandably tried to trick the gods and tried to steal the divine secrets which would have given him immortality. But he fell short, as did also the hero Gilgamesh; etc. Much of this mythology was reflected in the well-known story of the Greek figure Prometheus. He was the creator of mankind from red clay, and stole fire from heaven to animate and inspire his creatures—fire understood both as the external means of culture and as the rebellious spirit burning within him. In retaliation, mankind was burdened by the gods with many ills such as death, disease, and longing. Prometheus was punished with agonising pangs, and mankind too must bear their divided nature, always aspiring to heaven, always falling back to Earth and tormented by their glimpse of a happier state.*

The biblical prophets of Israel bring a new and different idea of revelation. It was not based on the idea that mankind had in some way managed to seize an element of divine power and Mystery-knowledge, but insisted that we should wait for God in good time to reveal himself to those whom he chose. History is always many-sided, however. The extended encounter between Israel and the cosmic wisdom of the Near East was not ultimately just a simple line drawn in the sand. The prophets' message about God was clear; yet it still had to be explained why human beings were always trying to pre-empt God, and claim their achievements as their own. Humankind had never looked purely to God, but had tried to master and understand, indeed to worship, the world. The themes of cosmic esotericism which lie behind the Day-

---

* In Greek myth, Prometheus is a figure who in many traits resembles the Jewish legend of Satan which developed out of the biblical materials. Though seen as a benefactor to humanity: he nevertheless tricks the gods. The punishments inflicted on human beings for accepting his gifts closely recall the story of Adam and Eve. Later he is unambiguously heroised, but in his early development is rooted in more complex and ambiguous attitudes: Forsyth, *Old Enemy* pp.86-7. For the Babylonian myth of Adapa ('Man'), who should have cheated the gods, Borsch, *Son of Man* pp.100-101.

star myths still demanded recognition. It is rather that although given a novel re-interpretation in the light of biblical monotheism, they persistently reappeared on the margins. The biblical message never proved wholly satisfying in itself. Officially rejected, works like the Books of Enoch were called apocryphal (or worse). Yet they testify strongly to the fact that even in Israel cosmic themes could never be fully banished by the Bible's message, though thrust aside to give the prophets' voices pride of place. It is actually in these apocrypha, and works like the Dead Sea Scrolls (the Essene writings), that a fuller diagnosis of changing spiritual conditions, and even (in our terms) changing consciousness, is directly addressed.

In Jewish religious thinking of the Essenes, the idea of the 'double revelation' takes up the myth of stolen wisdom from before the Flood associated with fallen heavenly beings. A subtle time-analysis also emerges. The knowledge which the 'rebel angels' gave to humanity is, in the eyes of this esoteric Judaism, not untrue but morally tainted. It belongs to the past, as does all luciferic wisdom, and should be rejected because God has since revealed himself. The Enochian tradition is confident that God has revealed to humanity all the knowledge needed for his existence, and moral advancement, *in due time*. Here it shows solidarity with the Bible tradition, which grew out of pastoral nomadism. The Bible lays great stress upon the life of 'waiting on God', which gradually became the idea of *history*, or everything in its time—whereas the Near Eastern mythology, as we saw, frankly admits to forcing the hand of the gods, eluding time at least in part. The knowledge that had been attained in this way is like the bright Morning Star which seems to be heralding the future and bringing the dawn; but as stolen knowledge it turns out to have no legitimacy, no solid base, and when the true dawn comes it collapses and reveals its inner vacuity. The 'double revelation' doctrine insists that the true version comes with time and the true God.

Cosmic wisdom was rejected 'as demonic knowledge coming from the betrayal of divine secrets (I Enoch 16:3) by the fallen angels,' writes Martin Hengel in one of the most penetrating studies: 'in effect this included all the wisdom of the pagans and the refined culture of the Hellenistic period.' The claim has been variously assessed by scholars. To some the Jewish esotericists seem to be just swayed by ideas they imbibed from the cultures surrounding them, and then claim that they would have got there anyway. Enoch especially, but also Noah and Abraham, were supposed to have received the true knowledge, so that the 'division in history was matched by two opposed streams of revelation and wisdom'.* But it is the subtle time-analysis which shows that something deeper and more interesting is really going on. An idea of changing consciousness, of deepened understanding, is suggested. The esotericists insist that they are the ones who have first penetrated to the real meaning behind those imperfect ideas, which had in past times purported to be the whole truth in themselves. It was their supplementary understanding which enabled older ideas to be combined with what is original and spiritually valuable in the biblical line of development that has been added on—the prophetic voices which were, in reality, the latest mode of oracular knowledge. At the heart of all this lay the idea of knowledge continuing and developing through time. Knowledge which had once sufficed needed to be given ever new forms. In a most remarkable way the concept of luciferic knowledge that is no longer applicable to the time uses the Daystar-mythology to symbolise the changing conditions of consciousness.† A past configuration which is no longer right for the present reveals its inner emptiness, and when the new form of truth that is needed blazes as the Sun, while Daystar/Lucifer and his cohorts fall from heaven, a new day dawns and God sweeps away the old order to make way for the new. But old truths are not so much abolished as reconnected with reality. (The Jewish 'Old Testament' in turn was treated in the same way by Christianity when it reconstituted it in the light of the 'New'.)

The Lucifer-mythology recognises the reach of the past into the pres-

---

* Martin Hengel, *Judaism and Hellenism*, I (London 1974) p.243. The Hellenistic period is the last flowering of Greek culture and thought when Alexander the Great made it a world-power, much of which was simply adopted by the Romans.

† It may seem surprising to attribute such a complex purpose to mythological stories. But myths tend to exhibit 'life in some great symbolic attitude', and bring out fundamental issues in practical terms. Myths are more imaginative than rationality, and often play upon our divided loyalties. Thus, the myths imply that rebels are a part of real order, which does not simply perpetuate itself as the Bible for instance implies we should just obey God. Originally important gods famously tend to fade; conversely, even established ruler-gods like Greek mythology's Zeus came to power as rebel tricksters in their time, with the help of a Prometheus. Myths did (and do) make us look therefore at how we see the world and its orderers, and explored how we might acknowledge other viewpoints. Rudolf Steiner was powerfully aware that mythic and imaginative pictures of change and of changing spiritual allegiance are essentially far more profound than rationalist ideas of 'progress' etc.

ent-day order even when it is no longer valid as an expression of reality, i.e. it has become 'rebellious' to it, refusing to go like the Star exposed when true Day comes. Our loyalties are inevitably divided. For in one way or another our being human is inconceivable without the double consciousness which the concept of the 'luciferic' articulates. The false brightness is after all the ideal of perfect beauty, an image of the past made perfect, or youth with all its promise that we always recall. Some-one who lived wholly without awareness of the pull of other times, of other possibilities, of ideals or illusions, would not be recognisably human at all. We need our never-never lands like Arcadia, but we also need to realise how they can be harnessed to achievable goals and the vision of a real New Age such as the Bible predicts. Rudolf Steiner's rehabilitation of the 'luciferic' from its moralistic and truncated guise is an achievement in itself—one which enables us to evaluate in a pro-found way the contributions of Bible and classical thought to our con-sciousness, while also prodding us into the realisation that the scale of the issue is bigger than both—as big as the question of our humanity itself. It does so by bringing us to understand the spiritual reality inher-ent in the myths, affecting and shaping human lives.

The Lucifer-myth rightly insisted, says Steiner, that the Greek feel-ing for the ideal should be recognised as a longing that emanated from a point in the heavenly spaces, from the planet Venus, the rebel Daystar.* In other words, they succumbed to the lure of the 'star of infinite desire'. However, in Steiner's day it would have been hard to find evidence to corroborate his assertion. Progress in tracing the development of both Greek and biblical mythologies, on the other hand, now makes his assertion fruitful against the larger backdrop of Near Eastern and more far-flung Indo-European cultures. It has been realised that there is a neglected Greek myth which does indeed con-nect with the whole complex of ideas we are tracing, namely the story of Phaëthon ('Bright Shiner'). A son of Helios, the Sun-god, he wants to prove his divine origins by driving his father's chariot through the heavens. He is undoubtedly the dazzling Morning Star trying to usurp the place of the Day, and he suffers the usual fate, proving inadequate and falling out of the sky. Later poets have coloured the story with sentiment, but originally it seems clear Phaëthon stole his

---

* Steiner, *Wonders of the World* p.78: The Greek sou, says Steiner, felt how 'the Luciferic principle wafts through our earthly existence' and 'looked up to the star Venus and said, There is the wandering point in the heavenly spaces toward which Luciferic longings tend.'

father's chariot and tried to usurp his place in the heavens. His overweening pride and the disaster he caused led 'Zeus to destroy the world with a flood or fire, a tradition that goes back ... to Near Eastern flood myths'. He plunged into the cosmic river, called Eridanus or Oceanus—another important indication of a link to the flood or rising of a new world from the waters at a new cyclic beginning, as a truth integral and whole.* Likewise the Bible, not to mention Plato's Atlantis-myth, traces the luciferic situation back to the Flood. In fact, we palpably have here and in the Enoch-books, the authentic myth of Atlantis in the full scope of its significance.

It is the story of humanity's link to a past mode of consciousness, once possessed by an archaic humanity 'before the Flood', in which leading figures could gain oracular knowledge and be caught up to the gods. Such a mental world fits the Neanderthal culture of memory and oracles, and it long went on being sustained in the continuing role of oracles, far into the ancient cultures of the Near East and Greece. It was originally a true connection to the world, when consciousness was less ego-centric, more like a true dream. But it became increasingly estranged from humanity's real situation, and connected itself with tendencies linked to man's ever-growing egotism, fostering illusions of godlike grandeur and power. Such 'luciferic' tendencies were (and are) psychically destructive, as imaged in the picture of the star-thrown-down-from-heaven, the charioteer unable to keep control, etc. The whole world that had once corresponded to such a consciousness had long ago passed away, dissolved into the cosmic waters from which all worlds arise and to which they go—another picture vividly preserved in the myths. The Greek civilisation had found a solution to the dangers of the luciferic consciousness, by eternalising it purely as a picture, an ideal world, above human capacities to possess. From this vividly pictorial realisation in Greek consciousness they fashioned the world of their heroes and gods, Olympian, perfect—above and beyond the human in their perfection. A profound and beautiful civilisation was the result. But another path was to take past intuitions and adapt

---

* Forsyth, *The Old Enemy* pp.132, 133. It is not clear whether this Phaethon is really distinct from the Phaethon in Hesiod, *Theogony* 986-91 in whose story similar themes occur: he is the son of Dawn or Day by a mortal man who was rapt up to heaven, youthful and radiant and strong; but he is seized upon by Venus-Aphrodite and becomes a guardian-spirit 'by night' in her innermost sanctum. Now 'by night' Venus (being close to the Sun) is under the horizon, hidden and occult. Is this a complementary picture to his more familiar daytime mythology?

them ever and anew to the present conditions of man's consciousness, as characterised in the Enoch-books; the crucial underlying factor here was the Old Testament conception of a God who enters into history with his people, and reveals himself through individuals. The past is not eternalised, but becomes the basis for further prophetic fulfilment, pointing into the future. From this point-of-view, the luciferic 'rebel angels' were seen to have exerted a growing influence on humanity ever since the Flood, being worshipped by the pagans; in the Jewish (and Christian) line they had ever to be cast down, though the process came to be an increasing and inner-spiritual one, rather than connected to a cosmic catastrophe.

Robert Forsyth in his study reports on distant Indo-European parallels as well as Near Eastern ones. Evidently the story is therefore much older than the clash of biblical and Greek civilisation where it rose to fame. The story is in fact very old and widespread, with other analogues in Pacific-coast North America, Oceania and Africa.* But within the Greek world too it brings light. Once we have been helped to see the deeper pattern behind the Phaëthon story, the old familiar Greek myths also start to become transparent—not only the myth of Prometheus who stole fire from heaven, but of his son Deucalion who, Noah-like, survives the Flood, and after it effects the repopulating of the Earth at the command of oracular Apollo.

Prometheus hiding fire from heaven and bringing it down to Earth is still another version of the Bright Shiner falling as Day dawns. He is punished as a rebel, but has succeeded in bringing an ambiguous gift to humankind nevertheless. Promethean fire is the energy enabling metalwork. In fact, in essential traits Prometheus is a mythic double of the fiery smith-god Hephaestus (Lat. Vulcan), who in mythology was again thrown out of heaven by Zeus, the high god whose name comes from *Dyaus*, Day (or *Dyaus-pitar*, Day-Father, Latinised as Jupiter). The celestial symbolism pops up too in the fact that Vulcan-Hephaestus is a god who is married to Venus, whose star is as always central. Forsyth notes how Hephaestus once probably played the role of rebel which we still see in Prometheus.†

---

* Forsyth, op. cit., p.133 with n37, p.134n41.

† Forsyth, op. cit., pp. 86ff, 132ff for the myths of Prometheus, Deucalion etc. see Apollodorus, *The Library* [a handbook of mythology] trans. J.G. Frazer (Harvard and London 2001) pp.50ff.

As the flood-water subsides, Greek myth describes how the ark carrying Deucalion and his wife comes to rest on Mount Parnassus. The sole human survivors, they wonder how to repopulate the devastated world. They turn—of course—to an oracle: the near-by oracle of Apollo, which advises them to cast behind them stones, which are the bones of their mother, and these become people: the oracle stretches meanings as oracles do, since *laas*, stone turns into *laos*, people. But the pun is employed because it draws attention to a mythic reality. In contrast to the men previously made by Prometheus, the new generation is explicitly more solid and earthy in nature. The coherence of all these features in Greek myth and Jewish esotericism, with far-reaching world-wide parallels, suggests not mutual polemics but mutual derivation from widespread much older stories about the luciferic effect.

All this may seem irretrievably remote from Neanderthal prehistory. We shall see, however, that something was already crystallising in the history of our consciousness at the transition from the dream-consciousness of the Neanderthals to the explosive awakening of the Upper Palaeolithic that indicates the inner course of human development sketched out in the myth. It is epitomised in the extraordinary consequence of the discovery of images—the luciferic solution par excellence. The fascination and the power of images is immensely potent. Rudolf Steiner most remarkably describes the struggle to transcend the potency of images, often associated with the Old Testament denunciation of idolatry, as rooted in the time of Atlantis and its end.*

What he refers to in such a disjunction is surely what we witness in the extraordinary birth of images that came with the transition to the Upper Palaeolithic, with its extraordinary world of pictures, carvings, ornaments and symbols. The consciousness that was once united closely with the environment attains to an inner freedom of expression that bursts out in pictorial art, doubtless also affecting language and thought. How that happened we must now try to evaluate in detail. The myth of Atlantis, in the more authentic forms we have uncovered, will furnish a link to the history of consciousness and its temptations that can furnish a brand new understanding of the Upper Palaeolithic breakthrough, its triumphs and its illusions.

---

* Steiner, *Cosmic Memory* pp.64-6.

Chapter 8

BREAKTHROUGH OR BETRAYAL?
ASSESSING THE 'EXPLOSION'

Since 2010 we have known of an encounter that had a permanent effect on our biological identity. Long-running arguments between once standard schools of thought about human evolution came to an end: it emerged that modern human beings were neither intrinsically superior to their rivals, the Neanderthals, nor had they cast them off as a more primitive version of themselves. A new kind of thinking was necessary. It turned out there had been more than one kind of advanced human, two major species though they interacted and interbred. Modern-day humans such as ourselves have features inherited from the Neanderthals—including (as we now know) aspects of their extremely capacious brain-case.[*]

The evidence for their mutual involvement came with a new sequencing of their presences in the Near East, especially in Israel. This showed that the Neanderthals were not older forms that gradually turned into Moderns, and neither were the Anatomical Moderns the crown of human development but an older type. They streamed 'out of Africa' and encountered the Neanderthals. And was it accidental that relatively soon after the Neanderthal encounter, Cro-Magnon modern humans stood on the verge of the famous Upper Palaeolithic 'explosion', when the evidences of creative human behaviour, with symbolism, cave-art, social organisation etc., all requiring a high level of intelligence, rather suddenly flood the archaeological field? The Darwinian model could not explain it. But something, or someone, intervened. 'The mystery is where, how, and why the change took place. There are no answers to be found in the vast bulk of hominid time on the planet.... On the classic time line, there is only one figure left between *Homo erectus* and the fully formed humanity of Cro-Magnon: the ancient enigma, Neandertal Man.'[†]

---

[*] See above, pp. 33f. The research done at the Max Planck institute in Leipzig led to the first estimate of a 1%-to-4% Neanderthal inheritance in non-African peoples, and already mentioned the specific areas of cognitive function and cranial features as well as the ribcage and metabolic features.

[†] James Shreeve, *The Neandertal Enigma. Solving the Mystery of Modern Human Origins* (London and New York 1995) pp.22-3. In a remarkably similar passage,

For it is extraordinary that in the time following on the Neanderthal encounter, human beings with their joint heritage began a trek across the continents that took them over to farthest Asia and literally to the ends of the Earth—and even that did not stop them, since somehow they spread over the sea into Australia too. Everywhere they went they took with them their visual symbolism, sophisticated implements and the other signs of an understanding of the world akin to our own: or, more correctly, there has been a continuity of development from that Upper Palaeolithic breakthrough that includes all of us, living virtually anywhere on the planet. While the Neanderthals were still considered nasty and brutish, it was inconceivable that they could have contributed significantly to such a decisive moment in humanity's history. But minimalising the Neanderthals has had its day. Especially if we combine what we know of them with Rudolf Steiner's perspective, we may recognise another possibility. Steiner can give us a radical understanding of the sudden diffusion with its drive toward the East, and of the real character of the Upper Palaeolithic 'breakthrough'. In fact we have reached the juncture at which we can define exactly how the Neanderthals fit into his evolutionary account.

Neanderthals: Who, Whence and Whither?

We have argued that the Neanderthals fit much that Steiner said about the 'Atlantean' phase of human development. The evidence suggests that they lived out of patterns of the past (memory), but also that they had begun to balance the conflicting tendencies within society by an appeal to what we would call reason. This pertains to Steiner's Atlantean fifth stage, puzzlingly called that of the Original Semites. He cautions, however, against assuming anything like modern thinking. 'This ability was only in preparation among the Atlanteans.' They still felt that their thoughts flowed into them from outside, and belonged to the order of the cosmos, not their own heads. They

Clive Finlayson recently looked back on the often paradoxical debate over modern human behaviour, but now, he comments 'we are faced with a new paradox, which I will call the Neanderthal Paradox': just at the key moment for Modern Humans, the Neanderthals surely confront us as a relevant factor in what has otherwise become a puzzle. 'Is this puzzle real or is it the result of a massive underestimation'—a refusal to see, in other words, what the Neanderhals really contributed? Thus Finlayson, *The Smart Neanderthal* (Oxford 2019) p.10.

were spiritual entities which directly 'influenced their will. Thus in a manner of speaking their will was determined from outside.'

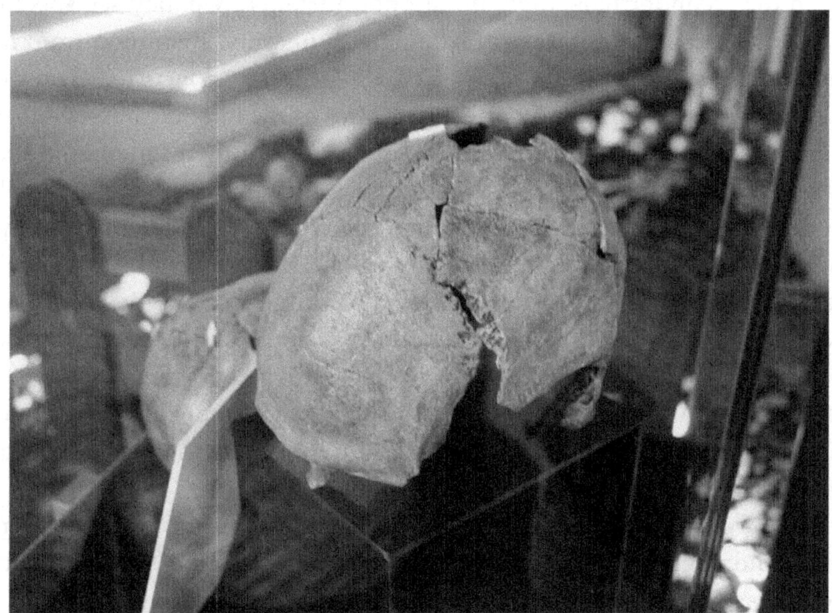

Neanderthal round skulls

The internalising of thought such as we know 'required enormously long periods of time'. But it was still a prehistoric achievement, though it belongs relatively to 'later times, after the decline of the Atlantean way of life'. That was in the period of what we call the last Ice Age, when a new phase was bringing a consciousness that felt the power of direction to lie within. One particular group, stemming from the 'Original Semites', then played a part in bringing over the Atlantean achievements in thought to the peoples who were subsequently coming forward in their evolution.*

Steiner very consistently presents the last Ice Age as the final moment, the end and indeed the transition out of 'Atlantis'. The myths of the Flood are a reminiscence of that time when the melt-waters streamed from the glacial covering of the Earth as it warmed. The period to which he refers, after Atlantis proper, would therefore coincide with the Middle Palaeolithic, the time of the classic Neanderthals in Europe, followed by the Upper Palaeolithic. Steiner's narrative continues with mass migration eastwards, a stream from Africa meeting another in the

---

* Steiner, *Cosmic Memory* (New York 1971) pp.59-60.

Middle East, right across into Asia and basically into all other parts of the world now inhabited by Modern Humans.*

In the Neanderthals, then, if I am not mistaken, we may see specifically that late manifestation of Atlantean humanity, stemming from its fifth 'race', to which Steiner refers. We have already concurred with the tendency to recognise the cognitive abilities of the later Neanderthals as being by then on a par with ours, emerging out of their mnemonic and musical earlier mentality. The Original Semites are the stage of development of Atlantean evolution, when the conflicting tendencies of the previous 'races' were balanced—the essential aspect of what we call reason. The ability to work together in a complex society is bound up with that development, as were new forms of spirituality going beyond the old oracular knowledge. It need no longer be an embarrassment to our special status if we admit that the Neanderthal connection may have contributed decisively to our modern humanity in this regard. For that is what Steiner boldly asserts.

The best documented evidence suggests that the scene of the encounter between the Anatomical Moderns reaching Europe from Africa and the Neanderthals was the Near East, centred in what we now call Israel. There in fact we may find the key to the designation 'Original Semites' when applied to a type of human being originating far back before the times of historical Semitic civilisation. Like several other designations that have been assigned to Atlantean peoples, it has a meaning drawn from a future connection that we can still discern: the specific group of these Atlantean humans that was involved in the Near Eastern encounter gave their name to the Atlantean fifth stage. (The Atlantean 'Mongolians' similarly appear to have received their name from the Neanderthal affinities of the Mongoloid type in the Far East—cf. above, p. 227.)

It is in Israel and the Near East also we see the best evidence of peaceful co-operation. And of course interbreeding. With their mixed genetics humans then embarked on their fantastic migration, populating our inhabited world.

Researchers have of course noted the significant coincidence of so many critical events: the encounter with the Neanderthals in the Near East; the Upper Palaeolithic 'Explosion' of cultural manifestations or 'cognitive revolution'; the subsequent migration of humans, carrying their symbols and social complexity across the world; and

---

* Described in Steiner, *Gospel of St. Matthew* (London 1965) pp.25ff, and in many other places in his lectures.

then, not least enigmatic, the extinction of the Neanderthals. Obviously their 'sudden' vanishing in terms of the fossil record, with its scattering of evidences, will hardly denote an overnight event. Even so, the latest reconstruction suggests an overlap of only a few thousand years.[*] Their unforeseen extinction remains, however, unexplained. Steiner is at his most intriguing in his understanding of this extraordinary concatenation of factors, offering a viewpoint which is less naïve than the popular triumphalist version of the human rise to prominence.

He distinguishes those human beings at the end of Atlantean times, i.e. emerging from the last Ice Age, as falling into three categories. The original humanity which still bore the old cosmic consciousness form one group. The still persisting 'Atlantean' population, which we have learned to call Neanderthal, form another. And then there is the small new population which they help to engender, of those who were able to go forward and to grasp the cosmic wisdom now in the form of thought. These latter were the bearers of a gift to future humanity. But Steiner tellingly remarks: 'The second group of human beings was doomed to gradual extinction'[†]—and this extinction of the greater part of the Neanderthal population is just what we find. Steiner has some interesting remarks to make on the end of the Atlantean humanity; still perhaps the most contentious aspect of the modern study of Neanderthal prehistory. It is connected with the luciferic tendency we have recently explored. But first we must pursue the migration and future role of the 'third group'. Steiner mentions that the specific portion descended from the fifth-Atlantean population was led through particular circumstances to a special place, namely a remote part of Central Asia, from which they were destined to exert an influence on subsequent spiritual evolution. The importance of Central Asian shamanistic ideas for the evolution of the civilisations which subsequently dominated the Orient is nowadays a matter that can be validated in cultural-historical terms and is central to Steiner's presentation of broader post-Atlantean spiritual history.

The core group were subject to a policy of deliberate isolation, but we should say at once that this was not racial exclusivity but the result

---

[*] The 2013 London conference brought forward evidence that the Neanderthals' extinction was completed earlier than had previously been thought.

[†] Steiner, *Cosmic Memory* p.63. Speculation upon the causes of Neanderthal disappearance is vast and so far inconclusive.

of a kind of religious teaching. In due course it will be necessary to consider its nature more exactly, but the religious inspirer who brought their teaching is asserted to be the real character behind the legendary stories of Manu, son of the Sun. In Indian myth he is the one who first taught humanity to worship with sacrifices, ordained the cultus of the gods and gave laws to men after the Flood. The great code of Hindu law is still preserved in his name. Manus are an integral part of the Indian mythology of the cycles of time, called Manu-cycles (*manvantaras*), not only the 'Atlantean' story of the Flood and the rescue of humankind. We noted that parallels from Polynesia to Egypt suggested a prehistoric source. It was the sun-Manu, says Steiner, who inspired the eastward migration of humanity. We may be glimpsing here the first known 'journey to the East', to the rising Sun, of which the Manu appears symbolically as the human child.* Steiner referred to prehistoric spirituality as 'pre-religious' in its sheer actuality. But here we have a first instance where spiritual significance is given to a goal-to-be-attained, evidently as a kind of 'promised land'.

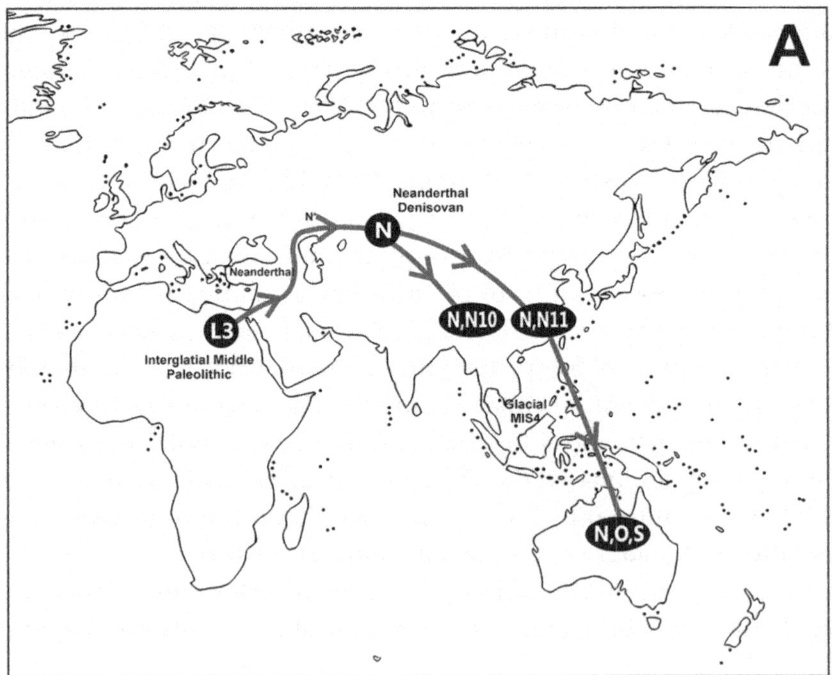

---

* The anthropologist Claude Lévi-Strauss stressed the role of mankind's tendency to journey toward the rising Sun as a consistent factor in human development: Lévi-Strauss, *Tristes Tropiques* (London and New York 2011) pp.122-3.

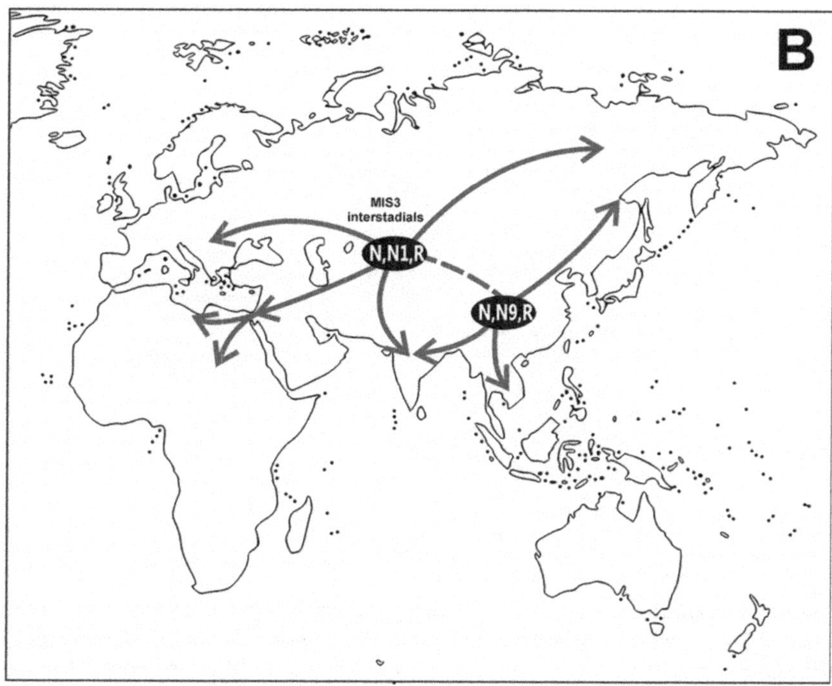

Modern conclusions about the spread of humans across the globe are based on a migration 'out of Africa', an encounter with the Neanderthals and a re-dissemination from a centre in Asia. The oldest religious-shamanic conceptions followed the same route. An identical track was described by Rudolf Steiner, who found its inner meaning conveyed in the myths about 'after the Flood'. Orthodox prehistory has been bedevilled by notions of a supposed Indo-European dispersion conquest of which no real archaeological traces have ever been found. The more compatible 'Palaeolithic continuity' model has recently gained much ground. Maps showing human dispersion 'out of Africa' and subsequent distribution. Modern reconstruction (2015) of human migration out of Africa (A) and secondary dispersion out of Central Asia (B), based on genetic analysis: by R. Fregel, V. Cabrera, J.M. Larruga, K.K. Abu-Amero, and A.M. González.

In addition to the specially selected group which followed the religious mission of the Manu, Steiner describes the dispersal of other populations, not all of which kept going as far as the steppes and formidable deserts of Central Asia.* Before we can understand events in detail, on the other hand, we need to clear up a discrepancy—more apparent than real—which seems to contradict what we have said about the agreement of Steiner's descriptions with the modern anthropologists' idea of prehistoric human migration.

---

* Steiner, *The Spiritual Foundations of Morality* (Vancouver n.d.) pp.34ff.

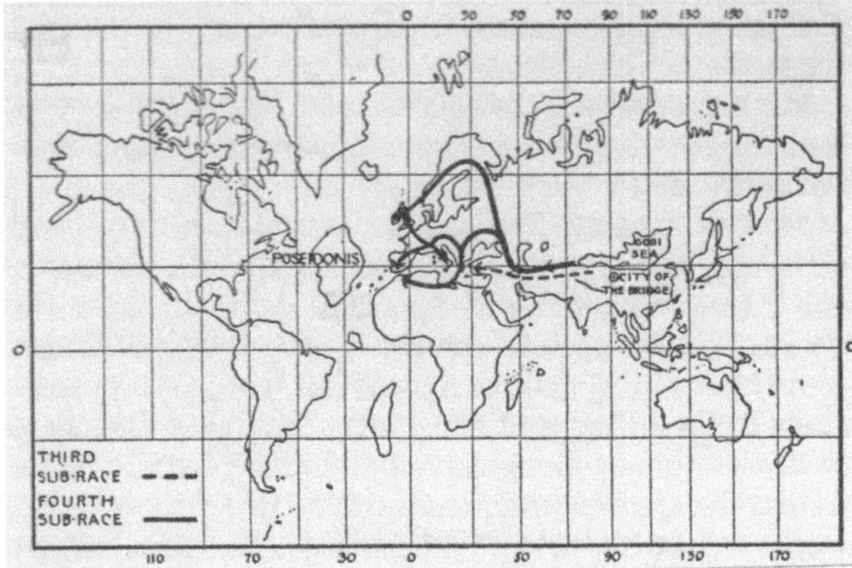

Some Theosophical accounts of post-Atlantean human dispersion form a partial antici-pation of Steiner's ideas. Atlantis is here called Poseidonis following the terminology of A.P. Sinnett. Map from Albert Schwarz, A Vade Mecum to Man: Whence, How and Whither, *1914 (shown in Powell,* Solar System *p.284).*

The Big Fish

The breakthrough modern interpretation of human descent speaks, of course, about a dispersal 'out of Africa'. It is the Anatomically Modern (though not necessarily behaviourally modern) ancestors of modern humanity who streamed through the Near East on their way to occupying the globe; the side-stream which evolved into the Neanderthals remained limited in their dispersion to Europe and a little more, and died out—or at least, they survived only as a genetic trace within the humanity that prevailed and continued. So it seems to researchers now, with the focus which we have adopted; but the Neanderthal experience may have looked different, and not without its deeper justification, and it may be that it survives in that mythic 'geological record' as Steiner calls it,* of the mythic record on which we have already touched.

---

* Steiner, *Wonders of the World* (London 1963) pp.112ff.

In Indian mythology, Manu ('Man') is the founder (and lawgiver) of a new humanity after the Flood: he himself escaped the waters with the help of the saviour-god Vishnu in his fish-incarnation, Matsya. The picture of the fish which grows bigger and bigger in the waters suggests imagery of the womb and the growing embryo, finally bringing humanity to birth. Out of that mythological world of Atlantis before the Flood, human-ity achieves birth and populates the world. Here Manu is accompanied by the seven Rishis. Illustration to Shatapatha Brahmana *(c.1840).*

When we turn to the myths about the Manu and his role, however, we find nothing about a migration or a mission to reach Central Asia, on which Steiner laid such stress as the historical reality underlying the tale. The account of Manu-son-of-Vivasvat (the Sun) is the story of a sort of Indian Noah who is saved from the universal Flood by the inter-vention of the great god Vishnu. The latter is also, though on a more exalted plane, a Sun-deity who in the mythology repeatedly descends from his transcendent level of existence to help or preserve mankind. On this occasion he takes the form of a fish, and this appearance is counted as one of his many 'lowerings' or *avataras*—of which Krishna will be the most famous and most loved. Yet so far from being uniquely Indian, his story here is in fact just a variation on the Flood-mythologies on which we touched previously in their Greek and Semitic guises. We suggested that it was closely linked with 'oracular' knowledge, which Steiner considers to have been the dominant mode of mental organisation in the Atlantean phase. In it, any problem in the world-as-we-see-it, or a viewpoint that has become 'rebellious' to our sense of world-har-mony, is put before a god: the answer to our problem is a stretching of the mind, an unrecapturable divine-creative moment which makes new sense of things. It seems as though illusion fell bodily away, and from the creative-unconscious waters the world rose anew at a god's bidding. Later versions were moralised as the overcoming of error but sometimes the more ambiguous mythic forms were richer in insight and tragic depth. An original, unrecapturable harmony, a challenge to god, a taste of divine fullness of life, and reintegration into reality though with a sense of descent or loss—all these are themes explored. The Indian Vishnu is certainly connected with cycles of creation, with the cosmic-originative waters, with illusion and insight. The special form of his myth has much to tell us, moreover, about the reason why the Flood-symbolism has been so well preserved (or constantly renewed) from archaic times.

The threefold nature of a functioning human society in its rudiments goes back to prehistoric times. Oracles are a characteristic form of one sphere in particular—the sphere which in modern times is that of the individual. When a specific problem needs to be solved, for the sake of the social whole, an individual must come up with an answer; but in pre-egocentric times the oracular experience finds the answer in an ambiguous manner, involving personal-divine linkage but in a mode eluding human possession. Rather events regroup. All three spheres have a basis in human nature, but what is the basis of the 'oracular' sphere? The *fish* which rescues Manu furnishes us a profound mythical answer.

Vishnu-as-fish plays the role of Noah's Ark, bearing its human cargo across the Flood-waters and landing him safe on the landward side. What does this mean? We need to envisage events on our three superimposed levels. First comes our suggestion about modern divided consciousness, torn by longing for unattainable experience like that of childhood, shut off as by an intervening catastrophe like a flood or by a 'fall'—such 'luciferic' consciousness we proposed is rooted in a reminiscence of prehistory, also preserved in Flood and Daystar mythology. Secondly, we followed Steiner in linking the dream-like content which was then cut adrift, in its original form, with larger rhythms including the shifting of the continents. So, scientifically: 'Atlantis' was divided from the Eurasian continents, where new developments will take place after the 'fall from heaven' which marked the Cretaceous-Palaeogene divide. Thirdly, on the subsequent historical scale, a new stage of life and culture may be characterised as starting with the dispersal of modern people across the world. On the simplest level, therefore, the myths mean that humanity was still being guided in its dispersion by *oracular* knowledge. Therefore they saw themselves as 'rescued' from a Flood-catastrophe and stepping into a new world-order, envisaged in the imagery of the old myths. But what is the sense of the fish-story specifically in this reality, as Steiner claims?

We need to ask how even dim memories of the prehistoric past consciousness of humanity have been handed down in the myths so (relatively) near to modern times? The fish-imagery actually is the key. For the Vishnu-fish does not just arrive to rescue the wisest and best of mankind; it involves him in several stages of engagement that reveal its true nature. At first the fish is tiny, and cries out for help to Manu until it can become big enough to save him! He keeps it in a bowl, and it starts to grow. When it gets too big for the bowl he digs a pond for it. When it becomes too great for the pond he puts it in the Ganges. When it becomes too great for the river he grows frightened of its continuing expansion, places it in the sea, and by clinging to its 'horn' he is the sole person to be saved from the Flood. The expansion motif is akin to the way Vishnu on another occasion laid out and organised the world in his 'dwarf'-avatar by his increasingly huge strides.* The underlying symbolism is undoubtedly that of the embryo in the 'waters' of the womb, and the way the tiny fertilised egg grows until it can no longer be contained there, but is 'saved' and indeed saves us all by populating the world. The child then defines the scope of man's action and thought by his 'strides'. The embryo that grows in the womb, the baby that grows up and lays claim to the world through his strides—these are the basic images of Vishnu.

Sometimes the myth rather includes a ship, other times not. Even so, Rudolf Steiner recognised the mythical meaning in the elaborate measurements of the biblical Noah's Ark, indicating that they allude to the measurements of the human body. And the presence of the creatures in the ark incorporates into the myth the Near Eastern symbolism of man's relation to the animals as their synthesis, which is so prominent in the Paradise-myths.† In some versions, notably the Babylonian 'epic of creation', the creature in the water becomes a monster, a dragon of the deep: it represents the animal-nature which

---

* The fundamental versions of the myth from the *Vishnu-Purana* and *Matsya Purana* are translated and commentated in W.D. O'Flaherty, *Hindu Myths* (Harmondsworth 1975) pp.180ff. As a dwarf, Vishnu is granted only a few strides of Earth by the demons, but by those strides measures the whole world.

† Steiner, *Occult Signs and Symbols* (New York 1972) pp.12ff.

must be overcome by the human in the birth-process. The saviour god Marduk has to defeat the dragon-nature, explicitly called a *ku-bu* or embryo, and out of its body he then shapes the humans' world of Earth, sky and water. (It is a recurring feature of Babylonian mythology that the gods have both animal and human forms, and often seem to fight their own animal-nature.) The creation-hymn was recited at yearly royal festivals, but it was also done at the request of the midwives as a charm in the case of a difficult birth. Then as on other occasions, its meaning is oracular rather than merely celebratory: it focusses on the 'determining of destinies' which ensure the well-being of the king and the state by recreating them.* The tendency of the resolution to become more difficult as time went on, so that the disturbing nature of the animal-bearer is increasingly stressed, undoubtedly reflects the ever further strengthening of the conscious ego-identity, making it harder to integrate fully by the old oracular methods.

The Flood-mythology, at any rate, shows up clearly as based not only on cosmic reminiscences, but at the same time indicates the little ocean a pregnant woman carries around inside her, and the perilous journey of the child from embryo to birth to become human. It is well-known that dreams often seem to recur to the foetal state. A mother-to-be who dreams of her baby is essentially, one might say, an oracular woman such as still spoke at Delphi: there Apollo is the saviour-god who conquered the dragon Python, but the dragon's mate Delphyne ('dolphin', like a fish yet human-like and intelligent) continued as a mythic presence in the prophesying. Steiner stressed that the god had to take on her nature and make it prophetic.†

---

* N.K. Sandars, *Poems of Heaven and Hell from Ancient Mesopotamia* (Harmondsworth 1972) includes the 'creation epic'; for the determining of Marduk's destiny, which is also that of the reigning king, see Tablet III. For the quasi-identity of the king with the saviour god, which coincides with much that we have said about the oracular state, see the good summary treatment in F.H. Borsch, *The Son of Man in Myth and History* (London 1967) pp.90-96.

† Steiner gave a detailed interpretation of this Greek version of the oracular mythology of Delphi: Steiner, *Christ and the Spiritual World* (London 2006) lecture 3 and see Karl König's development of his approach: *Seals, Dolphins, Salmon and Eels* (Edinburgh 1984) esp. pp.86-7. 'Fish' in ancient times of course covers dolphins and other creatures we now specify as water-mammals. Delphi, it will be recalled, was the navel of the world and the place which gave birth to a new humanity after Deucalion's Flood as well as Apollo's oracle-site. The Greek Flood myth (like the myth of Phaëthon) seems partly to have lost its link to cos-

Unmistakably, however, the nature of the creature-in-the-waters reflects the dominant characteristic of the embryo. Its most obvious feature is its already large and developed head, the embodiment of our individual human nature. The problem of the need for a large head was only decisively solved through the Modern Human method of birth, where the limbs are still too diminutive for use. These still animal-like 'dragon' limbs must be subsequently metamorphosed for earthly life-conditions. The large head is an 'ark' enshrining the human, transcending the animal and dominating it, like the hero-gods who often are said to enter into the dragon to conquer it. To Steiner's clairvoyant extended consciousness, such pictures are in truth an expression of the past evolutionary stage (his Atlantis), when the morphology of humanity's head-dominated form was 'separated out' from the reptilian animal forms which were physically realised at the time on Earth during the Cretaceous. That evolutionary history may not be repeated biologically, as Haeckel wished to suppose, but remains suggestively in the aura around human beings, as in that part we call their consciousness or sub-consciousness. The image of our embryo-head is the basis of the oracular mythology. To understand the handing-down of its mythology, we need look to nothing other than the fact of the increasing dominance of the head in original Modern Humans. That is the imagery which surfaces in the activity of their heads, in the making of their myths. Perhaps we are back yet once more at the meaning of the Neanderthal skull and the more-than-modern capacity of their brains, which was handed on to Modern Humans from the time of that Near Eastern encounter. That the Manu-myth tells of a Flood and humanity emerging thence in the form of a fish means: a new kind of human coming to be born, in whom the stony skull is a prominent feature but one that is overcome through a saviour-god/man. The trek toward the Sun was a matter of long-term dispersion, during which new humans would literally inherit the Earth, by being born from that fusion of the god- and the animal-nature.* We seem close to the idea

---

mology and world-renewal, and the oracular dragon/Delphyne myth likewise has become almost a vestige. The Near Eastern and perhaps remoter prototypes suggest more the original wholeness of thought. The prehistoric originals evidently allude to the changed manner of birth 'after the Flood': Erik Trinkaus argues that the Neanderthals still had a long gestation, and that the risky but advantageous 'premature birth' after only nine months was first developed by Modern Humans—the 'fish' mythology would be a spiritual imagination of that evolutionary step: see Trinkaus and Shipman, *The Neandertals* pp.382-3.

* In the Delphic Flood-myth, the stony men who are produced in conjunction

that begetting such a mixed new population was mythologised as a mission, in fulfilment of the oracular vision, combined with that religious duty to overcome the power of images which Steiner also attributed to the Manu.

Though the story has been drawn into the orbit of the Indian religion of Vishnu, we can see in fact that Manu was once more of a saviour-figure himself, though of an ambiguous kind rather as Phaëthon or Prometheus was. In the story, the fish needs saving (as any embryo-child needs protracted care), before the cyclic renewal that follows the catastrophe (Flood, or labour pains before a birth). The alarming growth energy of the fish recalls certain other Indo-European myths about the fiery animal energy under the waters reminding us of Phaëthon when he plunged into the ocean-river, and connected with the idea of a final great Flood. Such energy could also give birth to a hero of divine or semi-divine power, resembling the potent male seed in the watery womb. The appalling plunge to Earth is thus ultimately a reminiscence of our fall into birth—a version of that 'being thrown' into the world which Mircea Eliade recognises as one of the most ancient starting-points of religion. The luciferic themes here begin to show their venerable antiquity.

The Lucifer-mythology of the fallen stars finally makes sense as far more than just a natural phenomenon. It is not just the sight of shooting-stars nor the 'dimming of the planet Venus in the dawn sky. In order to view that phenomenon as a rebellion, one must first have a rebellion plot firmly embedded in one's own ... consciousness.'[*] For the luciferic dislocation is ultimately rejection of reality—a rejection of birth itself, and a fixation upon that primal horror of our tumble into earthly existence, and a nostalgia for a heavenly state that is lost. Some part of us holds instead to the mythic fullness which we intuit before and, beyond. The Bible version that deals with it, of course, serves to remind us of our duty: to put that primal loss behind us and, when we arrive in the world, start our engagement with things and with people. If we do not, we will be taught that lesson forcibly. Yet a part of us refuses to do so, thereby complicating and deepening our self-consciousness, bringing us valuable gifts, so that we brave our punishment nonetheless.

---

with the Earth-mother by Prometheus' son Deucalion, the Greek Manu or Noah-figure, repopulate the world. The structure of thought is consistent with regard to the inner history of the oracle itself, claimed mythically to have been taken over by the light-god from a female Earth-spirit.

[*] N. Forsyth, *The Old Enemy* p.134n41.

The notion that we humans received such a source of rebellious energy from a 'trickster' god is already found among peoples who have preserved prehistoric culture such as the Bushmen of Africa—more properly, San people. They believe that such energy is linked to ascent into heaven and descent into the underworld. They represent it too with cave-art representational symbolism,[*] and the fact that the Neanderthals have eventually been shown to have practised it probably implies that they too would have known the 'trickster' mythology of the ambiguous gift. The ladder-symbol found in the Neanderthal cave-paintings in Spain probably indicates cosmic levels, of ascent and descent, as in shamanism. Their oracular pattern of life would probably incline them to birth and Flood symbolism too, as would the prominence of their large heads and their dream-life.

Steiner's application of the Manu-story to the original dispersal of 'post-Atlantean' humans thus indicates to us that the popular phrase 'out of Africa' must not be interpreted too naively. It does not mean that a hugely long column of African-Modern Humans came marching through the Near East, stopping to mate with a few Neanderthals before streaming on eastward! Rather, it probably means that Anatomically Modern Humans were moving to occupy new territories when they expanded out of Africa, and in the Near East they probably also expanded their numbers again—with the help of the Neanderthals. This would lead to an interest in further movement, which it would now appear was given an ideological motivation. The 'migration' was therefore less a march, more like a series of overflows. At the root of it was a new burst of population growth especially of the new mixed type. This process went on into the period of the cultural 'explosion'. 'Recent genetic evidence suggests that populations all over the Old World underwent dramatic "explosions" around the beginning of the Upper Paleolithic in Europe,' notes James Shreeve, utilising the results of researchers employing 'a new method of analyzing mitochondrial DNA data to reconstruct the demographic history of *Homo sapiens*. In almost every population examined, they found statistical evidence for explosive growth just at that exact time.'[†]

---

[*] D. Lewis-Williams, *The Mind in the Cave* (London 2002) pp.138ff.

[†] James Shreeve, *The Neandertal Enigma* (London and New York 1996) p.304. The research into the population figures was led by Pennsylvania State University and H. Harpending.

Thus the wave of people spreading out was closely associated with the other so-called 'explosion', when symbolism and technical refinement rapidly became widespread in human societies. And we can now add that a further element in that process was the encounter and interbreeding with the Neanderthals. How did the Neanderthals see it? And what role did they actually play? Naturally they can hardly have directly caused it; but their population numbers seem to have been already in trouble, and they perhaps actively fostered the idea of renewing themselves through the incoming population—as the oracular mythology suggests. There must have been tensions, no doubt some hostilities. But the other humans do not seem—despite the wish of some researchers to have it so—to have been roused to any determined aggression or resistance in response to them. On the contrary, all appear finally to have been happy to join in.

It would once have been deemed unnecessary to suppose that the Neanderthals *thought* anything about it at all, but nowadays we know that they were culturally and cognitively advanced. If they thought, of course, it was in myths that were rooted (if Steiner is our guide) in oracular knowledge. And that knowledge is by all later evidence a mythology of the world-Flood, of a grasping at heavenly knowledge dragged down dramatically into the waters here below, of a struggle with animal powers and an eventual safe delivery—a delivery in its first and most literal sense: a bringing to *birth*. Is it conceivable that the otherwise declining Neanderthals would have thought of themselves mythologically furthering the renewal of themselves and the world which the birth-symbolism implied?

That they would have thought of themselves as cosmically motivated seems to me absolutely certain, and their mission would have been to shape the world and themselves expand into it (like the Vishnu in the fish and in the cunning dwarf-avatar). How did they look upon the ever-growing *sapiens* population? They would be bringing their heaven-descended knowledge and spreading it—a central thematic in the fallen stars mythology. 'And the sons of God saw the daughters of men, that they were fair....' (Genesis 6:2). It might seem too narrow, even fundamentalist, to seize upon one such element from the biblical account. However, we know nowadays that the stories of the fallen angels after the Flood bringing knowledge belong to much older and more authentic myth (still adumbrated in the Enochic and apocryphal sources) than the theological versions. It is almost as universal as the Flood story, and extremely archaic. The Neanderthals with their cosmic knowledge might indeed have seen their drive to reproduce themselves less in a Darwinian than in such a mythic-cosmological perspective. Moreover, there are other, quite independent myths which say something oddly similar in different terms. Following Steiner's hint, we may look again to the materials which came to be told as Roman history but which are really myths of human and cosmic prehistory. They are certainly part of something much older and more widespread than the rise of ancient Rome.

The story goes that the little band of followers of Romulus were all ready to found Rome, and in most respects had the skills and manpower to do it: but there lies the rub, manpower—but no women. The city they could found would have no life, as one historian quizzically puts it, beyond the first generation of its citizens! No future growth could come about unless a drastic solution was found. The region in which the city was destined to lie was by no means uninhabited, however, and the would-be Romans came up with a highly effective stratagem. They staged a great celebration festival together with the Sabines of the area, and at a prearranged signal suddenly seized and abducted all the Sabine women. The short-term result was of course fierce hostilities between them and the Sabines generally—but the hostilities were quickly defused and the marriages formalised. As a result, the small number of proto-Romans was enabled to multiply and in a few generations grow into a populous, not to say powerful city and, in accordance with the promise of the gods, in due course rule the known world.*

Even ancient historians had their doubts about the tale as history. Its real significance is more fundamental, as Rudolf Steiner's remarks on early Roman 'history' suggest. The story has indeed proved to be a central strand in the rediscovery of the oldest mythology which unites Indo-European cultures across much of the world, from Scandinavia to India. Such common myths must necessarily be of great antiquity and predate the individual civilisations. In Norse mythology, the comparative mythologist Georges Dumézil has shown, a very similar structure underlies the story about the two kinds of gods who governed the world, the Æsir and the Vanir. The Æsir are the gods such as Odin and Thor, who embody divine knowledge and strength, and are predominantly male; they originally contended with the Vanir, but rapidly came to an accommodation with them and now share in a joint rule.

---

* The story is best known from Livy's *The City from its Foundation* I,9-13. It is usually referred to as the 'rape of the Sabines', but the Latin *raptus* signifies abduction rather than the rape . The story basically stresses the diplomatic outcome rather than highlighting violent means of appropriation.

The Vanir are predominantly, though not exclusively female such as Freyja, who embodies youth and vitality—and like her brother Freyr brings peace and plenty. Brother-sister or twin gods are also known elsewhere in this category. Human beings would have a stern existence if only the Æsir ruled the world. Under the dual rule they can prosper and multiply. In Ireland the same polarity appears, though differently coloured: the bright gods who are called the 'tribes of Dana' are the deities of Druidical high magic and also of warrior strength, but in legend they are only one of the peoples who in successive waves occupied the island. The hostility between them and the Fomorians whom they find already there is grimly depicted, but again a solution is found and the two peoples contribute to the whole. In archaic India, the brahmanical gods such as the magician Varuna, and Indra the warrior, are not alone in having divine status: they share it with the Ashvin gods, who spend their time not in heaven but among human beings, and bring them healing and prosperity. The mythology about them again stresses their first hostility against the heavenly deities, yet this is always resolved to bring out the necessity of both divine races for the life of the world.

The stories about the Ashvin make them twins, and also feature themes of eroticism as in the story of the brahman's wife whom they try to seduce—she contrives to get them to rejuvenate her husband, however, and still remains faithful to him! Motifs of sexual renewal, twin/brother-and-sister, powers of healing and plenty belong in the same world of ideas which underlies all these myths, Dumézil amply shows, though different cultural shadings also appear.*

Sometimes the stories tell of human tribes, more often of the gods who stand for their values and culture. At the root of the culture of the Indo-European peoples generally, then, before they split up into different branches across the world, we have a basic story of people or gods who represent cosmic power and heavenly knowledge, but who cannot govern the world alone but need to join up with others. These, who even when called gods, are an earthly-orientated and pre-existing population remarkable for their fertility. In some of them we detect the theme of the small group which by joining up with the larger population are enabled to expand and rule the world (cf. the dwarf Vishnu, who in a few steps traversed and took possession of the cosmos). In the stories there is persistent sexual innuendo, as we find too in the Semitic story of the sons of God taking human wives, in establishing the new

---

* Dumézil, *Mythe et Épopée* I II III (Paris 1995) pp.313-320.

world after the Flood. It is also noteworthy that at this juncture in his account of esoteric thinking on the merging of two humanities, Rudolf Steiner at first uncharacteristically concludes that his readership was not yet ready for the whole story: 'For the present it is not permitted to make public ...'. Later he gave a fuller account.*

Most recently, Chris Stringer and some fellow researchers such as Lucile Crété have inclined to a view that certainly accords with the mythical deposit from that phase, which we have identified in the tales of sexual encounter between the sons of God and daughters of men when they 'saw that they were fair'. Sexual attraction to the Modern Humans streaming out of Africa may have been so intense that Neanderthal/Atlantean humanity disrupted its ordinary breeding pattern so extensively as to start a decline and fall. If it were so exciting, say Stringer and Crété in a nutshell, 'interbreeding with our ancestors would have reduced the number of Neanderthals breeding with each other, leading to their eventual extinction.'† Or better said: they lived on only in the burgeoning new population which was already starting to spread across the globe.

A Birth of Images

It is worthwhile considering together the elements in the human story at the remarkable stage we have now reached, as we enter the Upper Palaeolithic. The mingling of the 'two species' takes place in the context of a manifold population explosion, a cultural transformation, and a diaspora which was unprecedented for any species. In Steiner's terms, an important component in the whole process was the surviving population-group from the 'fifth race' of the Atlantean type, which

---

* Steiner, *Cosmic Memory* p.66. Cf. *Outline of Esoteric Science* (New York 1997) pp.235ff.

† See the article by Stringer and Crété, 'Mapping Interactions of *Homo neanderthalensis* and *Homo sapiens* from the fossil and genetic records' in the journal *Palaeo-Anthropology* vol. 2022 no.2 (2022) available at https://paleoanthropology.org/ojs/index.php/paleo/article/view/130/775. The sexual asymmetry is a curious factor on another level. Mats Andersson at https://www.quora.com points out that the chromosome and mitochondrial evidence argues that of Neanderthal-*sapiens* offspring, only the *daughters* of Neanderthal men by *sapiens* women were fertile, so that intermarriage with Humans placed limitations on Neanderthal reproduction. Andersson dismisses other notions as fanciful, arguing (like Stringer and Crété) that there was clear natural attraction between the species.

Nearly all prehistoric rock art comes from the Upper Palaeolithic period—the time of the 'Explosion to modernity'—and was long believed to prove that a new phase in human thought and abilities began with the arrival of Homo sapiens. *Neanderthal underground chambers were strikingly devoid of such evidence. However, a number of cave images and symbols have now been found which are demonstrably Neanderthal, though they remain rare. The reason for the rapid diffusion of art, not its origination, is the question posed by the Upper Palaeolithic awakening. Neanderthal engravings from La Roche-Cotard, showing the ability to use abstract signs.*

'survived the great Flood', whose last phase we know nowadays as the last Ice Age. The people we now know as Neanderthal humanity. After the Flood, 'the fifth Atlantean race sent its people everywhere, and they founded new civilisations, civilisations capable of growing and becoming more advanced ... up to our own.'*

Modern science now knows in some detail about that massive dispersion, and it knows that in certain crucial respects the humanity which spread across the globe was fashioned by the Neanderthals' gift. It seems to the present-day researchers, however, that the new population surged 'out of Africa' and was affected only slightly by its Neanderthal encounter. In fact, it is still widely held that the Anatomically Modern 'African' type or *Homo sapiens* was the one which made the leap to modernity, and that in its case the population explosion was at the same time a cognitive explosion, resulting in the dominance of an intelligent, inventive, artistically and socially superior humanity continuous with our own achievement—the the-

---

* Steiner, *The Apocalypse of John* (London 1958) p.140.

ory I have called 'triumphalist'—while the Neanderthals were soon left behind as a footnote, indeed deleted altogether by some agency or other. That was plausible and appealing while the Neanderthals were routinely belittled and played down. The recent rehabilitation of the Neanderthals in scientific estimation, on the other hand, and the recognition of their high social and intellectual level, musically and otherwise, has at the very least complicated the issue considerably, and indeed the triumphalist theory of the cognitive leap has not fared well in the light of the latest palaeoanthropological research. I have just suggested that we may partly be facing a matter of perspective. The archaic mythologies which are the 'fossil record' of the history of consciousness may be right in their hint that the interaction of species originally looked different—if we once suppose that the Neanderthals had such a mythology in mind. They may have seen union with the burgeoning *sapiens* population as a way of preserving themselves.* From Rudolf Steiner we may also find a resolution of the riddle that biologically the contribution of cultural advancement appears to us not to come from the Neanderthals but from the Anatomically Modern peoples who swept into the Near East and on into the world.

Perplexing aspects to the problem can be distinguished: firstly, the 'cognitive leap' has become problematic. Once the chronology had shown that the modern anatomy long preceded the cultural achievements which were once supposed (following Darwin) to derive from it, so it follows that the leap would have had to be prodigious indeed.† But, secondly, the elements appearing together so brilliantly in the Upper Palaeolithic world have turned out not all to be so new after all, and not all confined to the one species destined for triumph. One discovery in particular has pulled a large part of the rug from under the whole construct. That is the discovery that the Neanderthals, despite every-

---

* About 20% of the Neanderthal genome survives in the modern human constitution—Finlayson, *The Smart Neanderthal* pp18-19. The understanding of Neanderthal and modern genetics has been particularly furthered by the work of Svante Pääbo, winner of a Nobel prize for his work (2022). Without pressing the point, the need to see evolutionary survival advantage from the genetic level is of course an idea that has been stressed by biologists such as Richard Dawkins.

† Thus Finlayson presses home the lack of real evidence for Modern Humans having advanced tool-making skills before they left Africa: *Smart Neanderthal* p.62.

thing once thought about them, did actually produce examples of cave-art as at the Spanish site of La Pasiega (with parallels in at least two other Spanish caves), and of decorative ornament (Cueva de los Aviones) long before that putative dawn called the Upper Palaeolithic—for of course the basis of all the superiority supposedly attributable to the proto-modern humans of that time is: their use of images, meaningful, symbolic, value-laden images. Why was it then that images become with relative suddenness an insistent feature everywhere that human expansion went? We are perhaps still too inclined to think (if we admit our self-centredness) that the pioneers of symbolic statement were revealing their aspiration to become like ourselves. We naively suppose that the cave-artists' and carving-makers' gesture is a sort of flag-waving 'Look what I can do!' But the concatenation of circumstances we have just mentioned, such as the explosion in human numbers and wider dispersal, may hold a more convincing explanation.

Even before their sudden decline the Neanderthals, in this respect still certainly a striking contrast, had lived in relatively finite groupings, held together rather literally by being in tune with one another. Gatherings for sharing memories and deepening emotions, musically and gesturally were at the heart of Neanderthal togetherness. When the humans who spread out of Africa and caught them up in their wave of expansion and increased population came through, however, such intimacies may well soon not have been enough to keep people in touch. Rival groupings could easily arise; there could be problems of recognising who was who, and the need to assert some overall control. Recognition by symbolic signs (art images, patterns of clothing, or style of weapons) had far greater power than the slow musical establishment of empathy.

Symbolism has a power, in fact, that is almost unlimited—even in our modern civilisations we have barely scratched its surface let alone exhausted its possibilities all these thousands of years later. Even if we confine ourselves to its simplest forms, it constitutes an infinite extension over mere practical implements. An artistically worked bone tool with an animal carving, to take the simplest of examples, is no longer just a tool but a focus of complex meanings: it tells first of all of its intended task as a designedly human activity, readily becoming (for instance) a sort of badge of the maker's profession showing that it is acknowledged and valued socially. By its standard of workmanship it invites learners to aspire to work in that manner too, or perhaps just to admire, even among non-craftsmen. Or, it may also set a standard that warns off inferior workmen. Well-made clothing can establish status and demand respect, or even fear as indicating influence, power and resources. Images imply immediately a hierarchy of importance, by showing which things are therefore considered interesting or significant—conversely, outsiders' images with an alien style show the need for caution in evaluating their acts and meanings towards one's own group. By establishing a style, an implement with meaning also assigns to anything else a contrary meaning, even if it does not have such significance for the others. Meaning instantly makes everything meaningful, positively or negatively. To give oneself or one's representative a particular look, in dress or otherwise by symbolic accessories (objects, headgear etc.), presupposes a self-consciousness i.e. knowing how one should appear to others to have most power or to win recognition.

One's symbols not only speak to others: identity is suddenly to be defined in terms of the meanings we put upon our self-presentation to others. Consciousness is drawn into that endless 'luciferic' whirlpool of considering how we appear in others' eyes, and our adequacy or otherwise to what they expect. We ourselves feel the need of an image of what we aim to live up to, never fully attained and sometimes fuelling our inadequacy. Amid competing meanings we need to make ourselves heard by giving valid signals or we shall lack social dominance. Ripples of signification spread around and within, wherever a sign can be made visible or carried and shown. Neanderthals had lived in a world where the landscape carried symbolic significance; the Upper Palaeolithic evidence clearly shows to expert researchers how human symbols were creating a more differentiated society, and indicating power over an inner world. (Cf. above, pp. 271f.) Interacting with early language it must have hastened the complexity and rhetorical power of speech by its ability to evoke iconic meanings, i.e. to stand for a set of values. The ability to show one's allegiance to the range of meanings of a particular culture thus makes a potent kind of primitive mass-communication possible.

Perhaps now the future tense, probably absent from Neanderthal speech, comes into its own as declaring what one wants to happen becomes more important. A goal-orientated approach must have become necessary when larger numbers of people and the need for travelling to new localities became the norm. Though modern individualism is still far over the future horizon, people who could find solutions and encapsulate them in a pictorial or symbolic statement, which shows everyone immediately what is being proposed, must have been a *sine qua non* of the new expansive phase of human existence. Emblems of authority, insignia of specific skills, statements of a relation to specific goals and the ability to communicate them—these are the features which made the Upper Palaeolithic world, and which make it so surprisingly recognisable to us.

The power of symbols has been central to all human societies ever since, and never more so than today. Human history is the history of symbols. From the Pharaohs to Queen Elizabeth I the nature of political rule has been held up before the subject population in portrait, coinage and insignia; in reformed government the power of the flag or national emblem continues; in religion the Cross or the Crescent, to say nothing of the *eikon* succeeding to the idol in the temple, stained-glass windows or figures in a shrine are reminders that it too is inconceivable without symbolic images. In recent secular societies their power is even greater, where consumerism powers the State and the technologically multiplied image controls almost every aspect of life, showing us what it is we want, and what we want to look like not only in fashion but in our body-image, what we find erotic, how we want to live and where and what accessories we need, what life we want for our children, and how we wish to be represented in death (down to the nouveau riche Russians memorialised in the ownership of their designer shoes and expensive watches). The idea of one's person being *represented* (from smartphone images to appearing on television/social media) constitutes ultimate desirability in these societies. Orwell was prescient but he did not anticipate how intense would be the longing for Big Brother to be watching. The fascination of the image is so great that it readily usurps or absorbs identity bringing a sort of psychic inflation. Although control through State-coercion persists in certain societies, control through images (most explicitly, by aggressive advertising) has in many advanced countries made it largely redundant. The unleashing of image-power is therefore an event of huge and ambivalent significance. For we must of course mention that the potency of visualisation can equally lead to the highest achievements in the visual arts, in thought, and to the noblest idealisms of self-sacrifice. When the Upper Palaeolithic dispersion carried across the world the secret of the power of images, it was indeed opening a Pandora's box. Its power for exaltation and slavery flew out and will never be put back.

All this only reinforces the importance of that rather mythological-looking concept of Rudolf Steiner's, when he spoke of the 'luciferic' in a universal and evolutionary sense. For the image which liberates the meaning from a memory and the idea from the literal repetition of the thing, as Steiner explains, opened humanity even at that prehistoric stage to the surging currents of the luciferic tendency, where vision is cut loose from reality. To have discerned the prehistoric beginnings of the image and its impact on human development is indeed one of the most startling achievements of his 'evolution of consciousness'. For the dislocation of the image from the actuality is not merely, as we have seen, an inert process of abstraction. The discrepancy between consciousness and its concrete situation is a time-discrepancy, characterised by that twofold combination of precocity and retrogression which the Lucifer-myth so brilliantly expressed. In Steiner's anthroposophical interpretation, to fall into illusion in this way is actually to slip over into an alternative reality, or world construed in a way that contradicts our 'real' earthly world. The urge to rush ahead to a premature mental completion, which then tries in self-fascination to perpetuate itself, until it is undermined by the pressure of factors excluded from consciousness is, for that time, to give way to an alien rhythm: it is indeed a rhythm not belonging to our phase of evolution or the pattern of forces which belongs to our Earth.

The cosmic phenomena reflected in the ancient myth of Venus shining upon our Earth, then its luminosity being overwhelmed as Earth turns to the Sun, are a genuine expression of cosmic relationships which show different planetary rhythms. The faster rhythm of Venus (and its inverse direction of revolution) establish a contrasting cosmic 'individuality', a coherent otherness which usurps our earthly consciousness: in its slide into illusory self-fulfilment, only to fall back painfully to Earth. Of course, image-forming is not necessarily a luciferic lapse, in so far as it connects with things and an intuition of their essence; yet it is the faculty that most readily does tempt us to slide into that other reality, with its contrary rhythm, and to become fascinated by the image in itself, which then comes between us and the reality it represents. The discovery of the power of images, which already prompted the 'leap' of the Upper Palaeolithic cultural expansion, meant that the openness of human beings to the luciferic constellation, at once rushing forward yet holding on to the past, was inordinately increased. And therefore the inspiring message of the Manu, whose prehistoric mission is the truth behind the myths of escape from the all-engulfing Flood, is said by Steiner to have consisted in teaching a way of overcoming the fascination of the image—a kind of spiritual discipline. He taught those whom he singled out to master the power of images, and so he will be the one who inspires all the later, 'post-Atlantean' civilisations.[*]

Mind Control and the Mysteries

The explosion to modernity witnessed in the Upper Palaeolithic is an outbreak of images, and their effectiveness in the new situation is likewise undeniable—we ourselves are the ultimate outcome of the results. But how did image-making become with relative rapidity a power to be reckoned with? For that matter, how did images become possible to human minds? No more than the development of memory could 'just happen' can it have been that people just started to retain meaning in the form of a picture, whether mentally or in pigments or carvings, which 'referred' to realities beyond. Whence came this faculty which shows itself in indicative clothing, in carved implements whose artistry serves to make a statement, but par excellence in the art which left Picasso stunned, the dazzling iconography of the Upper Palaeolithic cave-paintings deep underground?

The cave-paintings give a clue. The extraordinary difficulty of access of their locations, for painting or viewing, already suggests the supreme importance attached to them, their spiritual eminence so to speak. But all older efforts to interpret them were mere guesswork until the researchers reached out for help to those who still know their language. The realisation that such people exist was a new beginning in human prehistory-studies. 'We do not have to indulge in the blind (and often wild) guessing that is frequently associated with the study of rock art,' candidly remarks Lewis-Williams, as he sets out to 'present key features of shamanism that are relevant to Upper Palaeolithic art.'[†] When shamanic practitioners of surviving archaic cultures, in this case the San

---

[*] Steiner, *Cosmic Memory* pp.64-70. These ideas are, first of all, the inspiration behind the Indian spiritual disciplines which later became *yoga* etc. In the old Indian mythology, the world itself had been conjured up in the form of magical images (illusion, *māya*) by the creator Varuna. The goal of the seer was to penetrate through the illusion to the reality, the unseen or transcendent truth. The image separates us from the reality it represents; to know the reality is therefore to be united with it. The goal was later given philosophical expression in the celebrated 'thou art that' of Shankarāchārya (ninth century AD).

[†] D. Lewis-Williams, *The Mind in the Cave* (London 2002) p.135 and see 136-179 for his case-studies first from the African San or Bushmen, secondly from native North America.

people (or Bushmen), were shown reproductions of prehistoric art they understood its message immediately, for its images (says Lewis-Williams) 'encapsulate the very essence of San religion'. The prehistoric people shown in the images also naturally shared those thoughts. They were practitioners, they explained, whose knowledge enabled them to stir up 'their supernatural potency, to cause it to "boil up" in their spines until it explodes in their heads and takes them off to the spirit realms—that is, they enter a state of trance'. The other associated images in the prehistoric pictures, and the strange 'geometric' signs, they explained, depicted the stages and turning-points of the trance journey. The principal occasions for shamanising (or 'trans-cosmological travel') are usually highly esoteric, so that 'only a few people may be present'—though it is also typically associated with a 'trance dance', a great event which 'is the central religious ritual of the San', when the men, women and children of the whole society are present in a ritualised (though also partly informal, even playful) way. This social event is likewise depicted in modern-day San rock-art and there appear to be scenes here too with very ancient parallels.* The esoteric character is typical of Mysteries—as is also, in fact, the relationship to all-inclusive gatherings and structured social activity of a festival nature.

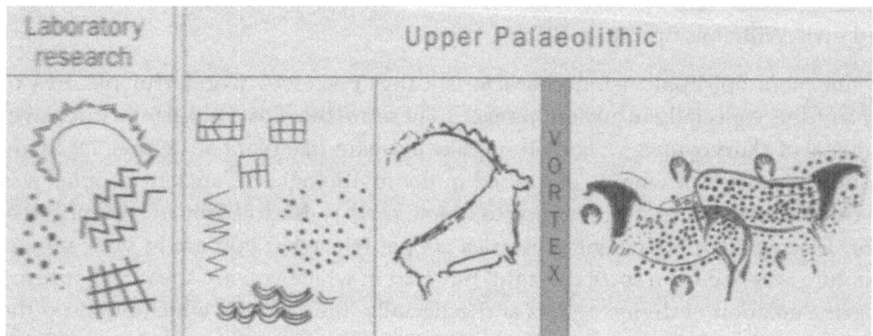

Cave-paintings from the Upper Palaeolithic are normally accompanied by mysterious signs, which became intelligible in light of the shamanic interpretation, now widely accepted. Such entoptic shapes appear in states of altered consciousness among modern subjects, closely matching prehistoric forms. Rudolf Steiner related patterns of dots, spirals etc. to experiences of entering and leaving the body. Institute for Research into Rock Art, Johannesburg.

The San shamans recognised many specific details of the ancient depictions including the geometric signs, which are not decorative

---

* Lewis-Williams, op. cit., p.138, pp.139-40 with illustrations.

or extraneous, nor indeed abstract but an integral part of the spir-it-journey and actually 'seen' in trance. Lewis-Williams is particu-larly fascinated by these 'entoptic phenomena' because they can be related to the activity of the optic nerve and the cortex of the brain. In certain states (of ecstasy, trance) 'the pattern in the cortex is per-ceived as a visual percept. In other words, people in this condition are seeing the structure of their own brains.'[*] Rudolf Steiner has very similarly described how at first in the unfolding of spiritual perception one sees one's own nervous system and its activity, often in spiral-forms or in plant-like wavy forms—an experience which he stresses is not objective but simply reflects one's own activity. He related certain archaic religious visions of an ecstatic nature, involving numbers of spirit-beings, to the seeing of the nerve-end-ings of the spinal cord. Patterns of dots also arise from these expe-riences.[†] All these configurations are prominent in cave-art, either independently or incorporated imaginatively into representational forms such as antlers, horns, body-markings and other shapes.[‡] Steiner also points out that the dot-formations, still half-consciously experienced in intermediate states of awakening, etc. underlie mod-ern ideas of atomism.[§]

---

[*] Lewis-Williams, op. cit., p.127.

[†] Incipient spiritual vision characteristically 'perceives wonderful pictures of plant-life, especially in ancient times.... These are the most rudimentary achieve-ments of clairvoyance ... something like a dream-like plant structure, yet living and real. Much of what is described in the mythologies of ancient peoples was seen in this way' (with reference to Norse myth)—Rudolf Steiner, *Apocalypse of St. John* pp.48-9; in ancient Mysteries people felt 'what you are in your inmost nature is veiled from spiritual sight' but 'that in which you are sheathed' appears as the creation of divine beings and especially 'the currents which build up the spinal cord with its twenty-eight lower pairs [out of thirty-one] of nerve fibres' (with reference to Zoroastrianism)—*Mysteries of the East and of Christianity* (Lon-don 1972) pp.43-5; these forces flow through us except 'where they are held back and reflected, ceasing to function and it is only the forces of the world of spirit which stream through. These are the forces that form, for example, the optic nerve ... and are actually reflected in the twelve cerebral nerves and thirty-one pairs of spinal nerves.... We become aware of a part of the elementary world [of forces] by holding it back.'

[‡] Illustrated in Lewis-Williams, op. cit., p.128 and plates pp.113-120.

[§] Steiner, *Macrocosm and Microcosm* (London 1968) pp.134-6; waking conscious-ness is experienced as a 'contraction' of our spiritual nature into these nerve-end-ings as 'consolidated' points (p.133).

All researchers are agreed that these 'entoptic' apparitions are a signal that the mind is in a state where it can visualise its own activity, and that combining such perception with further visionary content results in charged—or one might even say supercharged—images. Already in prehistory, shamans clearly knew how to bring the mind into such highly receptive states, allowing perceptions to become saturated 'ideas' which carry information about the way they relate to the social and spiritual situation. Such states equally work to *define* them—isolating them from the flux of experience and thereby fashioning a rich and complex cosmos of meanings. Lewis-Williams makes the essential point that the cave-images and their shamanic analogues conjure up a whole inner world.

Thus the Mystery-way of evoking them was significantly different from the configuration we have described as *oracular* knowledge, which accords so well with our knowledge of the Neanderthal culture. (The cave at Bruniquel so strikingly apt for cave-painting, but untouched, shows that the Neanderthals there were behaving quite differently.) In oracular consciousness, the moment of perplexity and openness in the momentary meeting with the god is perhaps a point of similar ideogenesis, but one from which the human petitioner immediately falls back. In shamanism, however, we have a consistent practice and discipline directed at bringing to visionary perception the content and power of the mind, through techniques of concentration, heightening of attention, trance, etc. issuing in symbolic images. Though we have become blandly familiar with forming ideas through long cultural history, we still talk about the way they come first through 'inspiration' as new insights. We have also become blandly accustomed and forgetful of the way that they reshape our perspectives and determine our attitudes, for good or ill. Though they were long denigrated, shamans are nowadays acknowledged by religio-historical research as the originators of the inner world of representations we nowadays call our mind. Some researchers at least are starting to grapple with the implications which follow.

The cave-paintings originate, then, in the context of shamanic procedures. This knowledge for the first time enables an expert like Lewis-Williams to enter into the process of their formation. It is therefore quite wrong, he points out, to think of them as images painted onto the rock. Rather, they belong to a reality through and beyond it, whence they are captured on the surface which, screen-like, makes them visible—indeed features of the rock-face are often incorporated:

> It seems that the walls of rock shelters were thought of as a 'veil' suspended between this world and the spirit-realm. Shamans passed

through this 'veil', sometimes following the painted 'threads of light', and, on their return, brought with them revelations of what was happening in the world beyond. We do not have any evidence to suggest that *only* shamans painted, but it seems likely it was the shamans themselves who painted images coming through into the world of the living and visions of the transformations they experienced in the spirit world …. The painted images of another world made sense because of their location on the 'veil', the interface between materiality and spirituality. The rock wall on which paintings were placed was not a *tabula rasa* but part of the images; in some ways it was the support which made sense of the images. Art and cosmos united in a mutual statement about the complex nature of reality. The walls of the shelters thus became gateways that afforded access to realms that ordinary people could not visit—but they could glimpse what it was like in that realm as painted images filtered through and shamans … described their journeys.[*]

Thereby ordinary people could also receive an intimation, secondary though it necessarily had to be, of the power which possession of such images gave to the trance-masters. They could gather round and express their awe and joy in dances and festivals. They could honour the revealed meanings and rally round them in loyalty or craft them into implements or carry them as signs.

The belated recognition of the shamanic reality behind prehistory's most famous productions has hugely enriched contemporary anthropology. It also furnishes confirmation of the immense antiquity of the spiritual discipline and mythology of shamanism. The question as to how ideas and symbols could be perpetuated over immensely long periods of time confronted us earlier, and found an answer in the suggestion that the basic images are found in our own constitution—the subconsciously perceived foetal head/skull immersed in fluid, ever renewed in the large-headed state of the embryo in the womb, dramatises itself in the oracular mythology of the Flood/ falling stars, ark or fish/monster and the rebirth of mankind. Whenever we grasp any new situation, or when we stand uniquely aware, our mind may evoke features of the mythology. Modern researchers have seized, similarly, on the neurological signs which indicate trance-induction. They are right to stress that we carry the elements of the complex within us. But even long after the eighteenth-century Enlightenment, science has kept the focus of human evolution too relentlessly pinned on the evidence of emergent rationality, and so on its neurological

---

[*] Lewis-Williams, *The Mind in the Cave* pp.148-9.

basis. At last, however, has come recognition that consciousness embraces much more than this, and needs to be incorporated in the picture.

Lewis-Williams correlates the experiences that generated shamanism with the most significant component parts of the shamanic hierarchical universe. The multi-level universe, he suggests, is not just a fantasy but corresponds to experiences at the limits or transforming-points of consciousness. Experiments can induce similar sensations. At the upper extreme of shamanic testimony are experiences of weightlessness, flight, the distention of consciousness to include wide perspectives, in short the 'heavenly' dimension; at the other pole, tunnels (in which 'iconic images' may appear) are often described in near-death situations when 'subjects often experience sensations of inhibited breathing, distorted vision ... difficulty in moving ... frequently interpreted as being underwater. Both underground and underwater travel are widely reported shamanistic experiences.'* Between them lies the earthly world. A shamanic 'gateway' typically opens a connecting-passage between these three realms, and the structure is still preserved in popular pictorial notions of this world, heaven and hell. Lewis-Williams bravely asserts that this structure is part of our heritage as human beings.

When dealing with issues of consciousness, modern theories tend to turn unhesitatingly to neuro-physiology because of our favoured scientific assumptions. Rudolf Steiner, however, gives a less narrow view of consciousness and consequently a more illuminating account of the factors involved in the visionary states, in which the nervous system is only one functional element. He brings to the question his stress on overall structural features and their social correspondences, which enables us to appreciate much more profoundly what these archaic Mysteries were about. He does mention, nevertheless, many of the same characteristic stages such as conditions where all movement is inhibited, or in which exhilarating heavenly visions are experienced.† To understand Steiner's perspective, we need to

---

* Lewis-Williams, op. cit., p.145.

† He says for instance of the initiation-process: 'A moment comes when a man confronts his physical body, whose hands he can move during waking life ... whose eyelids he can open and shut and so on, but now he feels as though his whole physical body were petrified, as though it were impossible to move the eyelids, legs, hands, etc.'—Steiner, *Mysteries of the East and of Christianity* p.18; it is compared to Mystery-language of lying in the mire etc., 'where the spirit reveals all of life to be death. One is then no longer in the world, but under it, in the Underworld. One has descended into Hades'—*Christianity as Mystical Fact* (New York 1997) p.6; a tableau of images is another early experience, and 'what we obtain from this is an

pay attention to the concrete reality of the shamanic techniques, typically involving intensive rhythmic activity, and altered breathing/hyperventilation. Consciousness, in Steiner's view, is not a camera-like activity of the nervous system only, but a function of our total way of being, involving also the rhythmic (breathing- and blood circulatory) system and even our limbs with their distinctive mode of placing us in the world.

The Mystery-symbolism must from this perspective be related to elements from our own threefold constitution and the social analogues of our individual make-up. He bases his view of this specific syndrome in the rhythmic processes of breathing and the closely related blood-circulation, which he sees as generally much more central to the way we inhabit the social and even the natural world than most contemporary thinkers.* When we breathe in we feel a heightening of consciousness, connected no doubt to the increased oxygenation of the nervous system: we talk about 'clearing the head', even of being 'heady' at such moments, and here certainly breathing touches, even presses up into the nerve-system activity of representation. Techniques of breathing can deepen the awareness of the connection, and bring our activity into consciousness in the ways we have previously touched on, thus opening the path to expanded or visionary experiences. (Our usual modern path, however, has been exactly the opposite. We have consistently limited our attention to those aspects of idea-forming that enable a link to our ego-representation, pushing the fuller activity into the penumbra of the unconscious. Hence the illusion that ideas 'just come' to us, and also that our 'ideas are our own'—that 'we just make them up'. The preference for more abstract formulation, though leaving us freer, has also progressively concealed much of the image-nature of primary thought. Steiner's original philosophical investigations were aimed at drawing attention to

---

enhancement of our own inner activity, the active experience of one's own subjectivity.... A tremendously strong feeling of happiness is united with this imaginal knowledge. It is this subjective feeling of happiness which has inspired all those religious ideals and descriptions ... where life beyond this world is pictured in glowing terms. They are the imaginal result of this experience of happiness' (with reference to Islamic religious experience)—*The Evolution of Consciousness* (London 1966) pp.23-4. Such characterisations of stages on the path are of course part of the texture of Steiner's life-long work, and could be far more extensively documented.

* A starting-point for the exploration of these themes in Steiner's work can be recommended in his *Foundations of Human Experience* (New York 1996) pp.39-44; there are of course from him many descriptions of the interacting systems of human existence and their social extensions, with a basic exposition being provided in *Toward Social Renewal* (London 1999).

our own hidden activity, and to pursuing Goethe's ideas of the 'perceptual power of thought'.)

At the other pole of our middle-system experience, breathing activity can bring us into a felt proximity with the metabolism and limbs, and to exertion of the will. In an effort of will the breathing may become heavy or strained and we have the sensation of overcoming resistance, and this experience may be heightened into the kind of shamanic vision of inertia and spiritual 'submersion' we noted above. But it should be noted too that our human absorption of such experiences, whether of a high or of a submersive nature, is mediated constantly by the rhythmic and breathing-system which actively responds to them. Our pulse-rate increases in response to our inner sensations, or is slowed by our meditative mood, and our breathing both leads and is modified by the feelings aroused. The Mystery-techniques are practised amongst an esoteric group of fellow-initiates who sympathise intensely with the trance-process, or often conduct a pupil step-by-step through its stages. We say of such close groups that they 'breathe together'; they will certainly be much aware of changes in breathing, as a sharp intake of breath may signal surprise (sudden heightening of awareness), or otherwise indicate to those who know the exact point reached by the trancer. Relaxation and return to normality will be accompanied by a release and an easier breathing common to the group. Similar modifications will also be observable in the ritual dances and special activities of the larger social group when Mystery-practices are opened up in part to the community at festivals.

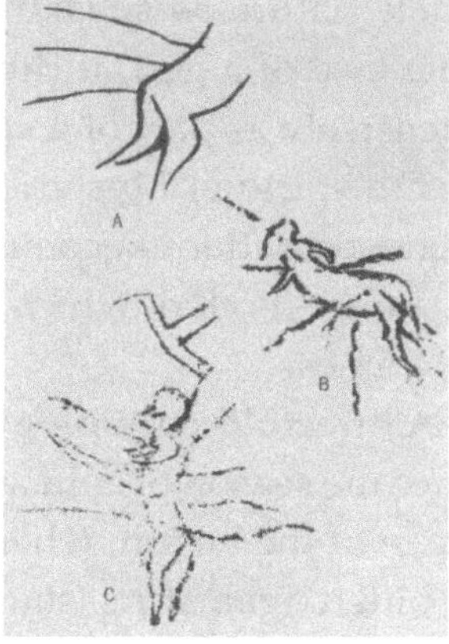

The scenes of death by multiple piercings in Upper Palaeolithic art are hard to interpret convincingly as realistic struggles. They seem rather to depict the mystical death and dismemberment experienced in dream-visions by the shaman, prior to rebirth on a higher level of being. Cave-paintings from Cougnac and Peche Merle, Quercy, France.

Thus blood-rhythms and breathing-changes accompany the inner experiences of the shaman and his mentors, and are a first intimate stage of social participation, though it remains for most people on the shadowy edge of consciousness. Breathing in and out is in fact our basic 'symbol' of the constant interchange between inner and outer in human experience, and changes in our pulse of the way we absorb such experiences. It expresses our need to take in life-substances and also ideas and challenges to our will, to come to a formulation when they are defined in their meaning for us, then to breathe them out and make our mark in turn upon the world. 'Inspiration' as an expression of the way an image or meaning enters into us and becomes part of us contains the element of transcendent origination but also, less noticed, an indication of our human way of coping with the impact of any powerful experience. Breathing further shades into utterance, starting with involuntary cries, or speech where the breath carries song-like sounds. Sung speech worked primarily on the emotional and empathic level. Rhythmic speech is a potent way in which we take hold of experiences and embody them in our own terms, and in which we socialise them and control the way they affect us ('interpret them').

As well as rhythmic action (such as drumming) and yogic breath-control, mantric utterance is one of the fundamental instruments of Mystery-activity. Its function is less descriptive than geared to shaping our inner response to fundamental features of our experience, utilising the power of the basic elements of language which emerge from our whole nature as speaking beings. Unlike the oracular, unrepeatable utterance of a god, it provides a formula which can be worked with intensively as part of a spiritual training—the decisive shift from oracles to spiritual Mysteries. In a mantram repeated ever and again, a linguistic pattern becomes itself like a meditative image charged with meaning—both of which we find employed in Mysteries from the earliest times.

A threefold cosmology is also projected by the breathing-process, since the season of cold and drought can be analogised with the drawing in of the breath, when life is held is reserve, withdrawn into itself or pitted against resistance; the hot or fertile season is equivalent to the breathing-out into the cosmos when life can expand and one can feel a dreamy sense of merging into the warmth and fulfilment of Nature; human life runs its course in-between. Steiner pointed to this 'cycle of the year as a breathing-process of the earth' as a meditative process which can be renewed in modern practice, having been the basis of the Mysteries from immemorial times.*

We suggested previously, following Rudolf Steiner's views, that when formally oracular society needed a higher-level framework in which to resolve the internal tensions which were increasingly generated by its growing complexity, it found intuitively a threefold structure. With that came a new type of hero who was less a defender, more a helper-figure and in some ways an organiser of society. Such a hero belongs to the sphere of co-operation or (in the broadest sense) economics, equivalent to Dumézil's function of fecundity out of the three domains. In historical times the symbolic imagery of the economic sphere will be predominantly agricultural, and it requires quite an effort on our part now to imagine our way back to an era when that particular fruitful domain was still far in the future. The fundamental economy of archaic society is hunting and gathering, with manufacture of implements, clothing and the resources of human fertility. Much of this can be related directly to the content of the cave-paintings. Earlier structuring of society, such as we may associate with the prevalence of oracles, is based on shared remembrances and on continuing patterns which had been successful (like the Mousterian craft-technique). Such knowledge was endued with a sense of divine authority. Alongside the newer complexities, we must of course remember such life continuing in all essentials, as also the tendency of increasingly powerful god-like leaders to dominate ever larger groups.

The emergence of the Mysteries, however, may be seen as a way of making fuller use of both newer and older tendencies, and as integrating them constructively. In fact through the picture-language of the Mysteries it first became possible to symbolise the value of activities and give them recognition as part of a system, in which they could be used most constructively to raise the level of life across the community. The Mysteries thus appropriated to human agents some of the power of the gods which had previously worked in as if 'from outside' especially in moments of crisis, and made use of it to give each sector its authorisation in appropriate symbols. The world of divine images came down into society through the shamanic ability to go into heaven or descend into hell.

---

* Steiner, *The Cycle of the Year as a Breathing-Process of the Earth* (New York 1984); also important in his life's work was the meditative 'Calendar of the Soul' (available in multiple editions and translations) which in a modern way restores our participation in the yearly process..

What the Mysteries effected was much more than the ability to evoke image-experiences. Hence it is inappropriate when sometimes the researchers evoke the analogy with psychotropic drugs. To use these to explain ecstatic states is most completely to miss the point, where Rudolf Steiner is most insightful. As significant as the images themselves was organising transmission of the techniques of their discovery in closed brotherhoods—and sisterhoods. There they could be safely taught and the ways of absorbing their potency could be developed—e.g. in the skills of cave-painting. Such an esoteric group was in a sense a microcosm of the larger social order that could be developed through the promulgation of its content via festivals, communal dances etc.* The knowledge of the way potent images could be humanised was essentially the knowledge of how to fuse oneself creatively with them in specific contexts, and so attain to a real 'identity' whether as actual craftsman, hunter, youth, old man, wife, mother, etc.

The right images could thus be transmitted for each phase or sphere of life. But their efficacy depended on insight into the way people could actually transform themselves by adopting them, and ultimately identifying with them. (Contrast the free-for-all competition for desirable images today, and the confusion of identity that so often results. Steiner does consider, however, that new and individual ways of finding ourselves are now needed.) The Mysteries accordingly seized upon important moments of transition and change in human life, and shaped them into a coherent cosmic and social vision: puberty rites were especially significant, gearing the psycho-physical changes that happen with biological maturity into the process of socialisation, and arousing awareness of the human dimension of life/death, responsibility and sexuality. But other 'rites of passage' were also highlighted—such as births, deaths, marriages, the giving of gifts, the making of sacrifices: all these events mark an inward

---

* For an introduction see my *Transformations of Religious Experience. The Approach of Rudolf Steiner* (Alistair Hardy Society Publications, Lampeter 2006); also Welburn (ed.), *The Mysteries. Rudolf Steiner on Spiritual Initiation* (Edinburgh 1997). Esoteric societies continued to lead social development up to and including recent Freemasonry, cf. the interesting studies in M.D.J. Scanlan (ed.), *The Social Impact of Freemasonry on the Modern Western World* (London 2002). For classic Mysteries in the ancient world and their relationship to power and social structure from prehistory to the Roman Empire, cf. Reinhold Merkelbach, *Mithras* (Königstein 1984) pp.153ff. Elements of the relationship reach back into the prehistoric hunting bands which furnish the oldest core of Mithraic myth, pp.4-6 .

transformation, of a situation or a personality.* Each is the opportunity for a new metamorphosis, ritually enacted in the Mysteries as a second or higher birth.

The mythology of Mysteries is inevitably centred therefore on the figure of the Goddess through whom an inner transformation, symbolic rebirth, can be obtained on the level of the spiritual images which make up cosmic reality. The figurines of a goddess-figure are some of the most well-known artefacts of the Upper Palaeolithic. Though popularly categorised as a 'fertility' goddess, it is certain that her meaning was already much richer and more complex. The fundamental mythology of Mysteries is probably already present, even though it will attain to its classic expression only in the time when agriculture has made human interaction with Nature and the handling of transformation (growing, processing food, overcoming death through life in the seed, etc.) still more comprehensive. The Persephone-myth from Eleusis near Athens, home of the agricultural Mysteries of the Two Goddesses, will be perhaps its purest expression.† In connection with the Goddess is also found a hero who himself descends into hell and is subsequently transfigured, like Heracles. Death and rebirth, the rhythmic of withdrawal into the earthly depths and vital expansion into the sky-world of the gods, the receiving of knowledge concerning human destiny and responsibility, ecstatic visions and symbols: all these widespread themes point to archaic origins. They predate even agriculture which they fostered: it is argued by at least one researcher into the Greek Mysteries that the agricultural diffusion myth is grafted on to a still older core mythology.‡ Already in the

---

* The classic study by E. van Gennep, *The Rites of Passage* (London 1960) is still valuable; a modern anthroposophical perspective on transformation, life-stages and society is beautifully presented by B. Lievegoed, *Phases* (London 1979)

† Persephone is abducted into the depths of the Earth to the dead, but returns to life every spring; through initiation into her Mysteries a birth-in-death took place, evidently of Dionysus, though descriptions of it are (deliberately) riddling. Analogous myths are connected to other Mysteries involving goddesses such as Isis, Ishtar, etc. and gods whom they regenerate such as Osiris or Tammuz. For Mystery-patterns involving 'mystical death', descent into the Underworld, etc. in the Upper Palaeolithic, cf. D. Lewis-Williams, *The Mind in the Cave* pp.281-2.

‡ K. Kerenyi, *Eleusis* (Princeton 1967) pp.126-30. Demeter is not only the Earth which yields the crops: originally it is the human dead who are put into the ground, who are called 'Demetrians' (p.93).

Upper Palaeolithic the Mysteries behind the cave-paintings will have comprehended that in order to absorb and share external or social experiences, people need to be moulded and guided by the power of images, reshaped in accord with the breath-like rhythms of life and of the world, reborn from the fire through the Goddess.

Betrayal

To many involved in the study of human evolution, the breadth and complexity of mind shown in the Upper Palaeolithic 'Explosion' of art and social expansion initially seemed to indicate a pushing aside of the old and the advent of the new. The Aurignacian culture was considered to belong to a species with distinctive abilities both technically and mentally which had simply not existed before. A stream of discoveries only seemed to confirm the phenomenon, and theories were rapidly developed to explain why earlier types of humans, especially the much denigrated Neanderthals, had not been able to achieve anything of the same calibre.* It was rather as if the world had been waiting for *Homo sapiens*.

Better chronology soon showed, however, that these same humans had been evolving much earlier, oddly without showing off their supposed abilities. Neanderthals on the other hand were emerging as cleverer and more organised, though needing new approaches to understand their musical and compartmentalised consciousness. It still seemed, though, that the Neanderthals lacked images and everything that images made possible. All that changed with the discovery of cave-paintings in several locations across Spain that were too old to be from *Homo sapiens*, and a collection of body-ornaments from Los Avi-ones. The find was widely reported. 'The Aviones finds are the oldest such objects of personal ornamentation known to this day anywhere in the world,' says study co-author João Zilhão, a University of Barce-lona archaeologist. 'They predate by 20 to 40 thousand years anything

---

* Even Lewis-Williams, though sceptical of the 'illusion' that a sudden step forward occurred (*The Mind in the Cave* pp.69ff), feels the need to postulate an inability of the Neanderthals to deal with significant images, involving even the speculation that they could not dream or remember their dreams (pp.190ff); Steven Mithen has to speculate that the new species possessed brains that were 'connected up' neurally in a way that Neanderthals were not: criticised by Lewis-Williams, pp.110-111. It will appear, however, that all such theories are shown to be unnecessary by the more recently uncovered facts.

remotely similar known from the African continent. And they were made by Neanderthals.'*

These discoveries change everything. They sweep away any last suspicions that the Neanderthals, when they showed signs of advanced culture, were belatedly imitating their human cousins. They sweep away the need for elaborate theories that only *Homo sapiens* possessed the wherewithal for symbolic representation. They show that the Neanderthals knew the techniques and made use of the symbols which we know were based upon shamanic Mystery-activities. The 'ladder'-symbol prominent in the most striking of the caves is indeed a typical representation of transition through different levels of existence. Yet we know that the Neanderthals had a rich life centred in quite different aspects of consciousness, and the blank walls of Bruniquel contrast starkly with the plethora of painting-sites in Upper Palaeolithic times. So the difference is not one of ability, but of context and focus. Perhaps it is difficult to see how the acknowledged social complexity of the Neanderthals could have been sustained without images. Nonetheless, the Neanderthal painting-sites are few and represent a less outwardly visible aspect of their culture, whereas in the Upper Palaeolithic we seem to see the cat let out of the bag. So far from representing a unique modern-human breakthrough, a novel ability, is it conceivable that we must alter our thinking radically? The Neanderthal cave-sites were, some of them, re-used over time into the Upper Palaeolithic. Rather than being enigmatic exceptions, were these and similar centres where it all began? Is it possible that a technique which was limited and controlled within Neanderthal culture, and already ordered aspects of their social organisation, was taken up and ran rampant in the activities of the Modern Humans who made contact with them and whose population then exploded across the world?

The aspirations which fed into the mission of the great leader known in legend, the Manu of Indian myth, were certainly not the only factor which played into these confusing events. 'One can only understand what happened at that time if one knows that … from a certain quarter men had come into possession of knowledge and of arts which were not directly connected with the above-mentioned Manu and the task he rightly took upon himself to perform. This knowledge and these arts were for the first time devoid of a religious character. They came to

---

* The study was summarised and photographs of the cave-images reproduced in fine quality at https://www.nationalgeographic.com/science/article/neanderthals-cave-art-humans-evolution-science.

mankind in such a way that precluded everything except the fulfilment of self-interest and personal demands.'* It was connected, Steiner adds, specifically with the use of fire. When he subsequently wrote about it more explicitly, Steiner describes what took place as 'the betrayal of the Vulcan Mysteries'.†

The Vulcan-mythology is very close to that of Prometheus, or indeed Lucifer, and its basic images are a fiery-creative power falling out of heaven, plus a connection to the luciferic planet Venus (Vulcan's heavenly consort in the myths), both indicating close links to man's ability to define himself through arts and crafts. Metalwork was its later expression—but the 'use of fire' of course originates very much earlier. It was crucial, for one thing, in preparing the pigments for cave-art. The basic mythology is notably close to the old *oracular* myth of the end-and-beginning of a cycle; but in the Vulcan myth it is definitely elaborated into a discipline of craft-work, and so assumes the character of a Mystery in which the divine act of making and remaking by fire, and the fashioning of images, is put into the hands of human beings.* The nature of this particular Mystery brings it especially close, therefore, to the luciferic 'rebellion' against reality—and so perhaps to the temptation (and beyond that to the possibility) of a betrayal. It reminds us that evolution is not so simple, and much more ambivalent, than 'progress'!

Without the mind-controls represented by the Mystery techniques of absorbing and balancing inner forces, the advent of image-thought in human development was dangerously open to the tendency Steiner calls luciferic. As a result, individuals would have been less content to identify with roles as they were needed; or they could find themselves torn between conflicting roles, tempted to assume power beyond their real abilities, to fall short of the pattern community projected for them and to live in fantasies of roles which were destructive of actual achievement. Of course, we recognise here human society as we know it in all its hypocrisies, inadequacies, ambitions and inequalities. If the society-making machine itself falls into the wrong hands, the results already in prehistory could only be, as Lewis-Williams perceptively comments, 'a conflictual scenario of social divisions.... Art and ritual may well contribute to social cohesion, but they do so *by marking off groups from other groups* and thus creating the

---

* Steiner, *Cosmic Memory* p.66.

† Steiner, *Outline of Esoteric Science* (New York 1997) pp.244ff; p.249.

potential for social tensions.' It was these Upper Palaeolithic tensions, he argues, which set off the 'spiral of social, political and technological change' which has 'continued … throughout human history'. *

Such is most likely the driving force which propelled humanity out into the world, producing cultural and linguistic differentiation; but perhaps humankind would have fought themselves once again into self-destructive stagnation had it not been for that opening of a new dimension, signalling the need to resolve the problems thrown up by history on a higher level, made possible by the vision of the moral ego. The individual had gradually to become an active force in evolution. That furnished the starting-point for all the 'post-Atlantean' civilisations which followed. These all depended on giving every individual a stake in a complex project, based on the great religious visions and aspirations: a sense of what society is for, epitomised in the act of sacrifice. The conundrums that emerged are the ones to which we are still striving to give authentic answers. But that is material for another book.

The insistent insertion of destruction by fire alongside the Flood makes especially interesting one particular view of the Neanderthals' end. One school of thought would associate their decline with the massive eruption of the volcano at Campi Phlegraei in Italy at just around the requisite time—one of the largest ever European eruptions, which would have drastically affected conditions all across Europe. Might this be synchronised in the myths with the Phaethon-event, the fire falling from heaven of the Cretaceous-Palaeogene transition?

---

* Lewis-Williams, *The Mind in the Cave* pp.95-6; cf. p.284 (and generally pp.268ff).

Appendix

TOO MANY MAPS

In our early chapters we examined Rudolf Steiner's response to the remarkable Theosophical map of Atlantis to which he gave a qualified assent, as to the rest of Scott-Elliot's presentation. We saw many reasons for being positive, in the light of more recent palaeogeography. But even the casual reader of Scott-Elliot's book will observe that there are to be found there not just one map, but four. The existence of four maps, chronicling the changes of sea and land in the intervening time to our own, is hard to square with some features of the author's account of his Atlantean source. The question of their legitimacy would be a side-issue, were it not that a true evaluation of the first map is difficult, unless we know how it was come by, and what role it played in determining the rest of the work, including the provenance of the follow-up maps. The story has a certain fascination of its own. But we have serious reasons too for confronting those aspects of the Theosophical 'research paper' that became a celebrated book which are (frankly) on a level quite distinctly lower than the best.

Frustration

The underlying narrative behind *The Story of Atlantis* is presented fragmentarily—a fact of which Scott-Elliot, to be fair, is painfully conscious—and the patchy 'scientific' rationale for it is often left blurred and unsatisfactory. The situation is further confused by some pure fantasy-elements, scintillations of which are quite evidently derived from Plato—a source which is for the most part, as also in Steiner's case, ostentatiously ignored. Even when it is not, there are strange dislocations. Perhaps under pressure to envision what was asked for, the 'scryer' blended vestiges, half-memories and expectations. The 'City of the Golden Gates' is the most bizarre case: clearly based on the *Critias*, with its mention of the famous circles, the canals ('moats'), even the horse-racing, and the transition from the city's circles to a system on the rectangular plan irrigating the surrounding plain. But the Platonic capital has swollen to a veritable London with two million inhabitants or more, suitably well provided with parks and public amenities and a sort of diplomatic quarter.

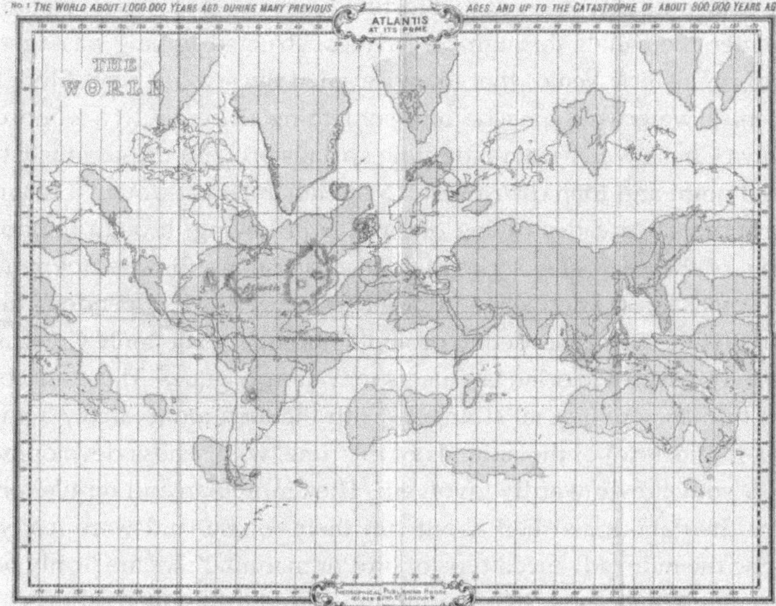

The map of Atlantis as originally published. The heavy grid showing latitude and longitude and modern continents recalls the 'British Empire' map often rolled out on school exercise-books, here of doubtful relevance. From Scott-Elliot, Story of Atlantis.

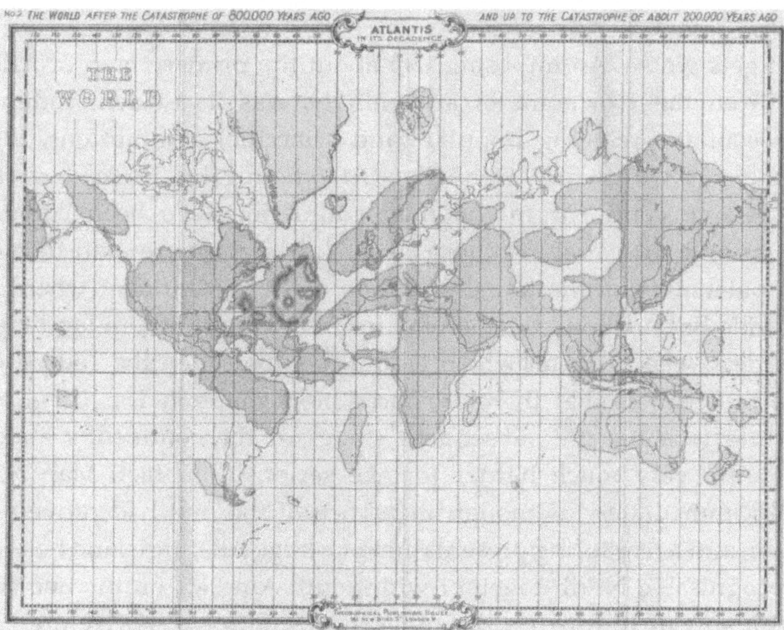

The additional maps in his book are hard to explain either from the account of Scott-Elliot concerning a globe and a parchment transcription, or from the facts of geography. They show dependence on the misguided recent theories of the Atlantic Ridge popularised by Ignatius Donnelly and others. From Scott-Elliot, Story of Atlantis.

Plato's Atlantic City was a sort of fortified harbour-port, where warships can enter the canals through cleverly constructed linking passages, a sort of prehistoric Venice that likewise ruled the seas. In Scott-Elliot the concentric waterways have become something else entirely: a wondrous supply of fresh water from mountain springs brought mysteriously to a central point and then distributed to the inhabitants through a magnificent series of cascades raying out and overflowing through the novel system of the *descending* series of concentric canals. 'It does not require much mechanical knowledge in order to realize how stupendous must have been the works needed to provide this supply': not only Greek and Roman but even Victorian engineers may simply gape.* These anachronistic human activities sit strangely with the mention (scarcely a page-and-a-half before) of the animals of the time, with whose development man is somehow inwardly involved: 'The amphibian and reptile forms which then abounded had about run their course, and were ready to assume the more advanced type of bird or mammal'; we are firmly back in the late Cretaceous. It is certainly an idea in occult teachings that man's existence is not a late fluke of evolution but an integral part of its story. Rudolf Steiner will have serious things to tell us on that front, but with these clashes we remain closer to the fantasy-film world of *One Million Years* B.C.

It gets worse. An imposing fact about the primary map of Atlantis, from the viewpoint of authenticity, was its evident independence of the reigning popular (and scientific) assumptions, ideas of the sort with which Donnelly's *Atlantis* regaled its readers. The impressive continent in the map could not have been conjured up on any such basis as those. It is startlingly more akin to modern paleomaps of the late Cretaceous, with its continent where the Atlantic Basin should be, the remains of a large southern land mass, a wide sweep of ocean which would now be called the Tethys, etc. The difficulty is that there are three more maps in *The Story*, purporting to show the subsequent stages of the continent's history. They are very much harder to take seriously. In fact, Map no. 2 looks immediately as though the shock of the first had proved too much, and a reassuring presentation of a large Atlantic island between a recognisable North-together-with-South America on the one side, and a sort of Europe on the other, now confronts us. We are back in the Donnellian-Platonic tradition, upgraded a little by the greater

---

* For Plato's layout, cf. the illustrated reconstruction in S.P. Kershaw, *A Brief History of Atlantis: Plato's Ideal State* (London 2017) p.103; cf. *Story* pp.44-46.

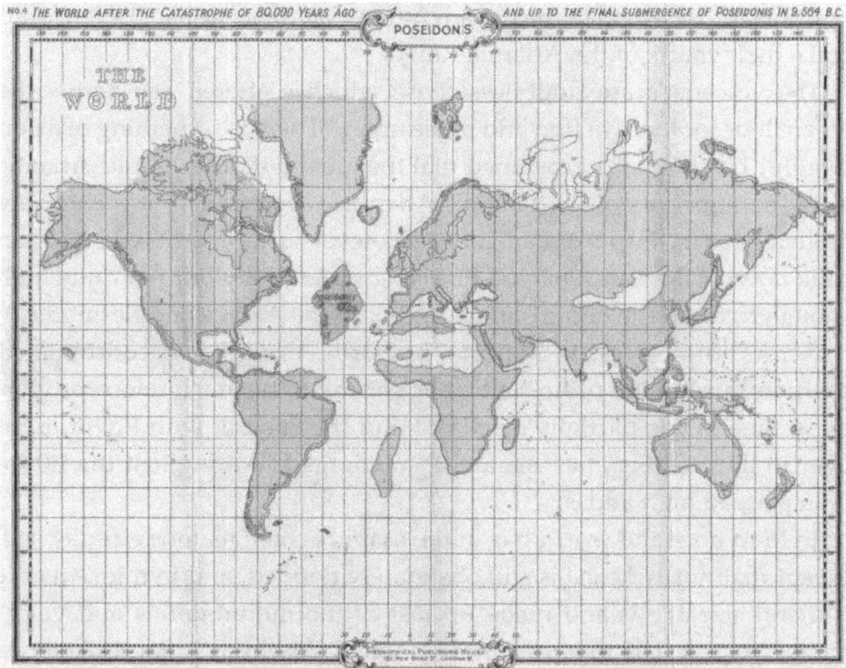

The final map shows the supposed survival of the old Atlantic continent as the island of Poseidonis (so christened by the Theosophist A.P. Sinnett), bringing it down to a date within the range of early history and the Atlantis-legend of Plato.

Atlantis of the first map. Elsewhere in Map no. 2's world strange and inexplicable things have happened. The sprawling southern continent has abruptly shrunk more or less to a modern Australia. Nearly all of Africa has arrived from somewhere, and attached itself in the north via Arabia to the Asian land mass. A new far-northern land formation has left much of Europe submerged beneath an unexpectedly large proto-Baltic Sea, while the Siberian steppes have surprisingly been encircled by water.

In Map no. 3 Atlantis has split into two disproportionate islands (Ruta and Daitya), but their detailed representation is unusual. We have noticed previously the almost complete dearth of relief indications on the *Story*'s maps, yet here we are most definitely compensated by the emphatically marked mountain chain which sweeps in a broad curve across Ruta. Nor is that all. The eastern half of the island within the arc of the mountain-range is dominated by two very large, solitary round peaks which from their shape and size more than hint at volcanoes. A sideways squint at the map, imagining away the lands around these various eminences, quickly reveals that the great mountain chain is positioned just where in the distant future *Challenger* and *Dolphin* would make their charts of the Atlantic Ridge.

The volcanoes are doubtless those which scattered ash across the ocean floor for them to find and muse upon. (There is something of a non sequitur here. One had assumed that the great outcrops in mid-Atlantic were the remnants of the Platonic Atlantis, whereas here it seems they originally soared up from an above-sea-level surface, and Atlantis then sunk down *with its mountains* still rising majestically but not quite high enough to reach the ocean surface except in the Azores and Bermudas.)

Across to the west in Map no. 3, North America has consolidated itself; in Europe the proto-Baltic has shrunk to nearer modern proportions; Siberia has partially dried out; but the Sahara, with Egypt, right over to the Red Sea, has remained submerged, leaving just the Ethiopian plateau high and dry.

In Map no. 4 the world has in the main become its modern-day self, except that north Africa is still obstinately flooded, and in mid-Atlantic is a still sizeable island made up just of the mountainous and volcanic region referred to above, now labelled Poseidonis. South America has assumed its long tapering shape, which presumably means that the Andes have now risen (though no such relief-indication exists on the map). When once Poseidonis goes down, we will have the world we know so well before us again.

In all this there is precious little discernible sense. There is no hint of Steiner's idea of the moving of the continents in response to the energetic life of the Earth. The continent that may be South America in Map no. 1, having broken away from the great composite land mass in the south-east, just seems to lose its shape and grind to a halt. Lands rise up or sink down without reason, appearing piecemeal in the places needed to make, bit by bit, the modern map of the world. Changes that would indicate lowering of sea levels globally from the greater land coverage in Europe and Asia still mysteriously leave north Africa submerged.

Names were no doubt added to the maps for publication, since the term Poseidonis for Plato's is no older than the 1880s: but the general idea of Atlantis shrinking in stages is itself closely bound up with that same adaptation of the story. It maintains a specious link with Plato and a disaster in something approximating historical times. In terms of its palaeogeography, as nowadays scientifically reconstructed, the continent's division into two island parts is not a step on the way to submergence altogether. The notion of the volcanoes left islanded in mid-Atlantic is a falsehood; the truth behind the evidence of vulcanism is the spreading sea-floor. The 'piecemeal destruction' scenario only emerges in the Theosophical interpretations of the late nineteenth century. In short, Maps nos. 2,3 and 4 cannot be older than Scott-Elliot or his immediate sources. And because Scott-Elliot does not tell us how he knows what he knows about the history of the continent, we cannot know whether he got his geographical information just from the maps. More likely, the maps have been concocted to include some things that were part of an existing body of ideas, just as they have included the fateful discoveries of *Dolphin* and *Challenger* and made them all together a jumble of their own. We are drawn back by them into Donnelly's misguided world, with the addition of a few—and of them some few that are interesting features admittedly, but as a whole a farrago. Moreover: by drawing in the first map, making it part of their sequence, they bring in question the value of its information too. Because we cannot access any underlying sources or source-distinctions, we must take all or nothing. So everything is compromised.

Everything is compromised—unless, that is, there were any reason to think that the maps were not all equal in value, but that the last three maps really were separate somehow from the first. Fortunately, there is good reason to support such a conclusion, though it does not appear in the *Story* as originally published. We must glance for a moment at its sequel.

Glimmerings

Scott-Elliot's *The Story of Atlantis* was published in 1896, to be followed in 1904 by another exercise in seeking fabled lands: *The Lost Lemuria*. The two short works were afterwards published together between single covers and reprinted many times. The second work concerns us in only limited ways, specifically where it revisits aspects of the earlier book and in doing so reveals new information about its methods.

The *Story* was at the very least reticent about its own basis—perhaps, in fairness, circumscribed by commitments to secrecy relating to particular esoteric matters or persons. It was left to A.P. Sinnett in the 'Preface' to assert the competence and diligence of the researchers:

> And to promote the success of their work they have been allowed access to some maps and other records physically preserved from the remote periods concerned—though in safer care than in that of the turbulent races occupied in Europe with the development

> of civilization in brief intervals of leisure from warfare, and hard
> pressed by the fanaticism that so long treated science as sacrilegious
> during the middle ages of Europe.[*]

Scott-Elliot added little more—until he wrote *The Lost Lemuria* (also with
maps). When discussing his later researches he took the opportunity to
make a somewhat defensive back-reference to the materials that had
been used to produce *The Story*:

> It was never professed that the maps of Atlantis were correct *to a single
> degree* of latitude, or longitude … still less must these maps of Lemuria be
> taken as absolutely accurate. In the former case there was a globe, a good
> bas-relief in terra-cotta, and a well-preserved map on parchment, or skin
> of some sort, to copy from. In the present case there was only a broken
> terra-cotta model and a very badly preserved and crumpled map, so
> that the difficulty of carrying back the remembrance of all the details,
> and consequently of reproducing exact copies, has been far greater.
>
> We were told that it was by mighty Adepts in the days of Atlantis
> that the Atlantean maps were produced …. But while guarding against
> over-confidence in the absolute accuracy of the maps in question, the
> transcriber of the archaic originals believes that they may in all import-
> ant particulars be taken as approximately correct.[†]

This is still infuriatingly vague. Evidently there had been objections
made to the maps in *The Story of Atlantis*, but we do not know what
exactly or from whom. The arch expression *'to a single degree* of latitude'
dismisses the problem with a hint of scorn, presumably meaning that the
author was not really quite sure how to give an answer. But he does not
throw the objection out. It seems the objection to the maps did not come
from someone with the viewpoint that the whole idea of such maps
was ridiculous; if it had, it would be rejected—but on the other hand
the arch expression would lose its point. It appears to have come from
some quarter claiming knowledge of the subject but apparently giving
the author of *The Story* a mild rap on the knuckles. Scott-Elliot seems to
intimate that he feels unfairly criticised, given the intractability of the
materials to hand.

Much else is even harder to understand precisely. There was a *terra
cotta* globe with basically clear relief formation, a *mappa mundi*. (Of what

---

[*] 'Preface', p.x.

[†] Scott-Elliot p.78.

size, it would be interesting to know, or with what indications of latitude, poles or orientation?) And there was a rather fine map on parchment—or rather on 'skin of some sort'. Alas, we do not know whether Scott-Elliot (or his assistants) would know in exact terms whether they were seeing parchment, vellum or an inferior substitute. 'Skin of some sort' sounds as though they did not consider it of the finest quality, or perhaps they thought it old and partly perished so that it would have been finer once upon a time. Before the perfecting of parchment manufacture (an elaborate business), writing on leather was a common way of making a copy which would endure—for legal or any other valuable purposes.

A point to be made here is that unless the map was genuinely on parchment or very fine skin, it is usually only possible to write or draw on the flesh-side. Parchment or vellum also requires to be stored under very strict conditions, usually pressed flat between boards, either singly or in quires. We ask in vain for any information here, but the implication is that no very elaborate form of preservation was employed. Again, that suggests it was not real parchment. Unless we can be sure that it was so, and all the indications suggest rather the opposite, it is likely that it was a map on one-side only. And indeed, Scott-Elliot mentions specifically 'a map'—and for that matter, 'a globe'; whereas the objection was to the 'maps' in *The Story*. Sinnett too refers to 'some maps' along with 'other records physically preserved', whose nature we would again be all too interested to divine. Given the statements of Scott-Elliot, it is most natural to suppose that Sinnett's plural 'maps' refers to the 'globe' (itself a map of course) and to the map depicted on 'skin of some sort'. Since there is no indication that the globe was held in a frame or stand to orientate it, it is again most natural to assume that the accompanying drawn map served the necessary purpose of showing north and south orientation, whereas an uncalibrated globe could be any way up. All this, if true, would account only for one map (in terms of content).

The very last sentences quoted from Scott-Elliot (above), as a sort of testimonial, refer of course fundamentally to the accuracy of the maps in the sequel, *The Lost Lemuria*, which do not directly concern us here. I have quoted him at length, however, because these sentences provide the only obscure hint of the process by which the 'records physically preserved' became printed maps in the book of 1896.

The 'transcriber of the archaic originals believes that they may in all important particulars be taken as approximately correct,' writes Scott-Elliot. What or specifically who does he mean? Could it be that his wife or a co-researcher had produced the 'transcription' and stated (to him) that it is to the best of his or her belief, accurate? Might the transcriber have been Charles Leadbeater, mentioned fifty years later as responsible for the maps by Jinarājadāsa? A careful reading says no. We must consider the style of *The Story*: a text which Scott-Elliot definitely composed, whether or not he was responsible for the content. It is written throughout in a stiff, formal style that avoids direct personal expression (with consequences we know all too well).* Within these stylistic parameters, it is virtually beyond doubt that 'the transcriber of the archaic originals believes' means that he himself stakes his reputation on the basic accuracy of his work. And that agrees with the tone of personal pique which we noted in the reply he makes to the objector(s) about his maps. The comments which he had just previously made about the way it had been possible to 'copy' the 'archaic originals', and then about the difficulty involved in carrying over the information, similarly sound like those of a man talking about the work he himself performed.

It is hard to be certain about the details, but the account of the map-transcribing for *The Story of Atlantis* which Scott-Elliot inserted in the later book offers a fairly consistent basis for conjecture. At least, the passage is accordant with the following conclusions. (1) The passage was inserted at the point where more maps were to be brought forward (depicting Lemuria), harking back to the original volume. The comments strike a markedly defensive note. Evidently objections have been made, not to the viability or usefulness of the project per se, but to the accuracy in some respect of the maps (pl.)—and so seeming to be based on actual knowledge. (2) In reply, Scott-Elliot recalls the circumstances in which the earlier maps were transcribed and to justify the former work catalogues the available 'originals', even though only in minimal detail. The difficulties of the operation are rehearsed, and Scott-Elliot seems to take a spirited stand on the accuracy of the 'transcriber'. (3) Scott-Elliot's

---

* He was an investment banker, and the bank-manager style which maintains a dignity and distance high above the poor applicant may have been his model—impersonal constructions on the analogy of 'It must be said immediately that the Bank is not in the habit of extending its facilities …' rather than 'I will not give you any further money' no doubt flowed easily from his pen. In the book he carefully avoids first-person assertions and substitutes such formulations as 'In making the statement therefore that … it must be understood' for 'When I say that … I mean': Scott-Elliot p.24 and *passim*, as they used to say. Sinnett in his Preface contrastingly writes in a far more relaxed manner, happy to identify himself as 'I' when appropriate, and to address his readers with armchair familiarity, even calling them his 'listeners' rather than striking a bookish tone.

account of the transcription involves two maps, one three-dimensional and one flat, which are best taken as correlated, and so appearing in the final copy as one map. Since he has staked his honour on the production of this map, but seems to feel compromised by the objector or objectors' comments about the 'maps', it is only fair to conclude that he felt defensive about work which he could not himself guarantee, but which has somehow tarnished his own efforts. The scornful reference to unreasonable expectations of accuracy *to a single degree* is a sweeping attempt to suppress the issue (it can hardly be taken as a serious literal admission of very minor points of variance). At any rate, his account is only able to justify the presence in the book of 'a map', or at the very most two if one copies the globe and one the parchment, so that the other two or three Atlantean charts remain problematic and may well have been at the bottom of the original objection. Scott-Elliot does not seem able directly to defend them except by a sweeping gesture.

Most likely there is only one real map. We cannot absolutely know which map it was that Scott-Elliot painstakingly took over from the globe and 'parchment' originals, but there is one in the sequence that seems far superior in intrinsic interest to the remaining three—and anyone who wants to lay claim to any or all of the rest is very welcome to do so. They clearly had no originals among the 'relics' that were transcribed. We are justified in treating the first map as on another level from the other three. Two serious questions remain to be solved. Firstly, can we glean from the writers' often guarded and defensive remarks an indication of its actual provenance? And secondly, can we ascertain its original meaning and function in that context? For the first (be warned!) we shall need a fine toothcomb, and for the second a broad historical perspective on maps generally.

The Maps—How, Whence and Whither?

We still have not exhausted the information grudgingly provided by Scott-Elliot and by A.P. Sinnett in his 'Preface'.

The latter gave vent to a little Gandhian remonstrance directed at the unspiritual qualities and history characteristic of Western civilisation, while basically accounting for the availability of the Atlantean relics—'maps and other records'—furnished to the investigators. He clearly believed that they had been preserved, during all the ages when Europe was rending itself with wars and Inquisitions, among the successors of those Egyptian priests mentioned by Plato, whose long keeping of historical records was the basis of their proud claim to knowledge of Atlantis.* Whence and how exactly they came into the researchers' sphere is not broached. Neither is it at all clear that Sinnett had seen them. That they were shown 'some maps and other records' sounds almost hearsay.

All that does seem clear is that the relics were loaned to them for the short-term only and under quite restrictive conditions. Scott-Elliot himself confirms Sinnett's expression by mentioning that 'among the records … there are maps … and it has been the great privilege of the writer to be allowed to obtain copies': quite probably Sinnett knew here only what Scott-Elliot told him, and Scott-Elliot is not saying much; he does not even reveal here that 'to obtain copies' meant literally transcribing from the original, and so he says nothing about the exact objects made available, only the final results. What he did do is acknowledge himself privileged, and this is the exact same wording he uses in *The Lost Lemuria*. 'In this case also the author has

---

* 'Preface' p.x. An alternative interpretation may be momentarily considered. The Atlantean investigators had been busy at their task 'for some years' prior to 1896. And from around 1890 Sinnett's household had been much frequented by Charles Leadbeater (not yet disgraced) who had recently arrived back from India with a young acolyte. Did he perhaps bring with him something much older—or at least supposedly much older—namely the proofs of ancient Atlantis preserved in the sanctity and serenity which is India?

A closer look at Sinnett's little diatribe, and a sense of its context, however, does not support such a view. Sinnett's claim is strictly a specimen of Atlanteology. Almost certainly he has in mind a passage from the *Timaeus* of Plato—for even if *The Story* is emphatically not based on Plato's tale, Sinnett doubtless considered that the truth of the new occult investigation was, in reality, identical with that behind the enigmatic ancient Greek account (hence 'Poseidonis'). India plays no part in its content. It will be recalled that the Egyptian priests address an early Greek lawmaker with mingled respect and condescension. He does not even know how significant his own country has been in world-history, they tell him. Thousands of year previous to his time, Athens had freed the world from the tyranny of a great Empire: but you Greeks do not even have any records of the event, because in your world instability rules and all records are periodically destroyed, whether it is by the rise and fall of cities and states, or natural disasters. The Greeks are mere children, and forget everything as they rise and fall with the times. But we Egyptians have for long ages kept records, and as a result our civilisation has endured unchanged and stable and with a knowledge of all that has happened, not least as regards Atlantis (*Timaeus* 23a-c). A pretty tale; but undoubtedly Sinnett considered the 'maps and other records' given to the Theosophical researchers as very similar (or more) to these 'records' of which Plato testified in the timeless repositories of Egypt .

been privileged to obtain copies ...' which suggests that there was a formality and a sense that a great responsibility was incurred—much indeed as if one were borrowing the *Mona Lisa* or the Magna Carta for some temporary private use. But who are the Museum authorities in the case? They kept firm control, we discover, over the copying process—again as mentioned in the second book. For in the same passage the writer mentions 'the difficulty of carrying back the remembrance of all the details, and consequently of reproducing exact copies'. So it seems that the researchers, or possibly even just the one 'transcriber', were not allowed continuous access or the use of facilities to make a copy on the spot, but had to carry back in memory what they had seen, and then commit it to paper! Small wonder that the transcriber was touchy about complaints that he might be out by a degree of latitude here or there.

Now these daunting restrictions are very similar indeed to the conditions imposed upon another Atlantean researcher—one of the most intelligent though not the least obsessive amongst them. He too was granted access, of a sort, to archaic and 'arcane' memorials of Atlantis, which he considered as possessing at the very least a considerable historical interest. His name was Lewis Spence (1874-1955).

Spence was a prolific author whose principal subject was folklore and mythology, on which he wrote some twenty books over the first half of the twentieth century. He was a Fellow of the Royal Anthropological Institute of Great Britain and Ireland, and rose to become Vice-President of the Scottish Anthropological and Folklore Society. He flirted with politics and may lay claim to have founded the Scottish National Party. Scottish folklore was prominent among his interests, but his contributions to understanding the mythology of South America, based on works like the *Popul Vuh* of the Maya, did more to make his name. However, he surveyed the traditions of ancient Babylon and Assyria, Egypt, stories of the German heroes and Celtic Druidism, fairy-lore and 'mystic Britain'. It is impossible not to be impressed by his scholarship and grasp of a wide range of sources, and his drive toward a synthetic approach grounded in cultural history going back to prehistoric times—see for example his *Outlines of Mythology* (1944). He was extensively read and admired, and many of his books are still being successfully reprinted or preserved as e-books today. Not all his colleagues were so convinced, however, when in the 1920s he began to base his theories of the development of archaic cultures around the myth of Atlantis.

The Problem of Atlantis (1924) and the *History of Atlantis* (1927) strive to integrate Plato's account of the ancient civilisation, now lost beneath the waves, with our knowledge of the Bronze Age and the rise of the high cultures of Mesopotamia and Egypt. He took up Ignatius Donnelly's amateur observations on cultural diffusion and turned them into well-documented hypotheses—whose presuppositions, unfortunately, few were inclined to admit. He has less to say about the evidence of flora and fauna, focussing on anthropological and comparative religious evidence.

It was only in 1943 that Spence chose to reveal that the sources of his certainty about the lost continent went beyond Plato and comparative folklore: they were those of the occultists, which he termed 'the arcane tradition'. (He was much more interested than the Theosophists, however, in harmonising everything with the *Timaeus* and *Critias*.) In *The Occult Sciences in Atlantis* he explained that he had deliberately refrained from using these sources in his earlier books, 'as I wished to make use only of such proofs as were capable of critical control' but that he has now decided to adduce them.[*] Rather oddly, perhaps, since he finds little in them of value in supporting his case for the Atlantean roots of the Bronze Age. Perhaps they were a gesture in part of desperation. At any rate, we may well let his contributions to folkore and mythological studies stand, independently of his Atlantean theories. What interests us here is his access to the archaic records of Atlantis held by an unnamed occult society to which he belonged.[†] We shall recognise similarities—more than slight—to the experience of William Scott-Elliot with the 'maps and records' he was shown.

Spence is considerably more forthcoming than Scott-Elliot had felt appropriate in describing his involvement with these sources, 'though it is, of course, impossible to divulge in what keeping they are, as very definite vows of silence are extracted on that particular score'.[‡] There is much that he can tell us, on the other hand, as it 'is well known to all mystics, initiated and otherwise' that:

> a very considerable body of traditional matter exists in the records of occult fraternities, generally known as 'The Arcane Tradition', which deals with occult history in its entirety. This is partly written, partly

---

[*] Lewis Spence, *The Occult Sciences in Atlantis* (London 1943) p.86.

[†] It is said to have been a 'European' society, and the guess is sometimes made that it was Rosicrucian. That, however, is unlikely.

[‡] Spence, op. cit., p.8.

conveyed by word of mouth. It does not, however, treat of the secrets of the Mysteries, which are incommunicable ... but merely with traditional circumstances, actual, probable, or, in some cases, merely mythical, and there exists no valid reason why this body of information should not be revealed and employed to fortify the purposes of occult argument and history. So much, indeed, is 'lawful'.[*]

So also appears to have thought someone among the guardians of the material given to the Theosophical circle researching Atlantis. The material laid open to Spence covered a wider range, though it was the evidence relating to Atlantis which of course particularly fascinated him. On this he was given oral teaching, 'my long training in Folk-lore making it comparatively easy for me to memorize its substance', and he was allowed to peruse a quantity of written material, 'which, so far as its Atlantean references are concerned, might be contained in about fifty or sixty pages of an ordinary printed book'.[†]

The records claimed tremendous antiquity (which is rather a matter of course with such reliquiae), but Spence was trained to analyse and evaluate source-materials and was by no means disposed to outright credulity. Scott-Elliot might believe that he was holding in his hand an artefact from 75,000 BC, but the Fellow of the Royal Anthropological Institute of Great Britain and Ireland scornfully declares of his 'source' that 'I have no hesitation in subjecting it to the same tests as the Scriptures and other traditional compositions have been subjected by the higher critics'. The result is indeed devastating (the Bible, no doubt rightly, got off lightly by comparison): 'a great deal of it openly calls for such criticism.... A certain, even a considerable, portion is of no utility and palpably absurd.'[‡]

The oldest among the records, he estimated, might be Alexandrian—which he read in a translation provided. Based on the language and the orthography of the bulk of the manuscripts in general, he thought them mostly late mediaeval to seventeenth-century. Some parts were in Arabic. The later content he thought worthless and unfounded. 'The older portion, however, appears to me to be much more precise on

---

[*] Op. cit., p.7.

[†] Op. cit., pp.8-9.

[‡] Scott-Elliot, *Story* p.17. 'We were told,' he adds subsequently, 'that it was by mighty Adepts in the days of Atlantis that the Atlantean maps were produced....' (p.78)—and evidently it was *no questions asked*. For the observations quoted from Spence, see *The Occult Sciences* p.7.

the whole, though manifest inaccuracies certainly occur, if indeed they do not abound. This older portion is clearly based upon more venerable material.'* On all of this we can of course form no independent opinion—which long deterred him from its use. His obsession with proving Atlantis probably still leads him to give the evidence a more than fair hearing, but his scholarship is probably sufficient to make it likely that some of the material was a few centuries old.

Scott-Elliot recalls in his case the 'considerable difficulty' in 'carrying back the remembrance of all the details, and consequently of reproducing exact copies'. Spence from his experience can only concur:

> These manuscripts, which are readily procurable at the time of initiation, are almost wholly withdrawn from the initiate subsequently; indeed, he is credited with much greater powers of memory than he usually possesses, and this has in many cases, I believe, militated against progress in occult study, a somewhat narrow and illiberal tradition existing against the publication or transcription of these books, which, after all, contain nothing of arcane value proper other than mere chronicle and allegory.†

An initial period of open access corresponds, then, to initiation at a certain grade into the esoteric Order to which they belong. More information about them, including the whereabouts of other copies, is something reserved 'until a higher grade has been attained than that to which I belong'. The initiate is expected to master their content and remember it in what seems to Spence an unreasonably short time, after which access is 'almost wholly withdrawn'—another tantalising obscurism, but perhaps indicating a situation like Scott-Elliot's, when it was possible to examine the materials under strict limitations, and 'carry away' information by memory. We might suppose then that when direct copying was forbidden to Scott-Elliot, it was at least after some initial period of access (without which it seems impossible that the transcribing could ever have been accurately done at all). Anyone who has ever set out to copy a map soon discovers that there is far more detail to take account of than could ever have first been supposed. The annoyingly fragmented account of his work is certainly more possible to understand if we fill out some details from Spence.

---

* Op. cit., p.8.

† Loc. cit.

It is true that Spence makes no mention of maps or *terra cotta*, which would anyway not have interested him for the special purposes of his book, and what little we get to know about the 'arcane' materials is very strictly filtered by its relevance to his own prior views—one version of 'higher criticism' that actually leaves very little 'of value' still standing, though we gather something of the content from his polemical comments.* According to Sinnett, the maps were accompanied by other 'records' which Scott-Elliot presumably also had to assimilate under pressure, and I think from yet another cryptic passage with a glint of temperament in *The Story* that he found it very hard to make much of them either:

> Drawn as they have been from contemporary records which were compiled in and handed down through the ages we have to deal with, the facts here collected are based on no assumption or conjecture. The writer may have failed fully to comprehend the facts, and so may have partially misstated them. But the original records are open for investigation to the duly qualified, and those who are disposed to undertake the necessary training may obtain the powers to check and verify.

No 'higher criticism' in this version: but does the sentence about 'may have failed fully to comprehend' and 'may have partially misstated' reflect just a general insecurity or, as I rather sense, a feeling that he could not understand much contained in them and in consequence may be (or is really being) criticised for 'getting it wrong' in some instances. The true way, he justly insists, is to follow the path of becoming 'duly qualified' in esotericism, since after all these documents and maps are intended as guides to developing the powers to confirm their truth by spiritual perception. It was within their guiding context, together with perspectives from Sinnett and Blavatsky, that Scott-Elliot's wife (or else, Charles Leadbeater) saw her visions. Scott-Elliot's confidence in those who 'obtain the powers' presumably indicates that he trusts her visions much more than the half-intelligible stuff loaned out under difficult conditions by an occult Order.

By inference from analogy with Spence's account, though, we must assume that Scott-Elliot was initiated into an occult society in order to be presented with the Atlantean records. (It must also have been a member of this occult society who was in a position to carp over the accuracy of the transcription.) It may quite conceivably have been Leadbeater who opened the way to his membership, since he was well-connected in such

---

* Spence, *Occult Sciences* pp.27-30, 32-4, 43-50, etc.

movements, although it is not possible to say what society that was. The relics cannot have been owned by the Theosophical Society,* however, or it is unthinkable that access to them would have been made so difficult when the advancement of Theosophy was the underlying aim of the Atlantean book and, as Spence complains, there was no real arcane reason for withholding them but only a set of old-fashioned protocols aimed at preventing them being copied and circulated. Movements like Theosophy were devoted to bringing such knowledge to the wider cultured world, whatever we may make of its efforts now. Although he did not know which other societies held copies of the arcane records, Spence did know that two such organisations did so. A particular one of them had the maps and globes, we surmise, but it was probably not the one to which Spence himself adhered.

The Mystique of Maps

The evidence of the last section may well appear to have been mostly negative. Yet it seems clear that the 'scrying' work was closely associated with versions of the Atlantis story that were being maintained by occult societies, a version which they supported by records and even artefacts such as maps purporting to originate from a fantastically ancient period. Their subsequent usage we may painstakingly glean from the researchers who gained tenuous access to them only for a brief time. We have been forced to subject the statements of the researchers to a forensic examination because of their vagueness or opacity. But we gather that written records and, in the case of Scott-Elliot, globes and maps were 'transcribed' as the basis of their new presentations; but the account of the transcription process makes it hard to account for more than one

---

* Spence is clear that the material he employs 'does not emanate from any popularly known occult society' (p.9), one of his usual synonyms for the Theosophical Society. Yet again there is an irritating qualification, 'Taking it as a whole': does this mean he is aware that some things were shown to the Theosophists? Later he refers to 'a centre which has achieved a wide popularity among occult students of the less critical kind' on Atlantean subjects, describing the result as disastrous though 'claiming to be founded upon the matter of the Secret Tradition'. Was this the Theosophical Atlantis? Disastrous means: not agreeing with Spence's own entrenched opinions. He also mentions a further development, emphatically not based on the arcana, but on 'an inventive capacity which either Jules Verne or Mr. H.G. Wells might well have envied' (p.10)—but how, whence or whither he does not say.

or at most two maps, whereas the book of 1896 contained four maps in all, one showing Atlantis and three showing the subsequent changes in geography that produced our modern world. The latter three maps contain problematic elements, including a sort of convergence on the then new-fangled idea of the Atlantic Ridge and mid-Atlantic volcanoes, and Poseidonis—a vestige supposed to link geological history to the quasi-history of Plato's *Timaeus* invented by A.P. Sinnett. Scott-Elliot has the ability to vex us one more time, for he comments that he obtained 'copies' of a set of the maps 'more or less complete'*—which is hard to square exactly with his later description of the difficult task of transcription from globe and archaic depiction on 'some sort of skin'.

Assuming that all the researchers are being honest according to their lights, and under their obligations—a view I still do hold—we must conclude that there was some irregularity about certain of the maps and treat the additional three as lacking the authenticity of the one (we cannot actually prove two) which the 'transcriber' took such pains to create. We should remember the extraordinary interest of the original map; we should remember Rudolf Steiner's general words of approval and the remarkable statements of his own allied approach; we should remember the assessment of Lewis Spence as an expert on ethnological documents from South America to Celtic Britain, that some of the arcane materials could be genuinely old, if hardly dating from Atlantean times or even (as Sinnett claimed) from ancient Egypt. If we remember these things, I believe it is possible to put the whole matter in a light which explains what possibly does lie behind the whole adventure, and furnishes us with something that is actually of value as a starting-point for our attempt to approach prehistory. We must start with the undeniable mystique of maps. We must all have pored over our first map as children, and 'lost ourselves' in the wonder of it. Let us build from there.

Let us join to it that the moment in history when the modern map more or less sprang into existence—from the incomparable hand of Leonardo da Vinci. Cesare Borgia was carving out territories for himself in northern Italy and one day in 1502 arrived at the conquered town of Imola, along with his agents Niccolò Machiavelli (the power-political justifier) and Leonardo (whose usefulness exceeded all bounds). It was an opportunity for the great genius to dazzle, for he had been developing instruments for measurement and a magnetic compass, and he

---

* *Story* p.15.

combined their use in the technical feat of an 'ichnographic' plan of the town—more or less what we would call a satellite map today—which he presented to his master.

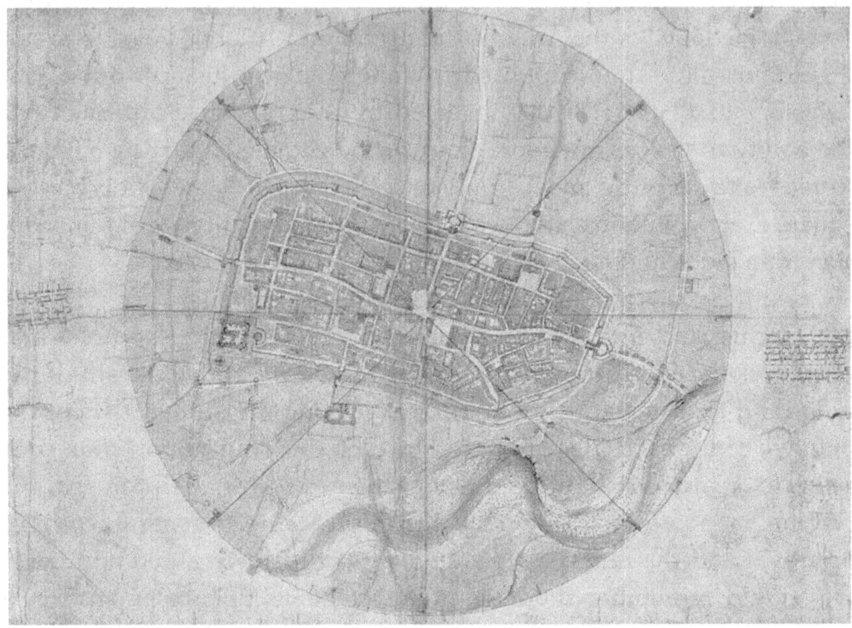

Leonardo da Vinci's map of Imola made for Cesare Borgia in 1502. It is projected on parallels from a point at infinity which transcends empirical consciousness.

By all accounts the Borgia was impressed not so much by the practical aid it gave to his occupation of the town, as by the apparently superhuman feat of its creation. He was right to be so. The plan shows the town from above, with every building as if seen from a 90° angle, directly beneath. To envision it Leonardo has freed himself from any particular observation point, and shows the town as if from an infinite number of theoretical positions, so that his consciousness can be present anywhere in the town with the same precision. He has stepped outside the limits of human perception from a single standpoint, which means some parts are visible and others are occluded, and sees more as a god would see. It was not the technical accuracy of the plan that was so remarkable, as the genius to envision such a freeing of consciousness from its limitations to enable the whole town to be shown. No one could actually stand at any of the high-altitude points from which we see the layout of streets and buildings. (How much we take things for granted when we look at maps today!)

Who would undertake to fathom the mind of Leonardo? But we do know that the Renaissance was inspired to a very great degree by the rediscovery of the writings called Hermetica, which were promoted by the Florentine Academy as the pinnacle in philosophy and religion alike. They are writings from Graeco-Roman Egypt in the first centuries AD—though they were originally thought much older. They do indeed draw upon the ancient wisdom-teachings of the cult of Hermes (Thoth), the wisdom-divinity of Egypt.* The Hermetica combine much of the best scientific thought of the Hellenistic era with a mystical approach to the world, seeing the order of Nature as a reflex of the mind of God. By his consciousness, man rises to a godlike ability to contemplate all things in the glory of God, and free himself from the standpoint of finite perception:

> Bid your soul travel to any land you choose, and sooner than you can bid it go, it will be there. Bid it pass from land to ocean, and it will be there too no less quickly; it has not moved as one moves from place to place, but it is there. Bid it fly up to heaven and it will have no need of wings; nothing can bar its way .... Leap clear of all that is corporeal, and make yourself grow to like expanse with that greatness which is beyond all measure; rise above all time, and become eternal; then you will apprehend God. Think that for you too nothing is impossible; deem that you too are immortal, and that you are able to grasp all things in your thought, to know every craft and every science ....†

Without being totally jocular, the whole passage amounts to a prophecy of Leonardo da Vinci. He surely had such concepts in mind when he conceived the impossible idea of his 'satellite' map.

---

* This admission marks a turnaround in scholarly attitude, seen most clearly in G. Fowden, *The Egyptian Hermes* (Cambridge 1986); see also the account by G. Quispel in Clement Salaman et al., *The Way of Hermes* (London 1999) which is also a valuable presentation of the Hermetica. Earlier, largely abortive trends in the same direction were initiated in the early twentieth century and strongly supported by the Theosophist-scholar G.R.S. Mead, *Thrice Greatest Hermes* (London 1906) which is now much out-of-date but still important in the field.

† *Corpus Hermeticum* XI,19-20 (trans. Scott). On the rediscovery and realisation of the importance of the Hermetica, see Frances Yates' fine study *Giordano Bruno and the Hermetic Tradition* (London and Chicago 1964) which includes a range of materials putting the whole subject in context. Passages from Hermes like the one quoted inspired many Renaissance panegyrics on the 'dignity' of man, i.e. his special exalted station, some of which are translated in S. Davies, *Renaissance Views of Man* (Manchester 1978). C. Zintzen gave the prestigious 'Teubner lecture' in 2000 on the *Menschenbild der Renaissance* (Munich and Leipzig 2000), in which the impact of the Hermetica in shaping the Renaissance view is powerfully brought out.

His brilliant application of them to the creation of a ground-plan, as if seen from heaven, reminds us that maps are the opposite of pictures. A picture shows us (albeit by 'representation') what something looks like: it is because we cannot see what Africa looks like that we turn to a map, or because we cannot see all the way across the desert following the dwindling track that we need a map to tell us what lies beyond our sight. Because we cannot look at once down all the streets in Imola, we need Leonardo's liberating town-plan. Another passage from the Hermetica again imagines soaring upward to a vantage point in the heavens, with the ability 'to grow wings and soar into the air! Poised between earth and heaven, you might see the solid earth, the fluid sea and the streaming rivers, the wandering air....' (*Corpus Hermeticum* V,5 trans. Scott). From such a height, the world is reduced to a map—mathematically, the 90° elevations used by Leonardo only meet, however, at a point infinitely high, transcending the possible, transcending the human condition. We see the world as a map when we are, empirically speaking, leaving it behind and it persists as an abstraction, reduced to 'land' or 'sea' or 'watercourses' or 'prevailing winds'. The reduction of the world to a map brings us to the bounds of actuality, looking one way at the factual with superhuman insight, yet already half out of framework and peering into the 'intense inane', 'to gaze on the things outside the cosmos (if indeed there is anything outside the cosmos), that too lies open to you' (XI,19 after Scott).

Earlier times were quite lucidly aware that such a vision of the world was a view from the perspective of after-death. The Hermetica were lost for many centuries, but a document called *The Dream of Scipio* from much the same Hellenistic-Roman time gave a more mainstream version of the same insights. In it Cicero tells a story to fix in his readers' minds the teachings he has tried to convey, which takes the form of a vision of the afterlife and a retrospective look back at the world 'below' from those same precipitous cosmic heights. It is too long to recount here, but we have the advantage of a great poetic moment which it inspired in the work of Chaucer. The poet's hero, Troilus, dies in the Trojan War, and his soul flies up while the elements of his being are converted back to their origins:

> His lighte goost ful blissfully is went
> Up to the holownesse of the seventh spere,
> In convers letinge every element;
> And ther he saugh, with ful avysement,

The erratik sterres, herkeninge armonye
With sownes fulle of hevenish melodye.
And doun from thennes faste he gan avyse
This litel spot of erthe, that with the see
Embraced is ....

Geoffrey Chaucer, *Troilus and Criseyde* Book V,1808-16

Through this and other channels, *Scipio's Dream* gave the mediaeval world its impression of the 'map of the world', Earth and sea beneath the erratic stars, the world in which we wander lost here below but there behold 'with ful avysement'—anticipating Leonardo's plan.

The same starting-point will help us with the history of actual world-maps, going back to ancient times. We make a sad mistake if we simply look back and find anticipations of our modern approach, and call that the history of map-making. The idea of mapping the world does not arise out of mundane human activities. Long-distance traders, even for instance on the Silk Road, did not need our kind of map. Each transportation was one in a series of relatively short relays; and on each one, it was sufficient to travel with those who had been before. For the unsure, it was more helpful to have a list of staging-points and numbers of days' travel, possibly little pictures of this distinctive mountain (bear left), or of that town by the river (watch out for the shaky bridge).* Territorial maps were the product of later, mainly imperial times. Generals required knowledge of the terrain to conquer and hold fast (avoid ambushes here; fortify that pass). The other practical reason for maps was legal—staking a claim to exact holdings and showing boundaries.† The Roman Empire eventually combined these two types into a comprehensive system for tax-purposes. But the earlier history of maps is based on quite different principles. We have to think along the lines of

---

* Such helpful listings played into map-making of course. See G.R. Crone, *Maps and their Makers* (London 1953) pp.23-4 for a Roman 'road map' of the Empire, more like a 'map' of the London Underground of today. A later Roman map (the basis of the Hereford *mappa mundi*?) showed routes and stylised pictures of mountains, rivers, towns etc. In the Hereford version, images from bestiaries or popular tales also crowd in, making 'the whole as much an encyclopedia of medieval lore as a map' (Crone, pp.26-7). A map of England from 1250 is still 'based on a straight line itinerary, from Dover to Newcastle' (p.27).

† Specially important in Egypt where the regular inundation of the Nile wiped out most markers. However, these plans do not lead on to wider application. Crone notes that these 'records were not it seems, combined to make maps of considerable areas on a smaller scale' (op. cit., p.15).

Scipio's Dream or Hermetic vision. Maps begin where we face that greatest unknown, the Beyond. Some of the oldest are those in the *Egyptian Book of Dead* (c. 1400 BC) which show the ways to be traversed by the soul to the realms of Osiris, and the beautiful fertile plots of land which the deceased will be given when he arrives there. These themselves stem from painted-coffin prototypes c. 2000 BC, like those found at El-Bersha which 'illustrate routes to reach the gateway of the Underworld. The upper route, in blue, is by water, the lower one, in black, by land. This collection of illustrated spells has in modern times been called *The Book of the Two Ways*. Linked with this is the day and night journey of Re, the sun god.'* A little later, the Sun's journey through the regions of the night is mapped out in splendid detail, again showing the world beyond human ken, which only enters the picture at dawn as the Sun passes through the gates of the horizon.

Even when we come to ancient Greece we find essentially the same standpoint. In Plato's (non-Atlantean) dialogue *Phaedo*, Socrates spends his last hours on Earth discoursing of the immortality of the soul, logically weighing conventional views and putting forward his (or Plato's) conception of the soul as its own self-sufficient reality, which need not be seriously affected by the death of the body. Nothing that we can prove about the soul suggests that it has physical-corruptible characteristics. Here again we stand in consciousness on the outer-edge of empirical, bodily things trying to grasp the infinite and self-sustaining. Then Socrates invokes a myth—a picture to help us. He looks in his mind's eye upon the world of men (not wholly unlike Troilus from his supernal heights), and portrays it as limited to a small hollow patch around the local watering-hole (the Mediterranean) where we dwell like frogs. There is, he points out, much of the globe which we do not know at all, mere vast tracts of terra incognita. Perhaps inhabited. The dead too must have some place in the world-picture, and so he brings in a picture which elaborates the old Greek idea of the Underworld and its sinister rivers—Styx, Lethe, Pyriphlegethon etc. Giving the account a scientific twist, he describes the Earth's recirculation of waters in the

---

* O.A.W. Dilke, *Greek and Roman Maps* (London 1985) p.16. Dilke fascinatingly notes other starting-points, such as the divinatory liver-map known from an ancient Etruscan bronze model. The world is a reflection of the celestial zones in Etruscan thought, which are determined by the orientation of the diviner as he contemplates the omens presented by the sacrificial liver. Signs observed there apply to the indicated region, so that the liver in effect becomes a map, and is marked as such in considerable detail (pp.18-19).

atmosphere, rivers and seas—and from there, underground until the waters rise again in rhythmic alternation. And in this process of recirculation, the souls of the dead are borne about until they are carried to the dark underworld prisons or to the regions of the Blest, or back into human rebirths.[*] The traditional map of the Other World thus forms the basis for the positioning of our living world in a 'map' of the whole Earth.[†] It was Plato's pupil Eudoxus of Cnidus who went on to refine the mathematics of the picture considerably, but the vision of the world which went on to become Greek geography had the same visionary roots we have seen from Egypt to Imola. Assessing the world we know becomes conscious mapping when we are setting out into the unknown.

Setting out was a divinatory act—dependent ultimately on the stars, those twinkling actualities of transcendence. To look beyond our own region means to set forth or set sail by the stars, and of course the stars are closely connected with the Beyond, the afterlife, the 'soul that rises with us, our life's star'. Turning the mental-leap to the idea of mapping into designs on papyrus or vellum must first take the form of mapping the stars. Only by their network of sightings and trajectories could ships take leave of the shore, watch it sink below the horizon and travel for weeks and months across the oceans. When the Renaissance kindled the spirit of exploration, mathematicians and astronomers had first to create the most accurate star-maps ever made, so that the far-flung territories stumbled upon by the first adventurers could be found again, measured and placed on the globe and claimed for nations and empires. The ships furrowed the immensities of the seas, but the minds of the map-makers traversed at once earthly distances and the spaces among the stars. Albrecht Dürer engraved the printed star-charts based on Tycho Brahe and Kepler's observations, and they were printed in Nuremberg in 1515. The great world-maps and atlases of Mercator and Ortelius followed later in the century.[‡]

So did the work of John Dee in Elizabethan England, who knew the great geographers and astronomers personally and himself made

---

[*] Plato, *Phaedo* 108-114.

[†] A map from the late Roman Empire, handed down by the Arabs in connection with the work of the geographer Ptolemy, still shows the inhabited world in its upper sections, and 'in the lower sector, Hades and its rivers': map reproduced, with description, as pl.29 in Dilke, *Greek and Roman Maps* (London 1985).

[‡] Gerard Mercator, *Nova et Aucta Orbis Terrae Descriptio* (1569); Abraham Ortelius, *Theatrum Orbis Terrarum* (1570). Cf, Paul Binding, *Imagined Corners: Exploring the World's First Atlas* (London 2003).

several brilliant contributions to mathematics and navigation.* In him the Hermetic intensification of consciousness, the concept of a science that would unite heaven and Earth or stars and earthly geography, the mysteries of the soul and the mastery of occult forces, were all as seethingly alive as they had been in Leonardo da Vinci. And he was fascinated by the feats of English navigation because he believed that they had rediscovered ancient Atlantis.

Through accidents of history and lack of understanding, Dee's prominence in the science of navigation afterwards fell from view, and many of his maps have been lost even though they were originally presented to Queen Elizabeth; some descriptions survive. We know that several of them showed 'Atlantis', i.e. the North American continent, and his fascination with it belongs to that whole mix of religious, scientific and mystical aspirations that combined with nationalist fervour and the 'cult' of the Virgin Queen. She had restored and purified Christianity, patronised arts and sciences, and brought in a new 'golden age'. In a sort of apocalyptic vision, Dee (and many others) saw her as destined to take that spiritual light to the ends of the Earth—that same Earth which her mariners were mapping and staking out its lands for her imperial rule. Dee's own Christianity was suffused with the Hermetic-cosmic piety which saw man as the instrument of divine powers on Earth and potential master of Nature's secrets. Such a new paradisal age might well see the restoration of lands long lost, whether those of the Lost Tribes of Israel (a major motive of exploration), or that mentioned by Plato.

Dee was able to envision Atlantis without the misleading speculation generated by those red herring-boats *Dolphin* and *Challenger*, and so need not think of a snaky island in mid-Atlantic. Plato's continent larger than Libya and Asia was amply met by the emerging information about the vast size of the new discovery. From Plato he would have deduced that it was once even larger and came closer to the European coast before the great disaster, whose 'biblical' overtones we have mentioned already. Dee's God might well have wrought such a change, being the God of Genesis who had told the waters

---

* Peter French, *John Dee* pp.173ff. He cites modern expert assessments that Dee 'was perhaps *the* major guiding spirit behind the glorious saga of English expansion' and that 'Dee's handling of the mathematical sciences as a whole is masterful ... but ... his analyses of navigation and hydrography have never been surpassed' (p.177). For his connections with Gemma Frisius, Mercator and Ortelius pp.177-8.

to 'be gathered together in one place, and let the dry land appear' (Gen. 1:9; Exod. 14:21-2)—he could just as easily reverse the effect (Exod. 14:28). But quasi-biblical speculation and imperial ambition was almost certainly not the full story of Dee's interest in Atlantis. For it coincides exactly with the high tide of his involvement in those sessions of scrying with Edward Kelley, who saw visions in the crystal 'showstone' and enabled Dee to converse with Angels and 'Some Spirits' on matters of which *A True & Faithful Relation of what passed for many Years* was afterwards published by the scholar Meric Casaubon in 1659.

Researchers still find themselves 'put on the spot' by Dee's involvement in these sessions, which were intensive and extensive and took in all aspects of his thought. It is impossible to divorce his advanced mathematics from his clairvoyant researches, many of which concerned the working of numbers and correlations with spiritual powers. The accurate mapping of the world also came together with his mystical-imperialist ideas, as a recently recovered document shows. *Britanici Imperii Limites* or *The Bounds of the British Empire* was apparently presented to the Queen in 1593, though the second part on *Geographical Reform* was written earlier.

The project of finding an 'ideal' standpoint for a map which Leonardo brilliantly inaugurated at Imola is taken up by Dee, who is seemingly out to find the place on which to centre his map which will reveal the hidden symmetry that indicates the scope of Elizabeth's rule. He thinks he has found it, and prophesies that greater things are yet to come:

> And if these things are true which we have so far heard tell, those four places which I have named have their own geographical symmetry. But concerning these things, and others relating to them (which are known hitherto to have lain hidden under the shadow of your wings) many wonderful, surprising, secret, and very delightful facts will, if it pleases our august and blessed Empress, with God's will, be revealed within the next seven years.[*]

He was preoccupied with similar ideas, but now aimed at the appropriation of Atlantis and its resources, when he got together in 1583 with two experienced navigators, Adrian Gilbert and John Davis. These were the men most interested in his Atlantean project—and

---

[*] K. MacMillan, with Jennifer Abeles, *John Dee, The Limits of the British Empire* ('Praeger Military History': California 2004) pp. 4-5. These documents were rediscovered in 1976. See also French, *John Dee* pp.197-8.

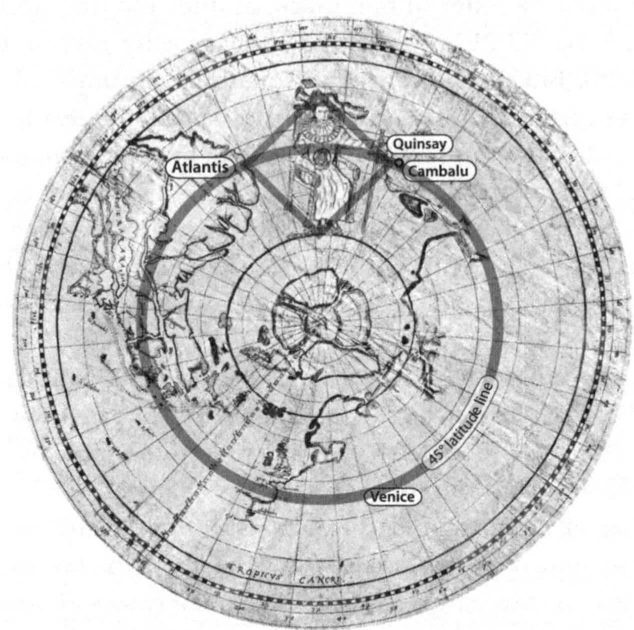

Map showing rediscovered Atlantis as a prophetic sign of the future British Empire under Queen Elizabeth I. Map from John Dee, Britannici Imperii Limites *(1593) interpreted by J. Egan, courtesy of Portland Museum.*

also amongst the most deeply involved in his divination sessions and scrying.[*] In that mystical way of joining two worlds, Dee wanted to persuade the Angels to support them in taking Christianity to the New World and so favour their planned conquest. Did he expect God to restore the Old Atlantis to its original size and glory as well? Did he get Kelley to peer into the dark backward and abysm of time, and see the world-map as it had once been? It would be not out of line with his mystical-mathematical geography. And he and friends form a rare instance of a circle which would be readily capable of translating visionary description into an actual map and of course a globe—by no means an easy feat, except for one of the world's mathematical finest, an associate of all the best cartographers in the world.[†]

---

[*] French, *John Dee* p.179 and n.4.

[†] As a matter of sheer speculation, it seems to me that the globe would offer the most ready way of modelling the described continents, perhaps in fresh clay onto an already finished sphere in what Scott-Elliot terms bas-relief. An impression might be taken thence in several sections and transferred to parchment/animal skin for the finished map. The use of the Mercator projection in 1896

There is no proof. Before Gilbert and Davis were due to embark, he had seized the opportunity to slip away to Prague and to advance his ideas at the Imperial Court of Rudolph II, taking Kelley with him. When Dee was away, his library and documents were outrageously dredged through for saleable items and the rest scattered or destroyed.* Into the latter category fell many of his esoterica. So if there had been such a map, it might well have perished as did many of Dee's other maps of Atlantis—or it might have found its way, along with a mass of other Hermetic and magical materials, into circles interested to delve into his inner world, or perhaps merely to make use of them as they could. Hermetic and occult societies continued to flourish and to preserve what they considered valuable heirlooms from the arcane tradition. If the

---

may well correspond to the practice of the original circle, though I have come to suspect that the final framework (which is very like the world-map that used to be rolled off for schoolboy geography lessons) is a last and not very intelligently applied afterthought. Greenland appears throughout the 1896 maps as a persistent and unchanging feature, possibly filled in from the grid-map; and I am not convinced that the ancient continents are very exactly correlated to the modern world. However, one must avoid correcting the evidence to find what one wishes.

The use of *terra cotta* for the globe fits well with the surmise of an Elizabethan origin. Though it was used in early times (cf. the well-known Chinese 'terra cotta army' of figurines from the third century BC), it fell out of use in the mediaeval time until it was revived as a new and exciting medium by the Italian sculptor Luca della Robbia. In the sixteenth century the Italian sculptors Pietro Torrigiano and Giovanni da Maiano brought the enthusiasm to England, and their work in the medium is extensively represented at Hampton Court, e.g. the well-known busts of the Roman Emperors. The material was therefore well established in England in Elizabeth's time. Of course the globe might have been earlier, and globes were manufactured as world-maps and star-maps from the fifth century BC. Lewis Spence thought some of his arcana were 'Hellenistic', and a well-known Roman copy of a Hellenistic (star) globe is famous as the Farnese Atlas. But such great antiquity for our objects is very uncertain. It is perfectly true, however, that a world-globe could have been made at any time since Pythagoras. The oddly prevalent notion that before Columbus everybody thought the world is flat is a pack of lies that was popularised by Washington Irving in a best-selling life of the explorer: see Carmen Blacker et al., *Ancient Cosmologies* (London 1975) pp.237 and 254 n.17. His 'evidence' for views expressed at Salamanca in 1486 is just made up.

* Just one opportunist, a certain Nicholas Saunder, stole more than a hundred volumes from Dee's collection of around 4,000 titles, but these have had the belated good fortune of ending up in the hands of the Royal College of Physicians, the only large cache of Dee-owned works to have been preserved together.

'records' which Lewis Spence was eventually to examine fall into that class, old enough to impress a moderately rigorous scrutiny, so possibly did the similar-sounding ones shown to Scott-Elliot. And where else might he have derived a map and a globe showing ancient Atlantis as the forerunner of America, before the opening up of the Atlantic Ocean? Certainly that map had nothing to do with the fashionable theories after Donnelly, or with whimsical fancies of Platonists in their studies.

The globe and map of Atlantis came to Scott-Elliot, we inferred, in the context of initiation into an occult society. Nothing else would explain the extraordinary difficulty of access—just as it is also described by Spence. Atlantis came to both as a part of a Hermetic *gnosis*, a stage of ascension, and the insistence upon memorising and upon a quasi-instantaneous assimilation of all the knowledge involved, before the texts were 'withdrawn', makes sense in that setting. The neophyte is to make an effort of self-transformation, of leaping by a spasm of mental effort to a new level of consciousness. If he can do it, he must do it now, after the esoteric preparation he has been given and in the face of the new vision the 'records' hold out to be attained. Such at least is the *beau idéal* of initiation, though often tainted with mystifications or reactionary features. Scott-Elliot was certainly aware that the material aids were there to stimulate inner apprehension, and that when that was achieved they fell away as an irrelevance. Yet the scryer's mystic perception was to be guided and stimulated by the maps and the traditional accounts, and the final 'story of Atlantis' was painstakingly put together from things seen and heard. The process was probably not so different from the sessions which generated the map in the first place. Some of the results, at least, were most interesting.

ACKNOWLEDGMENTS

Taking stock of my indebtedness, I come first to a lecture long ago by Dr Rudolf Lissau in Leeds stressing ideas of Rudolf Steiner from which my last chapter now takes its inspiration; it set me thinking. Not nearly so long ago Dr Judyth Sassoon (University of Bristol) kindly invited me to give a talk on Neanderthals in Stourbridge for which I already coined my current title; my recognition of her achievement as a Goethean scientist and palaeontologist far exceeds her modest citation in these pages. Of others I cannot rival, I must mention a visit from Professor Martin Lockley (University of Colorado) when we discussed his work on *How Humanity Came into Being*, which again made me think—of a book I could write. Since then there have been those patient friends who have scratched their heads at my ideas, but they became reality only with the keen interest and support shown by Sevak Gulbekian of Clairview Books, who has borne with my complicated requests, especially for pictures and maps, and last minute amendments and additions.

Other acknowledgments concern specifics: I must mention the generous permission given to me by Professor Christopher Scotese to reproduce details of his wonderful maps from the online 'Paleomap Project' and his pointing me to materials from his *Atlas of the Late Cretaceous*, of which I have gratefully availed myself. I thank Professor David Lewis-Williams for allowing me to reproduce drawings of the prehistoric scenes of initiation and signs, with their clinical parallels, from the Rock Art Institute of the University of Johannesburg. Jim Egan, of the Portland Museum, gave me permission to use his interpretation of John Dee's map showing the rediscovered Atlantis. Mike of Oxford Print Studio worked diligently and repeatedly at the Map of Atlantis to remove grid-lines and other excrescences, and to produce outline and selectively blocked and re-labelled maps to facilitate comparisons. The map of Neanderthal sites (shown in two parts) is reproduced by permission of Thames and Hudson from Clive Gamble and Christopher Stringer, *In Search of the Neanderthals* (1993). The map of European locations of Upper Palaeolithic art is reproduced under advice from Taylor and Francis Publishing group from *The Art of the Stone Age* (chapter Franco-Calabrian art by H. Breuil and L. Berger-Kirchner), copyright Methuen 1961.

The following images were reproduced unchanged under license as follows: the Sapiens-Neanderthal comparison from the Cleveland Museum of Natural History is licensed under the Creative Commons Attribution Share Alike 2.0 Generic (https://creative commons.org/license s/by-sa/2.0/deed.en) license. Author hairymuseummatt/ KaterBegemot. Homo Neanderthalensis from the Berlin Natural History Museum is licensed under Creative Commons Share Alike 4.0 International (https:// creativecommons.org/licenses/by.sa/4.0/ deed.en) license. Author Jakob Halun. Homo neanderthalensis from the Natural

INDEX